METHODS IN MOLECULAR BIOLOGY

For further volumes:
http://www.springer.com/series/7651

For over 35 years, biological scientists have come to rely on the research protocols and methodologies in the critically acclaimed *Methods in Molecular Biology* series. The series was the first to introduce the step-by-step protocols approach that has become the standard in all biomedical protocol publishing. Each protocol is provided in readily-reproducible step-by-step fashion, opening with an introductory overview, a list of the materials and reagents needed to complete the experiment, and followed by a detailed procedure that is supported with a helpful notes section offering tips and tricks of the trade as well as troubleshooting advice. These hallmark features were introduced by series editor Dr. John Walker and constitute the key ingredient in each and every volume of the *Methods in Molecular Biology* series. Tested and trusted, comprehensive and reliable, all protocols from the series are indexed in PubMed.

Deubiquitinases

Methods and Protocols

Edited by

Julie A. Maupin-Furlow and Mariola J. Edelmann

Department of Microbiology and Cell Science, Institute of Food and Agricultural Sciences, University of Florida, Gainesville, FL, USA

Humana Press

Editors
Julie A. Maupin-Furlow
Department of Microbiology
and Cell Science
Institute of Food and Agricultural
Sciences, University of Florida
Gainesville, FL, USA

Mariola J. Edelmann
Department of Microbiology
and Cell Science
Institute of Food and Agricultural
Sciences, University of Florida
Gainesville, FL, USA

ISSN 1064-3745 ISSN 1940-6029 (electronic)
Methods in Molecular Biology
ISBN 978-1-0716-2802-7 ISBN 978-1-0716-2803-4 (eBook)
https://doi.org/10.1007/978-1-0716-2803-4

This Humana imprint is published by the registered company Springer Science+Business Media, LLC, part of Springer Nature.
The registered company address is: 1 New York Plaza, New York, NY 10004, U.S.A.

Preface

Deubiquitinases (DUBs) and DUB-like enzymes are important proteases/hydrolases in cell biology [1]. These enzymes can generally or specifically remove ubiquitin and ubiquitin-like modifier proteins (Ub/Ubls) from substrate proteins, lipids, and chemical groups, thus, reversing post-translational modifications, altering signaling pathways, and replenishing the pools of free Ub/Ubls [2]. The C-termini of Ub/Ubls are ligated to the ε-amino group of lysine residues, the α-amino group of N-terminal amino acid residues, and less commonly, the thiol-groups of cysteine residues and hydroxyl- groups of serine/threonine residues [3]. As Ub/Ubls can be ligated to substrate proteins and sites on previously attached Ub/Ubl molecules, polymeric chains of Ub/Ubl can form with distinctive linkages. Ub/Ubl-signalling pathways are, thus, diverse in forming monomeric and polymeric Ub/Ubl tags on substrates that have widespread consequences on cell function, such as triggering the hydrolysis of substrate proteins by proteasomes, altering the cellular localization of biomolecules, impairing the active site of enzymes, and controlling protein-protein interactions. DUB/DUB-like enzymes can reverse these signalling events by cleaving Ub/Ubls from substrates and/or editing poly-Ub/Ubl chains [1]. Because of the diverse roles of DUB/DUB-like enzymes in cell biology, these proteins are often found critical in pathways associated with human health, viral infection, pathogenesis, plant biology, and other areas of biological significance.

This book aims to expand the understanding of DUB/DUB-like enzymes by highlighting advanced molecular tools and methods used in the field. Strategies are discussed on how to identify, biochemically characterize, and classify DUBs into groups including the USP (ubiquitin-specific protease), UCH (ubiquitin carboxy-terminal hydrolase), MJD (Machado–Josephin domain-containing protease), OTU (ovarian tumor protease), MINDY (motif-interacting with ubiquitin-containing novel DUB family), SENP (sentrin/SUMO-specific protease), and MPN+/JAMM (JAB1, MPN, MOV34 family) enzymes [1]. The first five DUB/DUB-like groups comprise cysteine proteases, whereas the latter group is comprised of zinc metalloproteases. This book also summarizes approaches used to discover DUB biological function and compounds/proteins that alter this function.

Chapters 1, 2, 3, 4, 5, and 6 are focused on methods used to characterize DUB activity and catalytic active sites. Chapter 1 describes structure-based approaches to characterize DUB catalytic sites by capitalizing on the finding that structural comparisons of DUBs in apo and Ub-bound forms can reveal conformational changes that lead to the catalytic activation of these enzymes. A series of activity-based probes (ABPs) that harbor C-terminal electrophiles that react with active site nucleophiles are found helpful in capturing cysteine protease DUBs in their Ub-bound form. The covalent complexes formed through this method are useful in probing DUB-Ub interfaces and identifying conformational changes associated with enzyme activity. Chapter 2 describes ways to study the Ub chain linkage selectivity of DUBs that cleave the synthetic fluorogenic substrate Ub-7-amido-4-methylcoumarin (-AMC). The hydrolysis of this monoUb-AMC substrate is compared to diUb-AMC substrates of different linkage types (*e.g.*, K29- and K33-linked Ub chains) by continuously monitoring fluorescence generated by a release of AMC, with DUB Ub chain preference corresponding to substrates with high levels of AMC released.

Chapter 3 highlights variations in the use of peptide-based fluorogenic probes to assay DUBs in purified form, cell lysate, and intact live cells. This chapter provides a cheaper alternative to the Ub-AMC probe by substituting Ub with the four C-terminal amino acids of Ub (LRGG) to generate LRGG-AMC. The chapter also reveals how to overcome cellular barriers and monitor DUB activity in live cells by conjugating the cell-penetrating β-hairpin peptide (RWVRVpGRWIRQ) to the four amino acid DUB recognition sequence (LRGG) and C-terminal fluorophore (7-amino-4-trifluoromethylcoumarin, AFC). Chapter 4 nicely describes current advances in analyzing DUBs from non-model animals of agricultural relevance. Ultrasonication-based tissue processing and bead-beating methods are outlined for researchers to generate cell lysate from these organisms, with the method found compatible with downstream ABP-labeling, purification, identification, and detection of cysteine protease DUBs. Chapter 5 makes use of recent advances in synthetic biology for DUB discovery. The researchers introduce amino acids other than canonical LRG at the P4–P2 positions in the Ub C-terminal tail (LRGG) using a Hybrid Combinatorial Substrate Library (HyCoSuL) approach and generate related substrates containing unnatural amino acids. The methods to produce and apply these Ub-based probes to detect DUBs in a cell lysate are carefully outlined in this chapter. Chapter 6 highlights one of the more challenging DUBs to characterize, Rpn11. Rpn11 is a DUB that removes Ub chains from protein substrates as they enter the proteasome for degradation. Like other members of the MPN+/JAMM metalloprotease group, Rpn11 contains Zn^{2+} in its active site. Here, the researchers outline a high-throughput assay in which the Rpn8/Rpn11 heterodimer is purified and assayed for DUB activity with a model substrate (Ub-GC-TAMRA) without the need to purify 26S proteasomes. Overall, Chapters 1, 2, 3, 4, 5, and 6 present methods helpful in identifying, classifying, and biochemically characterizing DUBs.

Chapters 7 and 8 feature new types of high throughput strategies for studying DUBs and the inhibitors of these proteases. Chapter 7 details how to profile DUBs and potential inhibitors in a cellular matrix using activity-based protein profiling (ABPP), with immunoblotting and mass spectrometry proteomics as readouts. To guide the reader on how to screen for DUB inhibitors that are potent and selective, the chapter nicely outlines the use of a standard ABPP method and then provides exciting new variations, including deep DUBome ABPP and high-throughput compatible ABPP-HT. Chapter 8 highlights how Matrix-Assisted Laser Desorption/Ionization-Time of Flight Mass Spectrometry (MALDI-TOF MS) can be used to perform a high-throughput assay of DUB activity. In this mass spectrometry-based approach, the in vitro cleavage of Ub chains into mono-Ub by the DUB is quantified and integrated with screening large and diverse compound libraries for drug discovery.

Chapters 9, 10, and 11 are focused on the discovery, classification, and characterization of novel DUBs/DUB-like enzymes. Chapter 9 provides a helpful guide on bioinformatic approaches based on evolutionary principles that can be used to discover and classify new DUBs. The researchers note several motifs that are useful in identifying DUB candidates from viral and bacterial pathogen genomes with high confidence. As the core set of bioinformatical tools needed to perform this analysis is freely accessible and easy to implement, this method has exciting potential for characterizing novel enzymes. Chapter 10 highlights how *Archaea*, one of three domains of life, can be used as tools for DUB/DUB-like enzyme studies. Using *Archaea* as an example, methods are outlined to synthesize, purify, and assay DUB-like enzymes that are active in organic solvents, high temperatures, and hypersaline buffers. Protocols for using *Archaea* as microbial platforms to synthesize Ub/Ubl isopeptide linkages on diverse protein targets are also delineated. Chapter 11 describes how to characterize the catalytic activity of the papain-like proteases, which play a vital role in the propagation of severe acute respiratory syndrome coronavirus 1 and 2. These papain-like proteases process the viral polyproteins essential for viral replication and remove Ubl ISG15 and Ub from host immune signaling proteins, thus, preventing host innate immune responses. The chapter clarifies how to use activity-based probes (ABPs) and fluorescence-based substrates to determine kinetic parameters of these papain-like proteases in their cleavage of the Ub/Ubl isopeptide bonds associated with disruption of the host signaling pathways.

Chapters 12, 13, and 14 are devoted to the challenging task of identifying DUB ligands and substrates. Chapter 12 provides an exciting approach combining next-generation sequencing with phage display (NGPD) to identify peptides that bind to DUBs. The steps and associated analytical methods needed to perform NGPD are outlined, and the initial evaluation using synthetic peptides in ELISA assays is included. As an example, the researchers have used NGPD to identify peptide ligands that bind USP11 in a unique pocket linked to DNA repair, thus, providing insight into biomolecular interactions that may reduce the resistance of cancers to current treatment options [4]. The next two chapters (Chapters 13 and 14) provide insight into strategies that overcome some hurdles when identifying DUB protein substrates. Proteasomes often degrade ubiquitinated proteins quite readily; thus, identifying these modified proteins can be challenging. DUBs can also display redundancy, complicating screening approaches that rely upon loss of function. In Chapter 13, the researchers identify DUB protein substrates and profile DUB specificity in *Xenopus* egg extract using ubiquitin vinyl sulfone treatment that inhibits DUB activity, which was combined with mass spectrometry. Since the DUBs were broadly inhibited, the DUB substrates were identified using quantitative proteomics. To pinpoint the DUBs responsible for removing the Ubl tag from the newly identified substrate proteins, the purified DUBs are added back to the samples to counteract the destruction of these proteins by proteasomes. Chapter 14 provides insight into an alternative strategy for identifying DUB protein substrates. Here, the researchers take advantage of the genome-wide resources available in *Saccharomyces cerevisiae* and identify DUB substrates that are stabilized upon systematic overexpression of DUBs.

Chapters 15, 16, and 17 provide some examples of alternative approaches that can be used to investigate DUB/DUB-like enzyme activity and use the DUB catalytic domain as a tool for cell biology studies. In Chapter 15, the researchers take a creative approach to examine DUB activity that does not depend upon the traditional fluorescently labeled artificial substrate. The method relies upon Atto532-tUI, a compound that can sense the release of free Ub from essentially any type of Ub-conjugate substrate and provide a fluorescence readout in real-time assays. A description of how to synthesize Atto532-tUI is also outlined. Chapter 16 describes a method to define SARS-CoV-2 Nsp3 activity using an ISG15-CHOP2-reporter. Nsp3 is challenging to characterize as it is a large multipass transmembrane protein but is of therapeutic interest as it cleaves the SARS-CoV-2 viral polypeptide and K48-linked polyUb and Ubl ISG15 from host cell proteins. Here, researchers guide how to isolate and characterize the biochemical activity of the full-length Nsp3. The ISG15-CHOP2-reporter used for the Nsp3 deISGylation activity assay consists of the Ubl ISG15 fused to a reporter enzyme. This fusion renders the reporter catalytically inactive. Following Nsp3 mediated cleavage of the isopeptide bond linking ISG15 to the reporter, the free reporter acts upon its substrate. Chapter 17 highlights how the function of Ub in endocytic trafficking can be investigated by fusing specific cargo to DUB catalytic domains. Here, the DUB fusions are examined for their impact on stimulated endocytosis of a transporter (MUP1) by fluorescence microscopy and flow cytometry. Endocytosis is also measured indirectly by cellular sensitivity to canavanine, a toxic amino acid analog taken up only by a surface-localized transporter (Can1). The assays of Chapter 17 nicely illustrate the utility of the DUB fusion strategy in investigating Ub in endocytic trafficking.

Overall, we hope that you find this book helpful when developing approaches to studying DUBs and related enzymes, expanding understanding of the involvement of these enzymes in cell biology, and creating new technologies and other advancements related to DUBs. These enzymes are emerging as central to controlling diverse cell functions, with the number of pathways regulated by DUB/DUB-like enzymes previously thought unimaginable. We thank all the authors who contributed to this book and the many pioneers who worked long hours to advance the field and make this work possible.

Gainesville, FL, USA *Julie A. Maupin-Furlow*
Mariola J. Edelmann

References

1. Harrigan JA, Jacq X, Martin NM, Jackson SP (2018) Deubiquitylating enzymes and drug discovery: emerging opportunities. Nat Rev Drug Discov 17:57–78
2. Cappadocia L, Lima CD (2018) Ubiquitin-like protein conjugation: structures, chemistry, and mechanism. Chem Rev 118:889–918
3. Li W, Ye Y (2008) Polyubiquitin chains: functions, structures, and mechanisms. Cell Mol Life Sci 65: 2397–2406
4. Spiliotopoulos A, Blokpoel Ferreras L, Densham RM, Caulton SG, Maddison BC, Morris JR, Dixon JE, Gough KC, Dreveny I (2018) Discovery of peptide ligands targeting a specific ubiquitin-like domain-binding site in the deubiquitinase USP11. J Biol Chem (this issue)

Contents

Contributors

HEATHER E. ARSENAULT • *Department of Molecular, Cell and Cancer Biology, University of Massachusetts Chan Medical School, Worcester, MA, USA*
JENNIFER A. BENANTI • *Department of Molecular, Cell and Cancer Biology, University of Massachusetts Chan Medical School, Worcester, MA, USA*
JUAN A. F. BONFANTE • *School of Veterinary Medicine and Science, Sutton Bonington Campus, Sutton Bonington, Leicestershire, UK*
YUN-SEOK CHOI • *School of Natural Sciences, Black Hills State University, Spearfish, SD, USA*
ROBERT E. COHEN • *Department of Biochemistry and Molecular Biology, Colorado State University, Fort Collins, CO, USA*
CHITTARANJAN DAS • *Department of Chemistry, Purdue University, West Lafayette, IN, USA*
PAUL DAVIES • *MRC Protein Phosphorylation and Ubiquitylation Unit, University of Dundee, Dundee, UK*
VIRGINIA DE CESARE • *MRC Protein Phosphorylation and Ubiquitylation Unit, University of Dundee, Dundee, UK*
AEVA S. DENISE • *Department of Cell and Developmental Biology, Vanderbilt University, Nashville, TN, USA*
KEN C. DONG • *Department of Molecular and Cell Biology, University of California at Berkeley, Berkeley, CA, USA; California Institute for Quantitative Biosciences, University of California at Berkeley, Berkeley, CA, USA; Howard Hughes Medical Institute, University of California at Berkeley, Berkeley, CA, USA*
PRISCILA DOS SANTOS BURY • *Department of Biochemistry & Structural Biology, University of Texas Health Science Center at San Antonio, San Antonio, TX, USA*
MARCIN DRAG • *Department of Chemical Biology and Bioimaging, Wroclaw University of Science and Technology, Wroclaw, Poland*
INGRID DREVENY • *Biodiscovery Institute, School of Pharmacy, University of Nottingham, Nottingham, UK*
MARIOLA J. EDELMANN • *Department of Microbiology and Cell Science, Institute of Food and Agricultural Sciences, University of Florida, Gainesville, FL, USA*
FARID EL OUALID • *UbiQ Bio B.V., Amsterdam, The Netherlands*
KEVIN C. GOUGH • *School of Veterinary Medicine and Science, Sutton Bonington Campus, Sutton Bonington, Leicestershire, UK*
RAPHAEL HEILIG • *Target Discovery Institute, Centre for Medicines Discovery, Nuffield Department of Medicine, University of Oxford, Oxford, UK*
NATHANIEL L. HEPOWIT • *Department of Cell and Developmental Biology, Vanderbilt University, Nashville, TN, USA*
THOMAS HERMANNS • *Institute for Genetics, University of Cologne, Cologne, Germany*
KAY HOFMANN • *Institute for Genetics, University of Cologne, Cologne, Germany*
HANNAH B. L. JONES • *Target Discovery Institute, Centre for Medicines Discovery, Nuffield Department of Medicine, University of Oxford, Oxford, UK*
BENEDIKT M. KESSLER • *Target Discovery Institute, Centre for Medicines Discovery, Nuffield Department of Medicine, University of Oxford, Oxford, UK; Chinese Academy for Medical*

Sciences Oxford Institute, Nuffield Department of Medicine, University of Oxford, Oxford, UK

RANDALL W. KING • *Department of Cell Biology, Blavatnik Institute at Harvard Medical School, Boston, MA, USA*

MAX S. KLOET • *Oncode Institute and Department of Cell and Chemical Biology, Leiden University Medical Centre, Leiden, The Netherlands*

JASON A. MACGURN • *Department of Cell and Developmental Biology, Vanderbilt University, Nashville, TN, USA*

ANDREAS MARTIN • *Department of Molecular and Cell Biology, University of California at Berkeley, Berkeley, CA, USA; California Institute for Quantitative Biosciences, University of California at Berkeley, Berkeley, CA, USA; Howard Hughes Medical Institute, University of California at Berkeley, Berkeley, CA, USA*

JULIE A. MAUPIN-FURLOW • *Department of Microbiology and Cell Science, Institute of Food and Agricultural Sciences, University of Florida, Gainesville, FL, USA; Genetics Institute, University of Florida, Gainesville, FL, USA*

SIGRUN K. MAURER • *Biodiscovery Institute, School of Pharmacy, University of Nottingham, Nottingham, UK*

ADAM T. MELVIN • *Cain Department of Chemical Engineering, Louisiana State University, Baton Rouge, LA, USA*

BINDU NANDURI • *Department of Comparative Biomedical Sciences, College of Veterinary Medicine, Mississippi State University, Mississippi State, MS, USA*

DIGANT NAYAK • *Department of Biochemistry & Structural Biology, University of Texas Health Science Center at San Antonio, San Antonio, TX, USA*

ALFRED N. NIJKERK • *UbiQ Bio B.V., Amsterdam, The Netherlands*

SHAUN K. OLSEN • *Department of Biochemistry & Structural Biology, University of Texas Health Science Center at San Antonio, San Antonio, TX, USA*

JONATHAN P. OWEN • *ADAS UK, School of Veterinary Medicine and Science, The University of Nottingham, Sutton Bonington, Loughborough, Leicestershire, UK*

JOAO A. PAULO • *Department of Cell Biology, Blavatnik Institute at Harvard Medical School, Boston, MA, USA*

ADÁN PINTO-FERNÁNDEZ • *Target Discovery Institute, Centre for Medicines Discovery, Nuffield Department of Medicine, University of Oxford, Oxford, UK; Chinese Academy for Medical Sciences Oxford Institute, Nuffield Department of Medicine, University of Oxford, Oxford, UK*

ALIREZA RAHNAMA • *Cain Department of Chemical Engineering, Louisiana State University, Baton Rouge, LA, USA*

CHRISTOPHER RAINVILLE • *Progenra Inc, Malvern, PA, USA*

VALENTINA ROSSIO • *Department of Cell Biology, Blavatnik Institute at Harvard Medical School, Boston, MA, USA*

WIOLETTA RUT • *Department of Chemical Biology and Bioimaging, Wroclaw University of Science and Technology, Wroclaw, Poland*

JOHN SANTELICES • *Department of Microbiology and Cell Science, Institute of Food and Agricultural Sciences, University of Florida, Gainesville, FL, USA*

LESLIE A. SHACK • *Department of Comparative Biomedical Sciences, College of Veterinary Medicine, Mississippi State University, Mississippi State, MS, USA*

ANASTASIOS SPILIOTOPOULOS • *Biodiscovery Institute, School of Pharmacy, University of Nottingham, Nottingham, UK; School of Veterinary Medicine and Science, Sutton*

Bonington Campus, Sutton Bonington, Leicestershire, UK; Vertex Pharmaceuticals, Abingdon, Oxfordshire, UK
DAVID E. STERNER • *Progenra Inc, Malvern, PA, USA*
KUMAR SURESH • *Progenra Inc, Malvern, PA, USA*
MARIA T. TSOUMPELI • *School of Veterinary Medicine and Science, Sutton Bonington Campus, Sutton Bonington, Leicestershire, UK*
GERBRAND J. VAN DER HEDEN VAN NOORT • *Oncode Institute and Department of Cell and Chemical Biology, Leiden University Medical Centre, Leiden, The Netherlands*
EVAN J. WORDEN • *Van Andel Institute, Grand Rapids, MI, USA*
GANG XIE • *Department of Molecular and Cell Biology, University of California at Berkeley, Berkeley, CA, USA; California Institute for Quantitative Biosciences, University of California at Berkeley, Berkeley, CA, USA*
ZHENGRUI ZHANG • *Department of Chemistry, Purdue University, West Lafayette, IN, USA*
MIKOLAJ ZMUDZINSKI • *Department of Chemical Biology and Bioimaging, Wroclaw University of Science and Technology, Wroclaw, Poland*

Chapter 1

Characterization of Deubiquitinase Catalytic State Using a Structure-Based Approach

Zhengrui Zhang and Chittaranjan Das

Abstract

The activity of deubiquitinases (DUBs) is tightly regulated in eukaryotes via various mechanisms. One of the regulatory strategies is substrate-induced catalytic triad rearrangement, where ubiquitin-binding helps the DUB adopt an active conformation for catalysis. The crystal structure of the apo form of such a DUB, when not bound to ubiquitin, reveals an inactive conformation of the catalytic residues, necessitating the structure of the ubiquitin-bound form to visualize the active state of the DUB. Comparing the apo and ubiquitin-bound structures reveals conformational changes leading to catalytic activation. To capture the deubiquitinase in its ubiquitin-bound form, a series of activity-based ubiquitin probes (Ub-ABPs) harboring C-terminal electrophiles were designed to react with the catalytic nucleophile of cysteine protease DUBs. The resulting covalently linked DUB-ubiquitin complex is amendable for structural studies to probe the DUB-ubiquitin interface and the potential conformational change of the DUB. Here, we present a detailed protocol for the generation and purification of ubiquitin carboxy-terminal hydrolase L1 (UCHL1) in complex with a Ub-ABP, ubiquitin-vinyl methyl ester (UbVME), and the subsequent structural analysis to characterize the catalytic state of the DUB.

Key words Activity-based probes, Deubiquitinase triad, Ubiquitin covalent complex, Catalytic state, Crystallization, Structural analysis, Conformation change, Substrate-induced rearrangement

1 Introduction

Deubiquitinases (DUBs) regulate various cellular processes by removing ubiquitin (Ub) from substrates, thus changing the physiological status of cells. Given the critical roles of DUBs in various Ub-dependent pathways, it is not surprising that their activities are often regulated by other cellular components. The structural analysis of DUBs has provided valuable insights into these regulations and the catalytic mechanisms of DUBs [1].

Among the regulatory strategies, substrate-mediated regulation, where Ub or the Ub chain promotes DUB activities, has been observed in some cases. For example, in USP7 (ubiquitin-specific cysteine protease 7) and UCHL1 (ubiquitin carboxy-

Julie A. Maupin-Furlow and Mariola J. Edelmann (eds.), *Deubiquitinases: Methods and Protocols*,
Methods in Molecular Biology, vol. 2591, https://doi.org/10.1007/978-1-0716-2803-4_1,

terminal hydrolase L1), catalytic triad rearrangements are induced upon Ub binding. The rearrangements activate the DUBs by causing local and distal conformational changes, leading to transforming an inactive triad arrangement to a productive one reminiscent of the classical papain-like active site [2, 3]. Another example is M1 di-Ub binding to OTULIN (OTU domain-containing deubiquitinase with linear linkage specificity), where the binding of the di-Ub secures the DUB in the active catalytic form [4]. Such a mode of regulation contributes to substrate selectivity and permits tuning of catalytic activity not possible with a preorganized active-site scaffold.

Characterization of the active state of the DUBs utilizing substrate-mediated regulation to promote enzyme activity requires structure determination of the DUB in the Ub-bound state and comparison of the structure with that of the apo DUB. This approach allows structure-based characterization of the conformational changes leading to the productive state. To understand the contribution of Ub to the DUB activity, structures of DUBs in complex with Ub or di-Ub chains have been solved [2–5]. However, such non-covalent complexes tend to exhibit weaker binding, making the co-crystallization highly unpredictable. Accordingly, the generation and crystallization of covalent DUB-Ub complex by linking the catalytic cysteine of DUB to the C-terminus of Ub in a mechanism-relevant manner have shown great success. In this case, a class of activity-based ubiquitin probes (Ub-ABPs), where an electrophile replaces the last glycine of Ub, reacts with the catalytic cysteines of DUBs [6, 7]. Upon binding to the S1-Ub site (which binds to Ub moiety prior to the scissile bond) of the DUB, the Ub-ABP orients the C-terminus of Ub close to the catalytic cysteine, presenting the electrophile for nucleophilic attack by the catalytic cysteine of the DUB. The resulting covalent complex, with a relatively inert thioether linkage tethering the DUB and Ub, is well-suited for subsequent purification, crystallization, and structural determination.

Two approaches to produce Ub-ABPs include a chemical means through solid-phase peptide synthesis [8] and a semisynthetic method combining recombinant protein expression and intein chemistry [6]. The latter approach is considered more viable for structural studies because it can produce Ub-ABPs on a larger scale with a more straightforward procedure and does not require expertise in protein chemical synthesis. To start with, a construct of Ub lacking the last glycine (Gly76), Ub_{1-75}, fused to the N-terminus of an intein preceding a chitin-binding domain (CBD) is needed. CBD is used for affinity capture by binding to the chitin resin. The first cysteine residue in the intein undergoes an N-S acyl shift producing a thioester linkage to the C-terminus of Ub [9]. The Ub can then be cleaved off with sodium 2-mercaptoethanesulfonate (MESNa) via transthioesterification from the affinity-captured Ub-intein-CBD species. The resulting Ub-MESNa can react with glycine-based electrophiles to produce the Ub-ABP of interest [6]. Here, we mainly focus on the production of UbVME (Fig. 1), which shows robust adduct formation with UCHL1 and could be used for structural determination in complex with UCHL1 [2].

UCHL1 is a Parkinson's disease-associated DUB expressed abundantly in neurons [10]. The crystal structure of the apo UCHL1 reveals a misalignment of its catalytic triad, with the catalytic C90 being 7.7 Å from the general base H161 [2, 11]. However, in the UCHL1-UbVME complex structure, the H161 reorients closer to C90 within 4 Å, suggesting that Ub may regulate the catalytic state of UCHL1 [2]. In this chapter, we describe the expression, generation, purification, and crystallization of UCHL1-UbVME and how to characterize the catalytic state of the DUB using structural information. The method described here may be applied to other DUBs to assess the potential conformational alteration of the DUB upon Ub binding, thus allowing a structural-based approach to determine the DUB catalytic state.

2 Materials

All the buffer, solutions, and media should be prepared using ultrapure water. Access to standard equipment used in molecular biology, protein purification, and protein crystallization (including PCR thermal cycler, gel electrophoresis apparatus, UV transilluminator, pH meter, orbital incubator, refrigerated benchtop centrifuge, nanodrop, fridge, freezer, fast protein liquid chromatography (FPLC) system, French press cell homogenizer, liquid nitrogen dewar, and microscope) is required.

2.1 Cloning

1. AccuPower PCR PreMix (Bioneer).
2. pGEX6P1 vector (Cytiva).
3. pTXB1 vector (NEB).
4. BamHI restriction enzyme.
5. XhoI restriction enzyme.
6. NdeI restriction enzyme.
7. SpeI restriction enzyme.
8. 10× CutSmart buffer: NEB, 1× buffer components: 50 mM potassium acetate, 20 mM Tris-acetate, 10 mM magnesium acetate, 100 μg/mL BSA, pH 7.9.
9. T4 DNA ligase.

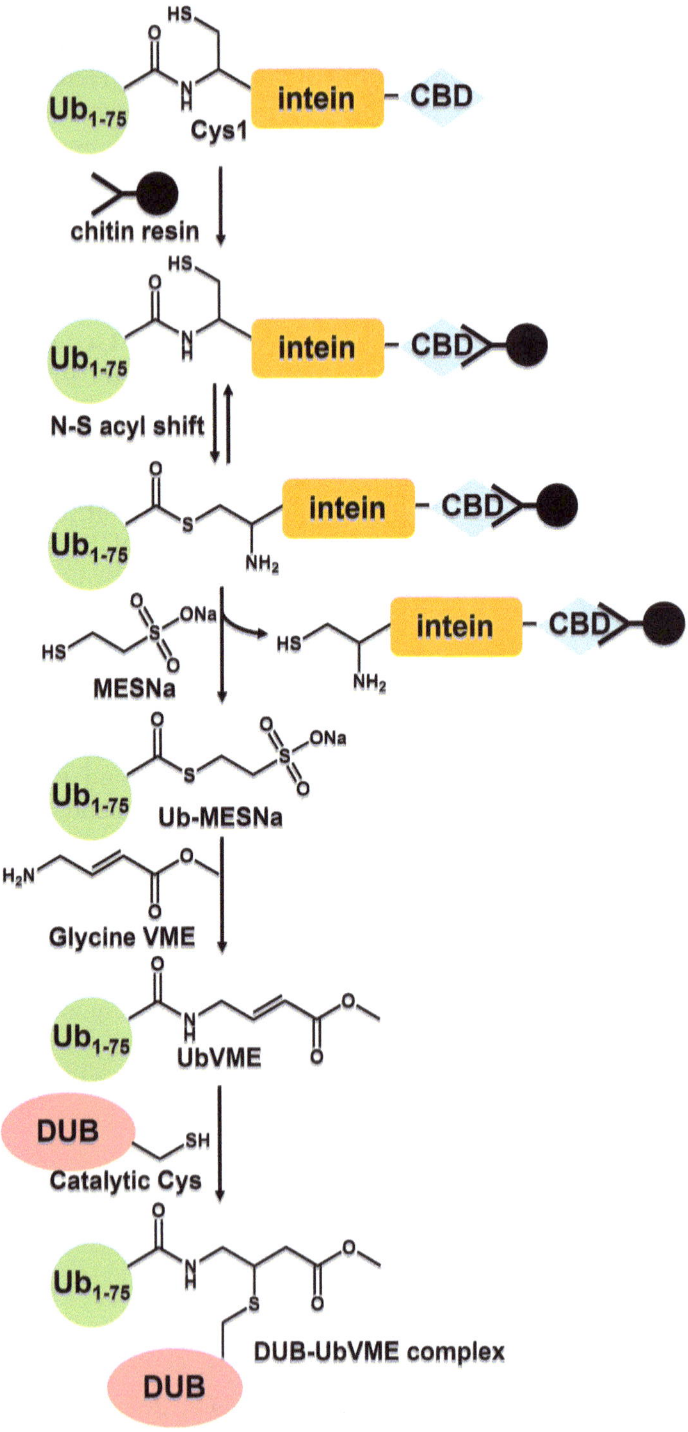

Fig. 1 Overview of DUB-UbVME complex formation. Ub_{1-75} fused to an intein and chitin-binding domain (CBD) is first expressed, where CBD is used for binding to the chitin resin. The N-S acyl shift occurring at the first cysteine residue of intein generates a thioester linkage between the Ub and intein. Sodium 2-mercaptoethanesulfonate (MESNa) can then release the Ub by forming Ub-MESNa via transthioesterification. Ub vinyl methyl ester (Ub-VME) is

10. 10× T4 DNA ligase reaction buffer: 1× buffer components: 50 mM Tris-HCl, 10 mM $MgCl_2$, 1 mM ATP, 10 mM DTT, pH 7.5.
11. *E. coli* DH5α competent cells.
12. Luria-Bertani (LB) media: 10 g NaCl, 10 g tryptone, and 5 g yeast extract in 1 L ultrapure water.
13. LB agar plate with ampicillin: 20 g agar and 100 mg ampicillin per liter culture.
14. QIAprep Spin Miniprep Kit (Qiagen).

2.2 Protein Expression and Purification

1. *E. coli* BL21 (DE3) competent cells.
2. LB media and agar plates containing 100 μg/mL ampicillin.
3. Erlenmeyer flasks (1× 500 mL and 6× 3 L).
4. 1 M isopropyl β-D-1-thiogalactopyranoside (IPTG).
5. Empty glass chromatography columns (50 mL and 200 mL).
6. 15 mL Microcon centrifugal filters (3 kDa and 10 kDa cutoff, Millipore Sigma).
7. 1× Phosphate buffered saline (PBS) with 400 mM KCl: 137 mM NaCl, 2.7 mM KCl, 8 mM Na_2HPO_4, and 2 mM KH_2PO_4, pH 7.4 supplemented with 400 mM KCl.
8. Glutathione Sepharose 4B GST-tagged protein purification resin (Cytiva).
9. GST elution buffer: 50 mM Tris-HCl, 10 mM reduced glutathione, pH 8.0.
10. 50 mL conical tubes.
11. PreScission protease (Cytiva).
12. GST dialysis buffer: 25 mM Tris-HCl, 300 mM NaCl, 1 mM DTT, pH 7.5.
13. 0.22 μm syringe filters.
14. 10 mL syringes.
15. HiLoad 16/600 Superdex 75 prep grade size exclusion chromatography column (Cytiva).
16. UCHL1 storage buffer: 50 mM Tris-HCl, 100 mM NaCl, 1 mM DTT, pH 7.4.
17. Ub-MESNa equilibration buffer: 50 mM MES, 300 mM NaOAc, pH 6.0.

Fig. 1 (continued) generated by reacting the Ub-MESNa with glycine VME. The catalytic cysteine in the DUB performs a nucleophilic attack and generates the DUB-UbVME complex, which mimics the intermediate stage of the deubiquitination reaction

18. Chitin resin.
19. Sodium 2-mercaptoethanesulfonate (MESNa).
20. Ub-MESNa elution buffer: 50 mM MES, 300 mM sodium acetate, 50 mM MESNa, pH 6.0.
21. 3 kDa and 10 kDa cutoff dialysis membrane.
22. 5 L dialysis buckets.
23. Bicinchoninic acid (BCA) protein assay kit.

2.3 UbVME Generation and Purification

1. 1 M $NaHCO_3$, pH 8.0.
2. N-hydroxy succinimide (NHS).
3. Glycine vinyl methyl ester [6] (*see* **Note 1**).
4. UbVME dialysis buffer: 50 mM sodium acetate, pH 4.5.
5. Mono S™ 5/50 GL cation exchange chromatography column (Cytiva).
6. Mono S buffer A: 50 mM sodium acetate, pH 4.5.
7. Mono S buffer B: 50 mM sodium acetate, 1 M NaCl, pH 4.5.

2.4 UCHL1-UbVME Generation and Purification

1. 1 M Tris-HCl, pH 8.0.
2. Mono Q 5/50 GL anion exchange chromatography column (Cytiva).
3. Mono Q buffer A: 50 mM Tris-HCl, pH 8.0.
4. Mono Q buffer B: 50 mM Tris-HCl, 1 M NaCl, pH 8.0.
5. UCHL1-UbVME storage buffer: 50 mM Tris-HCl pH 7.4, 150 mM NaCl, and 10 mM DTT.

2.5 Crystallization and Data Processing

1. VDXm 24-well hanging-drop crystallization plates with sealant (Hampton).
2. 18 mm siliconized glass cover slides (Hampton).
3. UCHL1 crystallization solution: 0.1 M HEPES, 2.4 M $(NH_4)_2SO_4$, pH 7.0.
4. UCHL1-UbVME crystallization solution: 0.1 M BICINE, 2.4 M $(NH_4)_2SO_4$, pH 9.0.
5. Mounted crystal loops (Hampton).
6. Magnetic sample mount (Hampton).
7. Universal pucks (UniPucks) and tools (MiTeGen).
8. Cryogenic dry shipper with shipping case (MiTeGen).

2.6 Online Tool and Software

1. ProtParam Tool on ExPASy server [12] (https://web.expasy.org/protparam/).
2. HKL-3000 (HKL Research).
3. Phenix [13] (https://phenix-online.org/).

4. Coot (https://www2.mrc-lmb.cam.ac.uk/personal/pemsley/coot/).
5. PyMOL (Schrödinger).

3 Methods

3.1 Expression and Purification of UCHL1

1. Clone the gene encoding UCHL1 (UniProt: P09936) into pGEX6P1 *E. coli* expression vector using BamHI/XhoI restriction sites. Verify the insertion of gene by Sanger sequencing.
2. Transform the expression plasmid into *E. coli* BL21(DE3) competent cells. Plate the cells onto an ampicillin selection agar plate, and incubate at 37 °C overnight. Resuspend a single colony from the plate in a 500 mL Erlenmeyer flask containing 100 mL LB media with 100 μg/mL ampicillin. Place the flask in an orbital shaker, and incubate at 37 °C overnight with shaking at 220 rpm.
3. Inoculate 60 mL of the overnight culture in 6 L LB media with 100 μg/mL ampicillin (6× 3 L Erlenmeyer flasks each containing 1 L media, 10 mL of the overnight culture in each flask). Place the flasks in an orbital shaker, and incubate at 37 °C with shaking at 220 rpm until the OD_{600} reaches 0.6–0.8. Induce the protein expression by adding 400 μL 1 M IPTG per L culture at 18 °C with overnight shaking at 220 rpm.
4. Harvest the cells by centrifugation at 8500 rcf at 4 °C for 7 min.
5. Resuspend the pellet in 100 mL ice-cold 1× PBS supplemented with 400 mM KCl. Unless otherwise mentioned, the steps below should be performed at 4 °C or on ice.
6. Lyse the cells by passing the suspension twice through a French press at 1500 psi. Remove the cell debris by centrifugation at 100,000 rcf for 1 h.
7. Load 5 mL GST resin to a benchtop 50 mL empty chromatography column. Equilibrate the column with 20 mL 1× PBS with 400 mM KCl by gravity flow.
8. Load the supernatant onto the column at 1 mL/min; keep this flow rate for the rest of the protein purification steps.
9. After all the supernatant passes through, wash the resin with 50 mL 1× PBS with 400 mM KCl.
10. Elute the GST-tagged UCHL1 with 50 mL GST elution buffer; collect the eluate in one 50 mL conical tube. The expected yield of GST-UCHL1 from 6 L culture is approximately 80 mg.

11. Add 50 units of GST-tagged PreScission protease to the eluted GST-UCHL1; dialyze the GST-UCHL1 with GST-tagged PreScission protease with a 10 kDa cutoff dialysis membrane twice against 4 L GST dialysis buffer. Each dialysis should take at least 4 h.
12. Remove the GST-tagged PreScission protease, and free GST from the tag-free UCHL1 by applying the dialyzed sample back to the 5 mL GST resin (pre-equilibrated with GST dialysis buffer). Collect the flowthrough containing tag-free UCHL1 in one 50 mL conical tube. Verify the protein by a ~25 kDa band on a reducing 15% SDS-PAGE gel (*see* **Note 2**).
13. Concentrate the sample to 5 mL using a 15 mL 10 kDa cutoff centrifugal filter. Filter the concentrated sample through a 0.22 μm syringe filter, and apply all 5 mL to a HiLoad 16/600 Superdex 75 size exclusion chromatography column pre-equilibrated with UCHL1 storage buffer (*see* **Note 3**). Perform the size exclusion chromatography run with UCHL1 storage buffer at 1 mL/min, and collect fractions every 2 mL. The elution volume of UCHL1 is expected to be between 55 mL and 70 mL.
14. Verify the fractions containing pure UCHL1 by a ~25 kDa band on a 15% SDS-PAGE gel. Pool the fractions together, and concentrate down using a 15 mL 10 kDa cutoff centrifugal filter to a desired volume (e.g., 1 mL).
15. Calculate the extinction coefficient and molecular weight of UCHL1 based on UCHL1 sequence using ProtParam tool on ExPASy server.
16. Determine the protein concentration based on the calculated extinction coefficient and molecular weight of UCHL1 using nanodrop. The concentration of UCHL1 is expected to be approximately 40 mg/mL for 1 mL concentrated sample.
17. Flash-freeze and store the protein sample at −80 °C until further use.

3.2 Expression, Generation, and Purification of UbVME

1. Clone a gene encoding Ub (UniProt: P0CG48) lacking the last glycine residue ($Ub_{1\text{-}75}$), into the pTXB1 vector using NdeI/SpeI restriction sites (*see* **Note 4**). Verify the insertion of gene by Sanger sequencing.
2. Express the $Ub_{1\text{-}75}$-intein-CBD fusion protein, and harvest the cells as described in **step 2** to **step 4** in Subheading 3.1.
3. Resuspend the cell pellet in 100 mL Ub-MESNa equilibration buffer. Lyse the cells and clear the lysate as described in **step 6** in Subheading 3.1. Unless otherwise mentioned, the steps below should be performed at 4 °C or on ice.

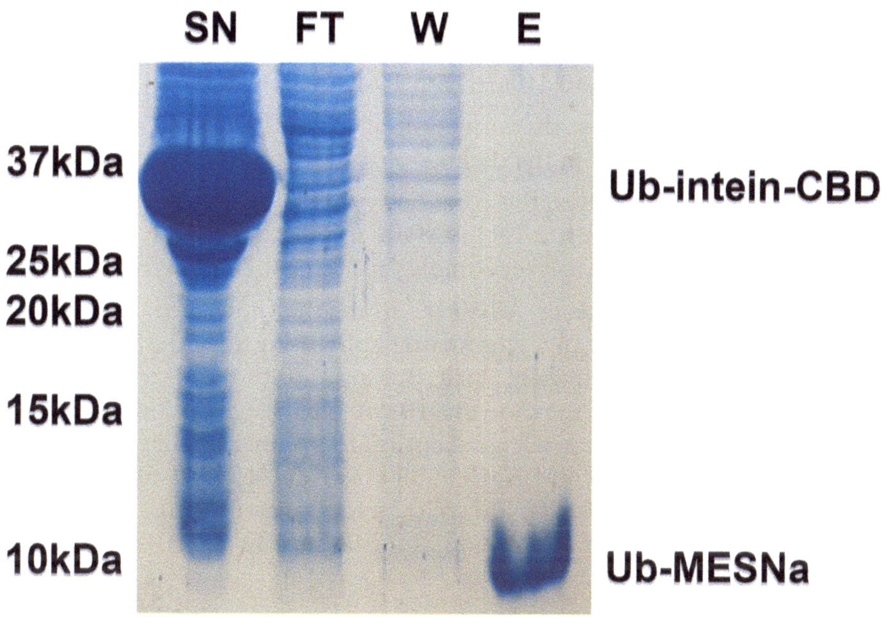

Fig. 2 Purification of Ub-MESNa. The expression of the Ub-intein-CBD in *E. coli* is confirmed by a thick band at ~35 kDa in the lysate supernatant (SN) on a reducing 15% SDS-PAGE gel. The absence of that band in the flowthrough (FT) and wash (W) samples indicates the efficient binding of the protein to the chitin resin. The elution (E) of Ub-MESNa is verified by the band at ~9 kDa

4. Load 50 mL chitin resin to a benchtop 200 mL empty chromatography column. Equilibrate the column with 150 mL Ub-MESNa equilibration buffer by gravity flow.
5. Load the supernatant onto the column at 1 mL/min; keep this flow rate for the rest of the protein purification steps.
6. After all the supernatant passes through, wash the column with 150 mL Ub-MESNa equilibration buffer.
7. Add 60 mL Ub-MESNa elution buffer. Cap the column and incubate at 37 °C overnight.
8. Next day, collect the flowthrough, which contains Ub-MESNa. Add additional 60 mL Ub-MESNa elution buffer to elute the remaining Ub-MESNa from the column. Pool the two eluted samples together (~120 mL in total). Verify the Ub-MESNa by a ~9 kDa band on a reducing 15% SDS-PAGE gel (Fig. 2). The expected yield of Ub-MESNa from 6 L culture is around 150 mg.
9. Concentrate the Ub-MESNa to 1.5 mL using a 15 mL 3 kDa cutoff centrifugal filter. The expected protein concentration at this stage is around 100 mg/mL.
10. Add 150 mg N-hydroxy succinimide (NHS) to 6 mL 1 M $NaHCO_3$, and then slowly add 120 mg glycine VME to the solution.

11. Mix 1.5 mL concentrated Ub-MESNa with the 6 mL 1 M $NaHCO_3$ containing NHS and glycine VME. Dilute the reaction to 10 mL with 1 M $NaHCO_3$, pH 8.0.
12. Allow the formation of UbVME overnight at room temperature in the dark.
13. Dialyze the reaction with a 3 kDa cutoff dialysis membrane against 2 L UbVME dialysis buffer for 8 h (*see* **Note 5**). Filter the dialyzed sample through a 0.22 μm syringe filter.
14. Separate UbVME from the unreacted species by cation exchange chromatography using Mono S 5/50 GL column. Specifically, load 3.5 mL of the dialyzed sample (around one-third of the 10 mL sample) every time on the column. Run a salt gradient of 10 column volumes (CVs) from 0 to 400 mM NaCl in 50 mM NaOAc, pH 4.5 (0–40% of Mono S buffer B), and collect fractions every 1 mL. UbVME should be eluted at around 200–300 mM NaCl (20–30% Mono S buffer B).
15. To verify the fractions containing UbVME, take 8 μL from each fraction, and neutralize the pH with 1 μL 1 M Tris-HCl, pH 8.0. Add 1 μL of 2 mg/mL UCHL1 to each sample, and incubate the reactions at room temperature for 30 min. Run these reactions on a reducing 15% SDS-PAGE gel. The fractions containing UbVME should migrate as a ~34 kDa band on the gel, attributing to the formation of UCHL1-UbVME covalent complex.
16. Pool the fractions containing UbVME together, and concentrate down using a 15 mL 3 kDa cutoff centrifugal filter to a desired volume (e.g., 2 mL; *see* **Note 6**). Determine the concentration of UbVME using BCA protein assay kit. The concentration of UbVME is expected to be approximately 30 mg/mL for 2 mL concentrated sample.
17. Flash-freeze and store the protein sample at −80 °C until further use.

3.3 Generation and Purification of the UCHL1-UbVME Complex

1. Dilute 7.5 mg UbVME in 1 mL 1 M Tris-HCl pH 8.0, mix well, and verify that the pH is nearly 8.0 with a pH strip. Add more 1 M Tris-HCl pH 8.0 if needed.
2. Add 20 mg UCHL1 to the solution.
3. Add ultrapure water to reach a final reaction volume of 10 mL, and incubate the reaction at room temperature overnight. Confirm the formation of the complex by a ~34 kDa band on a reducing 15% SDS-PAGE gel (Fig. 3).
4. Buffer-exchange the reaction into 50 mM Tris-HCl pH 8.0 using a 15 mL 10 kDa cutoff centrifugal filter (*see* **Note 7**).

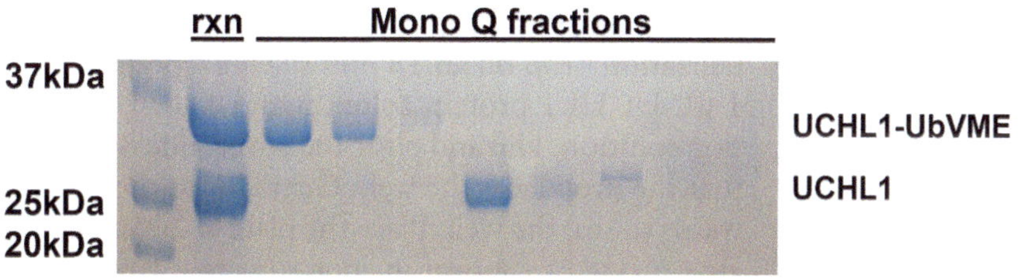

Fig. 3 Purification of UCHL1-UbVME using anion exchange chromatography. The reaction of UCHL1 with UbVME generates a mixture of UCHL1 and UCHL1-UbVME (rxn lane). Anion exchange chromatography (Mono Q) successfully separates these two proteins, with the UCHL1-VbVME eluted at a lower salt concentration

Concentrate the sample to ~10 mL, and filter through a 0.22 μm syringe filter.

5. Perform anion exchange chromatography using Mono Q 5/50 GL column. Load 3.5 mL of the dialyzed sample every time on the column. After loading the sample, run a salt gradient of 10 CVs at 0.5 mL/min from 0 to 200 mM NaCl in 50 mM Tris-HCl, pH 8.0 (0–20% Mono Q buffer B), and collect fractions every 1 mL (*see* **Note 8**). UCHL1-UbVME should be eluted at 10–50 mM NaCl (1–5% Mono Q buffer B), while UCHL1 should be eluted at 70–150 mM NaCl (7–15% Mono Q buffer B).
6. Verify the fractions containing pure UCHL1-UbVME on a reducing 15% SDS-PAGE gel (Fig. 3). Pool the fractions containing pure UCHL1-UbVME together.
7. Buffer-exchange the protein into UCHL1-UbVME storage buffer using a 15 mL 10 kDa cutoff centrifugal filter (*see* **Note 9**). Concentrate down to a desired volume (e.g., 500 μL).
8. Calculate the extinction coefficient and molecular weight of UCHL1-UbVME based on protein sequence using ProtParam tool on ExPASy server.
9. Determine the protein concentration based on the calculated extinction coefficient and molecular weight of UCHL1-UbVME using nanodrop. The concentration of UbVME is expected to be approximately 40 mg/mL for 500 μL concentrated sample.
10. Flash-freeze and store the protein sample at −80 °C until further use.

3.4 Crystallization, Data Processing, and Structural Analysis

1. Dilute UCHL1 stock to 20 mg/mL with UCHL1 storage buffer.
2. Crystallize UCHL1 by hanging drop vapor diffusion method using UCLH1 crystallization solution (*see* **Note 10**). Transfer 500 μL of UCHL1 crystallization solution to a well in the

24-well hanging-drop crystallization plate. Set up a 2 μL crystallization drop on an 18 mm siliconized glass cover slide with 1 μL UCHL1 protein solution and 1 μL UCHL1 crystallization solution. Flip and place the cover slide on top of the well so that the drop faces the well. Gently press the cover slide downwards to seal the well. Place the plate at 20 °C. UCHL1 crystals are expected to appear in approximately 1 week.

3. Dilute UCHL1-UbVME to 25 mg/mL with UCHL1-UbVME storage buffer.
4. Set up UCHL1-UbVME crystallization using the same approach described in **step 2** in Subheading 3.4. UCHL1-UbVME crystals are expected to appear in approximately 2 months.
5. To harvest the crystals, prepare crystal loops by attaching the mounted crystal loops of desired sizes (e.g., 0.1–0.2 mm diameter) on the sample mount cap using Epoxy. Allow a few hours for secure attachment of the loop.
6. Open the cover slide with tweezers and loop the protein crystal using the crystal loop. Immediately flash-freeze and store the looped crystals in liquid nitrogen for further data collection.
7. Collect X-ray diffraction datasets of the crystals (*see* **Note 11**).
8. Index and scale the dataset using HKL3000, and perform molecular replacement (MR) in Phenix using full-featured Phaser-MR with the deposited UCHL1 (PDB: 2ETL) and ubiquitin (PDB: 1UBQ) structures as models.
9. Carefully check the MR solutions in Coot to see if the protein structure agrees with the electron density. Manually adjust certain regions of the structure as needed. Refine the structure in Phenix using phenix.refine.
10. Repeat the manual adjustment process until the refined structures are satisfactory (*see* **Note 12**); open them in PyMOL.
11. Superimpose the UCHL1-UbVME structure onto the apo UCHL1 structure. Select and show stick annotations of the catalytic triad (C90, H161, and D176). Zoom in at the catalytic triad.
12. A significant reorientation of H161 (the general base in the triad) should be observed (Fig. 4a). Measure the distance between H161 and C90 using the "Measurement" tool under "Wizard" (Fig. 4b and c).
13. To explore other conformational changes around H161, select the H161 in apo UCHL1, and modify the surrounding residues by 4 Å. Show stick annotations of the residues. The reorientation of F53 should be observed.

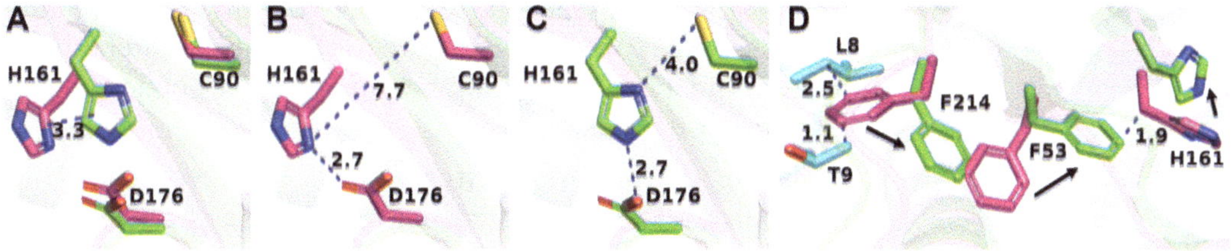

Fig. 4 Structural analysis of UCHL1 compared to UCHL1-UbVME. Apo UCHL1 is colored in magenta, UCHL1 in UCHL1-UbVME in green, and Ub in cyan. (**a**) In the UCHL1-UbVME catalytic triad, the imidazole ring of H161 shifts towards the catalytic C90 by 3.3 Å. (**b–c**). This conformational change produces a closer distance (4.0 Å) between H161 Nδ1 and C90 Sγ compared to that of 7.7 Å in apo UCHL1. (**d**) Ub binding induces this conformational change. Specifically, the L8 and T9 in Ub occupy the space for UCHL1 F214 in the apo UCHL1, forcing F214 to reorient and subsequently rearrange the conformation of UBCHL1 F53 and H161

14. Repeat the previous step with F53 to observe the reorientation of F214.
15. Repeat the process with F214. Ub L8 and T9 should be observed to occupy the space of F214 in the apo UCHL1.
16. Select L8 and T9 in Ub and F53, F214, and H161 in UCHL1. Show stick annotations of only these residues to demonstrate the mechanism of the structural rearrangement induced by Ub binding (Fig. 4d).

4 Notes

1. The synthesis of glycine vinyl methyl ester is not discussed in this chapter and can be found in previous literature [6]. If the synthesis cannot be performed in the lab, one may consider obtaining commercially available UbVME to generate the DUB-UbVME complex for structural studies.
2. The molecular weight of UCHL1 is very close to that of GST. Therefore, the efficiency of GST removal from the UCHL1 sample is difficult to assess solely based on SDS-PAGE. Yet, after GST removal, the total protein amount should be approximately half of that in the original dialyzed sample. Additionally, the subsequent SEC run can remove the remaining GST in the sample, as GST tends to dimerize in solution and thus elutes earlier than UCHL1. The SEC chromatogram can also be used to evaluate sample purity.
3. If the UCHL1 sample is used only to generate UCHL1-UbVME, it is unnecessary to perform an SEC run. The purification steps for UCHL1-UbVME can remove the impurities in the sample.
4. The SpeI restriction site is within the intein gene. Therefore, a DNA sequence (5′- GCATCTCCCGTGATGCA-3′) should

be added between SpeI site and Ub G75 in the reverse primer to ensure the completeness of the intein gene. Alternatively, ligation-independent cloning methods may be applied to generate the correct plasmid.

5. Dialyzing the reaction mixture for an extended time is not recommended, as unwanted precipitation has been observed in our previous experiences.
6. UbVME is stable in the buffer at pH 4.5, even after multiple freeze-thaw cycles. However, if needed, an additional buffer exchange step can be performed. UbVME is stable in commonly used Tris, HEPES, and MES buffers.
7. To buffer-exchange the reaction sample using a centrifugal filter, first concentrate the reaction volume down to approximately 1 mL, add Mono Q buffer A up to 15 mL, and concentrate the reaction volume down to 1 mL again. This step should be performed at least three times to ensure the proper binding of UCHL1-UbVME to the column and efficient separation of UCHL1-UbVME with unreacted UCHL1.
8. The setup for ion-exchange chromatography to separate DUB-UbVME and apo DUB varies by the nature of the DUB. To apply this to your DUB of interest, the choice of the column (cation exchange or anion exchange), buffer pH, salt gradient, and gradient length should be optimized. Typically, compared to apo enzyme, the UbVME complex is more positively charged.
9. Depending on the sample purity on SDS-PAGE gels, it might be necessary to perform an additional SEC run (with UCHL1-UbVME storage buffer) to remove impurities for successful crystallization trials.
10. The crystallization conditions for UCHL1 and UCHL1-UbVME were previously reported [2, 11]. To discover the suitable crystallization condition for your protein of interest, it is necessary to perform a crystallization screening experiment first.
11. The appropriate strategy to collect X-ray diffraction datasets varies based on the overall quality of the crystal and the technical specifications of the beamline. Consult experienced protein crystallographers if insights on data collection are needed.
12. Several metrics, including R-free, clashscore, Ramachandran outliers, side chain outliers, etc., should be considered to decide the validity of the refined crystal structure. Depending on the quality and resolution of the dataset, the values of these metrics can vary. It is recommended to consult experienced protein crystallographers for specific suggestions on data processing and structure refinement.

Acknowledgments

This work was funded by National Institute of Health Grants R01GM126296 (to C.D.) and T32AI148103 (to Z.Z.).

References

1. Sahtoe DD, Sixma TK (2015) Layers of DUB regulation. Trends Biochem Sci 40:456–467
2. Boudreaux DA, Maiti TK, Davies CW, Das C (2010) Ubiquitin vinyl methyl ester binding orients the misaligned active site of the ubiquitin hydrolase UCHL1 into productive conformation. Proc Natl Acad Sci U S A 107:9117–9122
3. Hu M, Li P, Li M et al (2002) Crystal structure of a UBP-family deubiquitinating enzyme in isolation and in complex with ubiquitin aldehyde. Cell 111:1041–1054
4. Keusekotten K, Elliott PR, Glockner L et al (2013) OTULIN antagonizes LUBAC signaling by specifically hydrolyzing met1-linked polyubiquitin. Cell 153:1312–1326
5. Reyes-Turcu FE, Horton JR, Mullally JE et al (2006) The ubiquitin binding domain ZnF UBP recognizes the C-terminal diglycine motif of unanchored ubiquitin. Cell 124:1197–1208
6. Borodovsky A, Ovaa H, Kolli N et al (2002) Chemistry-based functional proteomics reveals novel members of the deubiquitinating enzyme family. Chem Biol 9:1149–1159
7. Ekkebus R, van Kasteren SI, Kulathu Y et al (2013) On terminal alkynes that can react with active-site cysteine nucleophiles in proteases. J Am Chem Soc 135:2867–2870
8. El Oualid F, Merkx R, Ekkebus R et al (2010) Chemical synthesis of ubiquitin, ubiquitin-based probes, and diubiquitin. Angew Chem Int Ed Engl 49:10149–10153
9. Chong S, Mersha FB, Comb DG et al (1997) Single-column purification of free recombinant proteins using a self-cleavable affinity tag derived from a protein splicing element. Gene 192:271–281
10. Wilkinson KD, Lee K, Deshpande S et al (1989) The neuron-specific protein PGP 9.5 is a ubiquitin carboxyl-terminal hydrolase. Science 246:670–673
11. Das C, Hoang QQ, Kreinbring CA et al (2006) Structural basis for conformational plasticity of the Parkinson's disease-associated ubiquitin hydrolase UCH-L1. Proc Natl Acad Sci U S A 103:4675–4680
12. Gasteiger E, Hoogland C, Gattiker A et al (2005) Protein identification and analysis tools on the ExPASy server. In: Walker JM (ed) The proteomics protocols handbook. Humana Press
13. Liebschner D, Afonine PV, Baker ML et al (2019) Macromolecular structure determination using X-rays, neutrons and electrons: recent developments in *Phenix*. Acta Crystallogr D Struct Biol 75:861–877

Chapter 2

Synthetic Diubiquitin Fluorogenic Substrates to Study DUBs

Max S. Kloet and Gerbrand J. van der Heden van Noort

Abstract

Development of (semi-)synthetic methods to prepare ubiquitin (Ub)-based reagents has proven to be helpful in the classification of deubiquitinating proteases (DUBs). To study DUB selectivity for one or more of the eight possible poly-Ub chains, fluorogenic assay reagents have been reported relying on the appearance of a fluorescent signal upon DUB-mediated cleavage of the reagent. In this protocol, we describe the use of such an assay to profile the selectivity of TRABID, a member of the OTU family of DUBs.

Key words Fluorogenic substrate, Activity assay, Fluorescence, Deubiquitinating protease, TRABID

1 Introduction

The use of semi-synthetic methods, such as intein chemistry or genetically encoded orthogonal protection and activated ligation (GOPAL), and fully synthetic methods, such as linear solid-phase peptide chemistry, to prepare ubiquitin (Ub) [1], Ub-chains [2], and Ub-conjugates [3] opened the way to an arsenal of tools for the identification and activity monitoring of deubiquitinating enzymes (DUBs) [4–6]. Some DUBs show a distinct linkage selectivity concerning the recognition and proteolysis of only a subset of the eight distinct poly-ubiquitin linkages (M1, K6, K11, K27, K29, K33, K48, or K63 linked). Not only are some DUBs able to recognize specific subsets of Ub chains; they also do this using various modes of action [7]. To determine linkage selectivity of DUBs, one can follow the cleavage of all eight diUb (di-ubiquitin)-chains by the recombinantly expressed DUB, one-by-one in gel-based assays, exemplified by studies on the OTU DUB family [8] or the USP DUB family [9]. Other approaches to study DUB cleavage rely on fluorescence polarization (FP) [10], Forster Resonance Energy Transfer (FRET) [11], mass spectrometry [12], or chemiluminescence [13] as methods for efficient readout.

Julie A. Maupin-Furlow and Mariola J. Edelmann (eds.), *Deubiquitinases: Methods and Protocols*,
Methods in Molecular Biology, vol. 2591, https://doi.org/10.1007/978-1-0716-2803-4_2,

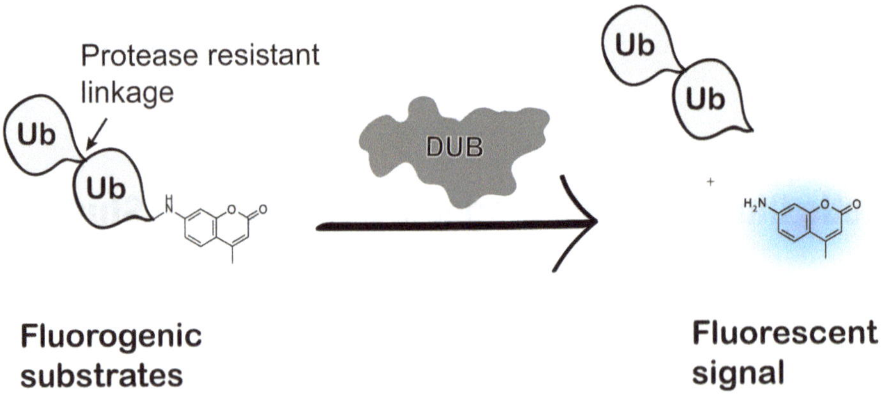

Fig. 1 Schematic representation of the diUb-AMC assay

When studying interactors of Ub-chains in cell lysate, including active DUBs, it can be beneficial to prepare Ub-chains that cannot be proteolyzed during the assay itself, for instance, by replacing the isopeptide linkage with a protease-resistant triazole linkage [14–16]. Further development of such non-hydrolyzable diUb molecules by introducing a reactive warhead or a fluorogenic group led to the next generation of diUb probes used to study DUBs, such as OTUD2 [17], UCH-L3 [18], SARS-CoV PLpro [19], and the COVID-19-causing SARS-2-CoV PLpro [20]. Non-hydrolyzable diUb-7-amido-4-methylcoumarin (-AMC) reagents can be used to interrogate linkage-specific reactivity of the (recombinantly expressed) DUB. Incubation of the DUB in the presence of each diUb linkage type in parallel can be monitored in real time using a fluorescence plate reader. If the DUB recognizes the specific diUb-AMC and cleaves the conjugated AMC moiety of the C-terminus, a fluorescent signal appears that can be directly correlated to the activity of the DUB (*see* Fig. 1). The non-hydrolyzable isopeptide bond mimic warrants that diUb-AMC cannot be converted to monoUb-AMC during the assay. Below we describe in detail how a typical diUb-AMC assay can be conducted. In this assay, we applied the diUb-AMC assay on TRABID (ZRANB1), a cysteine protease belonging to the ovarian tumor (OTU)-associated DUB family. TRABID is a protein implicated in Wnt–β-catenin signaling, and it shows a preference for K29- and K33-linked Ub chains [21, 22].

2 Materials

2.1 Assay

1. Assay buffer: 20 mM Tris-HCl, 150 mM NaCl, 2 mM DTT, 1 mg/mL CHAPS, 1 mg/mL bovine gamma globulin, pH 7.6. Add 24.2 g Tris-HCl and 87.6 g NaCl to 750 mL Milli-Q, and stir until fully dissolved. Adjust the pH to pH 7.6 by adding 1 M NaOH, and fill with Milli-Q to a final volume of 1 L to obtain a 10X stock: 200 mM Tris-HCl, 1.5 M NaCl,

pH 7.6. Dilute this buffer solution ten times by adding 100 mL 10X buffer to 900 mL Milli-Q, and supplement with 2 mM DTT, 1 mg/mL CHAPS, and 1 mg/mL bovine gamma globulin to obtain the assay buffer.

2. Non-hydrolyzable diUb-AMC substrates and monoUb-AMC: prepared as described in the literature [2, 17]. Dissolve the lyophilized fluorogenic diUb-AMC substrates in DMSO, and dilute in assay buffer to obtain stocks of 40 μM monoUb-AMC, K29 diUb-AMC, and K48 diUb-AMC (*see* **Note 1**).
3. Amino-methyl-coumarin AMC (CAS: 26093-31-2) from a commercial vendor. Prepare a parent stock by dissolving 2 mg AMC in 570 μL DMSO to obtain a 20 mM AMC solution. To prepare a standard curve, dilute 1 μL of this solution in 499 μL assay buffer to obtain 40 μM AMC as a stock solution.
4. Recombinant human TRABID (ZRANB1). Dilute in assay buffer to a concentration of 1 μM (*see* **Note 2**).
5. BMG LABTECH GmbH PHERAstar FSX plate reader. Use the PHERAstar with excitation wavelength (350 nm) and emission wavelength (450 nm); filter FI350–450.
6. Greiner bio-one sterile 96-well V-bottom microplates. Use to make serial dilutions of the fluorogenic substrates.
7. Corning 3820 black polystyrene 384 wells, non-binding surface, no lid, low volume, and flat bottom assay plates. Use for the assay.

2.2 Data Analysis

1. MARS software; BMG LABTECH PHERAstar FSX version 5.4. Use to measure fluorescence on the PHERAstar, and export measured data to Excel.
2. Microsoft Excel 365 version 2102. Use to convert measured fluorescent intensities to concentration formed AMC using the standard curve, and calculate the slope of this data.
3. GraphPad Prism 9 (1992–2021) LLC for Windows 64-bit version 9.0.1 (151) software. Use for the preparation of figures, standard curve, and Michaelis-Menten plot analysis.

3 Methods

3.1 Assay

3.1.1 Preparation of AMC Standard Curve

1. Add 40 μL of the 40 μM AMC stock solution to well A1 of a 96-well plate. Add 20 μL assay buffer to the wells A2 to A11 as preparation for serial dilution. Transfer 20 μL of well A1 to A2, and pipet up and down a few times to ensure homogeneity. Follow the same procedure for A2 to A3, A3 to A4, ..., A9 to A10 to obtain a descending concentration of AMC. Fill well A11 with assay buffer only.

2. Transfer 20 μL of the serial dilution of AMC (wells A1 to A11) to an assay plate using a multichannel pipet. Place the assay plate in the PHERAstar plate reader, and measure fluorescence intensity for 5 min.

3.1.2 Serial Dilution of diUb-AMC Substrates (Duplicate)

1. Pipet 40 μL of the reagent solutions (40 μM: monoUb-AMC, K29 diUb-AMC, and K48 diUb-AMC) to A1 (monoUb-AMC), B1 (K29 diUb-AMC) and C1 (K48 diUb-AMC) of a 96-well plate.
2. Fill the wells A2 to A11, B2 to B11, and C2 to C11 with 20 μL assay buffer.
3. Transfer 20 μL of A1 to A2 followed by pipetting up and down for proper mixing. Transfer 20 μL from A2 to A3 and so on until A10, to reach a serial dilution with 20 μL end volume at concentrations from 40 μM to 80 nM. A11 well serves as a control containing no fluorogenic substrate (*see* **Note 3**).
4. Repeat this procedure for rows B and C, or use a multichannel pipet to perform the serial dilution action on rows A, B, and C simultaneously.

3.1.3 Real-Time Monitoring of the Assay in the Plate Reader

1. Transfer 10 μL of each prepared concentration fluorogenic diUb-AMC substrate in the 96-well plate to a Corning black 384-well assay plate using a multichannel pipet equipped with 11 pipet tips. Perform this action in duplicate for every fluorogenic substrate concentration to generate two rows for mono-, K29-, and K48-linked diUb-AMC.
2. Set up the MARS software before starting the assay (excitation 350 nm, emission 450 nm) (*see* **Notes 4–5**).
3. Start the assay by adding 10 μL TRABID solution (1 μM) to each of the wells containing the Ub reagents in the assay plate (in duplicate) using a multichannel pipet (*see* **Note 6**).
4. Centrifuge the assay plate for 1 min (106 rcf) to ensure (A) proper mixing of the liquids and (B) removal of air bubbles in the solution.
5. Insert the assay plate in the PHERAstar plate reader, and record the fluorescent intensity for 45 min using the MARS software (*see* **Note 7**).

3.2 Data Analysis

3.2.1 Conversion of Fluorescence Intensity to Concentration Using Standard Curve AMC

1. Export measured fluorescence intensity (FI) values from standard curve to Excel using embedded MARS software.
2. Plot the obtained FI values in GraphPad Prism per concentration of AMC used. On the *x*-axis, plot the concentration AMC from 0 to 10 μM. On the *y*-axis, plot the FI values.
3. Select the linear regression function to draw a straight line through the inserted FI values leading to an equation belonging to the standard curve (*see* Fig. 2).

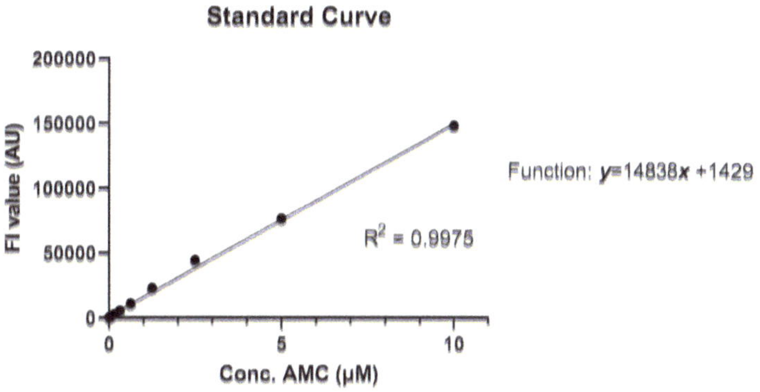

Fig. 2 Standard curve AMC plotted from 0 to 10 μM

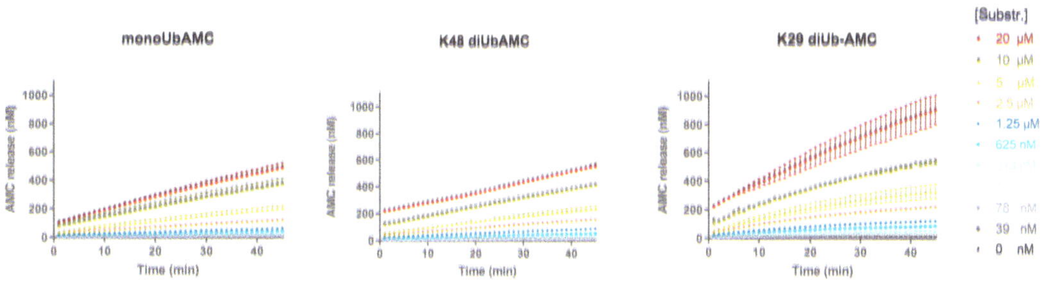

Fig. 3 AMC release at various used substrate concentrations

3.2.2 Plotting Concentration AMC Released at Various Substrate Concentrations

1. Export measured fluorescence intensity (FI) values per linkage type per concentration to Excel using embedded MARS software.
2. In Excel, convert FI values to concentration formed AMC using the formula obtained in Subheading 3.2.1.
3. Plot the obtained values formed AMC in GraphPad Prism per concentration of used substrate to obtain the AMC concentration (nM) released for each Ub linkage (Fig. 3). On the *x*-axis, plot the time in minutes from 0 to 45 min. On the *y*-axis, plot the concentration of AMC formed.

3.2.3 Preparation of Michaelis-Menten Plot

1. In Excel, calculate the slope of the [AMC] formed over the first 15 min per Ub linkage type, per concentration of substrate used (*see* **Note 8**).
2. Insert the calculated slope values into GraphPad Prism. The Michaelis-Menten function can be selected to draw a curve through the inserted slope values leading to a Michaelis-Menten plot (Fig. 4). On the *x*-axis, plot the substrate concentration from 0 to 20 μM. On the *y*-axis, plot the initial velocity V_I from 0 to 25 nM/min.

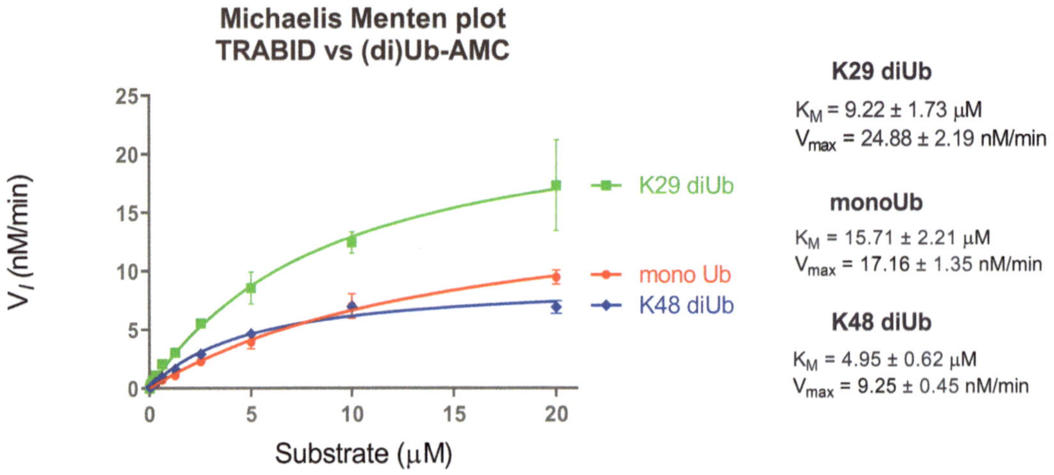

Fig. 4 Michaelis-Menten plot and kinetic parameters derived from the plot

3. Calculate the kinetic parameters such as maximum velocity (V_{Max}) and the Michaelis-Menten constant (K_M) by GraphPad Prism based on the inserted slope values per concentration of substrate used (*see* Fig. 4) (*see* **Note 9**).

4 Notes

1. MonoUb-AMC can be purchased from multiple commercial vendors.
2. We thaw the enzyme stocks on ice and dilute the working stocks on ice for use in the assay. Parent enzyme stocks are snap frozen by submerging in liquid nitrogen and stored at − 80 °C for longer storage periods. We used recombinantly expressed TRABID, kindly gifted to us by Dr. P. Elliott, Department of Biochemistry, University of Oxford, UK. TRABID is also available from multiple commercial vendors.
3. The prepared diUb-AMC substrate concentration series ranges from 40 μM to 80 nM. However, due to the 1:1 (v/v) dilution with an enzyme, the final concentration in the assay will range from 20 μM to 40 nM.
4. Parameters to consider when setting up the plate reader, besides excitation and emission wavelengths, are the correct range of wells to be assayed and the correct focal height and receiver gain to ensure the maximum amount of signal will be within the linear range of detection.
5. It is essential to properly set up the assay parameters in the MARS software before adding the enzyme TRABID to the assay plate. When TRABID is added, the assay immediately

starts, and the time lost to set up the plate reader will delay recording the initial crucial data points of the assay.

6. After addition to the assay, TRABID starts to cleave the amide bond between the diUb and AMC moiety, and the assay immediately starts. Therefore, add the enzyme solution to all wells as quickly following each other as possible, without delay.
7. Ensure to centrifuge no longer than 1 min (106 rcf) immediately after adding the enzyme and quickly enter the assay plate in the PHERAstar plate reader to minimize the time lost between the addition of the enzyme and the start of data collection.
8. Typically, the first 15 min of enzyme activity are used to calculate the initial slope of the reaction. It is important to have a linear increase (straight line) over this period of time. A larger or smaller interval can be selected depending on enzyme activity to calculate the slope.
9. Selectivity of TRABID for K29-linked diUb-AMC over monoUb-AMC and K48-linked diUb-AMC is observed in this particular assay, which is in line with reported studies [21, 22].

Acknowledgments

MSK acknowledges funding by NWO (VIDI grant VI.192.011). GJvdHvN acknowledges funding by NWO (VENI grant 722.014.002 and VIDI grant VI.192.011).

References

1. Tan Y, Wu H, Wei T, Li X (2020) Chemical protein synthesis: advances, challenges, and outlooks. J Am Chem Soc 142(48): 20288–20298
2. El Oualid F, Merkx R, Ekkebus R et al (2010) Chemical synthesis of ubiquitin, ubiquitin-based probes, and diubiquitin. Angew Chem Int Ed 49:10149–10153
3. Chatterjee C, McGinty R, Pellois J, Muir T (2007) Auxiliary-mediated site-specific peptide ubiquitylation. Angew Chem Int Ed 46:2814–2818
4. Dang L, Melandri F, Stein R (1998) Kinetic and mechanistic studies on the hydrolysis of ubiquitin C-terminal 7-amido-4-methylcoumarin by deubiquitinating enzymes. Biochemist 37:1868–1879
5. Hassiepen U, Eidhoff U, Meder G et al (2007) A sensitive fluorescence intensity assay for deubiquitinating proteases using ubiquitin-rhodamine110-glycine as substrate. Anal Biochem 371:201–207
6. Borodovsky A, Ovaa H, Kolli N et al (2002) Chemistry-based functional proteomics reveals novel members of the deubiquitinating enzyme family. Chem Biol 9:1149–1159
7. Lange S, Armstrong L, Kulathu Y (2021) Deubiquitinases: from mechanisms to their inhibition by small molecules. Mol Cell 82:15–29
8. Mevissen T, Hospenthal M, Geurink P et al (2013) OTU deubiquitinases reveal mechanisms of linkage specificity and enable ubiquitin chain restriction analysis. Cell 154:169–184
9. Faesen A, Luna-Vargas M, Geurink P et al (2011) The differential modulation of USP activity by internal regulatory domains, interactors and eight ubiquitin chain types. Chem Biol 18:1550–1561

10. Geurink P, Oualid F, Jonker A, Hameed D, Ovaa H (2012) A general chemical ligation approach towards isopeptide-linked ubiquitin and ubiquitin-like assay reagents. ChemBioChem 13:293–297
11. Geurink P, van Tol B, van Dalen D et al (2016) Development of diubiquitin-based FRET probes to quantify ubiquitin linkage specificity of deubiquitinating enzymes. Chembiochem 17:816–820
12. Ritorto M, Ewan R, Perez-Oliva A et al (2014) Screening of DUB activity and specificity by MALDI-TOF mass spectrometry. Nat Commun 5:4763
13. Gutkin S, Gandhesiri S, Brik A, Shabat D (2021) Synthesis and evaluation of ubiquitin-dioxetane conjugate as a chemiluminescent probe for monitoring deubiquitinase activity. Bioconjug Chem 32:2141–2147
14. Eger S, Scheffner M, Marx A, Rubini M (2010) Synthesis of defined ubiquitin dimers. J Am Chem Soc 132(46):16337–16339
15. Zhang X, Smits A, van Tilburg G et al (2017) An interaction landscape of ubiquitin signaling. Mol Cell 65:941–955
16. Zhao X, Lutz J, Höllmüller E et al (2017) Identification of proteins interacting with ubiquitin chains. Angew Chem Int Ed 56:5764–15768
17. Flierman D, van der Heden van Noort G, Ekkebus R et al (2016) Non-hydrolyzable diubiquitin probes reveal linkage-specific reactivity of deubiquitylating enzymes mediated by S2 pockets. Cell Chem Biol 23:472–482
18. van Tilburg G, Murachelli A, Fish A et al (2021) K27-linked diubiquitin inhibits UCHL3 via an unusual kinetic trap. Cell Chem Biol 28:191–201
19. Békés M, van der Heden van Noort G, Ekkebus R et al (2016) Recognition of Lys48-linked di-ubiquitin and deubiquitinating activities of the SARS coronavirus papain-like protease. Mol Cell 62:572–585
20. Shin D, Mukherjee R, Grewe D et al (2020) Papain-like protease regulates SARS-CoV-2 viral spread and innate immunity. Nature 587(7835):657–662
21. Licchesi J, Mieszczanek J, Mevissen T et al (2012) An ankyrin-repeat ubiquitin-binding domain determines TRABID's specificity for atypical ubiquitin chains. Nat Struct Mol Biol 19:62–71
22. Virdee S, Ye Y, Nguyen DP, Komander D, Chin JW et al (2010) Engineered diubiquitin synthesis reveals Lys29-isopeptide specificity of an OTU deubiquitinase. Nat Chem Biol 10:750–757

Chapter 3

Fluorometric Characterization of DUB Activity: From Single-Enzyme Reactions to Live Intact Cells

Alireza Rahnama and Adam T. Melvin

Abstract

Fluorescently tagged molecular probes capable of time- and concentration-dependent quantification of deubiquitinating enzyme (DUB) activity allow for precise characterization of both enzyme and DUB inhibitor. These probes are compatible with most plate readers allowing for rapid, facile fluorometric analysis of DUB activity. DUB activity can be measured in purified enzyme reactions, in cell lysates, or in intact cells depending upon the choice of the fluorometric probe. This chapter describes protocols and potential analysis tools to investigate DUB activity in these three scenarios.

Key words Fluorometry, Deubiquitinating enzymes, Lysates, Live intact cells

1 Introduction

Protein polyubiquitination is a tightly regulated process that governs several intracellular signaling pathways, including DNA repair, cell cycle regulation, and, most importantly, protein degradation. Deubiquitinating enzymes (DUBs) are an essential component of the ubiquitin-proteasome system (UPS), as they mediate the reversible process of polyubiquitination, allowing for the recovery of proteins targeted to the 26S proteasome for degradation. DUB functionality is to cleave the isopeptide bond formed between the C-terminal carboxylic acid of ubiquitin (Ub) and the amine group on the side chain of a lysine or ornithine residue on the target protein removing the poly-Ub chain and rescuing the protein from degradation before it is recognized by the proteasome [1]. DUBs are divided into five classes based on sequence and structural similarities, with the two most common being ubiquitin-specific proteases (USPs) and ubiquitin C-terminal hydrolases (UCHs). Several of these DUBs have been associated with a multitude of different human pathologies such as cancer [2],

Julie A. Maupin-Furlow and Mariola J. Edelmann (eds.), *Deubiquitinases: Methods and Protocols*,
Methods in Molecular Biology, vol. 2591, https://doi.org/10.1007/978-1-0716-2803-4_3,

neurodegenerative diseases [3], and cardiac disease [4] resulting in increased interest in exploiting them as therapeutic targets [5, 6].

Methods, such as Western blots, exist to evaluate protein levels qualitatively; however, these approaches lack the ability to quantitatively measure enzyme activity in both time- and concentration-dependent manners. The gold standard for evaluating DUB activity is fluorometry which requires the use of fluorescently tagged molecules that function as probes for a direct readout of DUB activity. The underlying principle of such probes is the recognition of an enzyme-specific sequence by a DUB which leads to the cleavage of the fluorescent molecule, releasing a fluorescent signal that can be measured and used to quantify the enzyme activity. The fluorescent signal indicates enzyme activity since as long as the fluorogenic molecule is bound to the probe, it does not emit a fluorescent signal. One of the most abundantly used fluorescent probes consists of a full-length Ub with a fluorophore (e.g., 7-amino-4-methylcoumarin, AMC) conjugated to the C-terminus (Ub-AMC) [7]. Other fluorescent molecules such as 7-amino-4-trifluoromethylcoumarin (AFC) [8] and rhodamine-110 [9] have also been incorporated in the design of such probes. A peptide-based variant substitutes the full-length Ub, with the four C-terminal amino acids of Ub (LRGG), allowing for a cheaper alternative to the Ub-AMC probe measure DUB activity, although at the expense of selectivity. These two probes (Ub-AMC and LRGG-AMC) help measure DUB activity in single-enzyme, one-pot reactions and in cellular lysates; however, they are not compatible with intact cells. Moreover, cell lysis is associated with the dilution of cytoplasmic and nuclear proteins and can lead to the dissociation of proteins necessary for DUB activity [10], which highlights the need for cell-permeable probes that quantify DUB activity in live intact cells. Recently Safa et al. [8] developed a cell-permeable DUB probe by conjugating a β-hairpin sequence motif (RWVRVpGRWIRQ) identified as a cell-penetrating peptide (CPP) [11] to the four-amino acid DUB recognition sequence (LRGG) and a C-terminal fluorophore (7-amino-4-trifluoromethylcoumarin, AFC). This chapter describes experimental protocols to perform time- and concentration-dependent studies of DUB activity (and DUB inhibition) using peptide-based fluorogenic probes in single-enzyme, lysate-based, and live intact cell studies.

2 Materials

Prepare all cell culture media (for both the adherent HeLa cell line and suspension OPM.2 cell line) inside a biosafety cabinet to ensure sterility. Prepare all buffers using deionized water, and sterilize the solutions inside a biosafety hood using 0.2 μm polyethersulfone

(PES) filters. All cells should be cultured in a 37 °C, 5% CO_2 incubator. Cell growth media and all buffers should be stored in a 4 °C refrigerator.

2.1 Enzyme Kinetics Assay

1. Assay buffer: 50 mM HEPES, 150 mM NaCl, 2 mM dithiothreitol (DTT), 0.02% Tween-20, pH 7.4.
2. Extracellular buffer (ECB): 5.036 mM HEPES (pH 7.4), 136.89 mM $NaCl_2$, 2.68 mM KCl, 2.066 mM $MgCl_2 \cdot 6H_2O$, 1.8 mM $CaCl_2 \cdot 2H_2O$, and 5.55 mM glucose.
3. Sodium phosphate buffer: 2.26 mM $NaH_2PO_4 \cdot H_2O$ and 8.43 mM $Na_2HPO_4 \cdot 7H_2O$, pH 7.4.
4. UCH-L1 is a commercially available DUB from the ubiquitin C-terminal hydrolase (UCH) family (*see* **Note 1**).
5. 7-Amino-4-methylcoumarin (AMC).
6. Ub-AMC (commercially available DUB probe) (*see* **Note 2**).
7. Z-LRGG-AMC (commercially available DUB probe).
8. DUB inhibitor PR-619 (*see* **Note 3**).
9. Dimethyl sulfoxide (DMSO). Store at room temperature.

2.2 Cell Culture and Lysate Generation

1. HeLa cells (*see* **Note 4**).
2. OPM.2 cells (*see* **Note 5**).
3. HeLa culture medium: 500 mL Dulbecco's modified eagle medium (DMEM), supplemented with 10% v/v fetal bovine serum (FBS). HeLa cells should be maintained at 37 °C with 5% humidified CO_2 and subcultured every 2–3 days depending on the cells' growth rate and confluency level.
4. OPM.2 culture medium: 500 mL RPMI 1640 medium, supplemented with 12% fetal bovine serum (FBS), 21.8 mM glucose, 8.6 mM HEPES (pH 7.4), and 1.0 mM sodium pyruvate. OPM.2 cells should be maintained at 37 °C with 5% humidified CO_2 and subcultured every 2–3 days depending on the cells' growth rate and confluency level.
5. Mammalian protein extract reagent (M-PER). Store at room temperature.
6. Trypsin. Store at 4 °C for short-term and at −20 °C for long-term storage.
7. 1X Phosphate buffer saline (PBS).
8. T-75 and T-175 cell culture flasks.
9. An incubator that can maintain 5% humidified CO_2 atmosphere and 37 °C temperature.
10. 15 and 50 mL canonical tubes.
11. 1.5 mL microcentrifuge tubes.

3 Methods

3.1 Measure DUB Activity in a Single-Enzyme Reaction (See Note 6) (Fig. 1)

3.1.1 DUB Probe Preparation

1. Prepare a stock concentration of 2.2 mg/mL Ub-AMC by adding 1 mL DMSO to 2.2 mg of Ub-AMC powder. This dilution will yield a stock concentration of 250 μM.
2. Prepare five secondary stock solutions of Ub-AMC at concentrations of 0.5, 2.5, 5, and 25 μM by serial dilution from the 250 μM stock solution in assay buffer (*see* **Note 7**).
3. Prepare stock UCH-L1 solution of 1000 nM in DMSO (*see* **Note 8**).

3.1.2 Fluorometric Analysis of DUB Activity

1. Add 70 μL assay buffer to each of the reaction wells A1–A12 (*see* Fig. 1 for plate layout for this step and subsequent steps) and 80 μL of assay buffer to reaction wells B1–B12 (*see* **Note 9**).
2. Add 10 μL of 1000 nM UCH-L1 to each of the reaction wells A1–A12. This distribution yields a final concentration of 100 nM UCH-L1 in the total volume of 100 μL per reaction well (*see* **Note 10**).

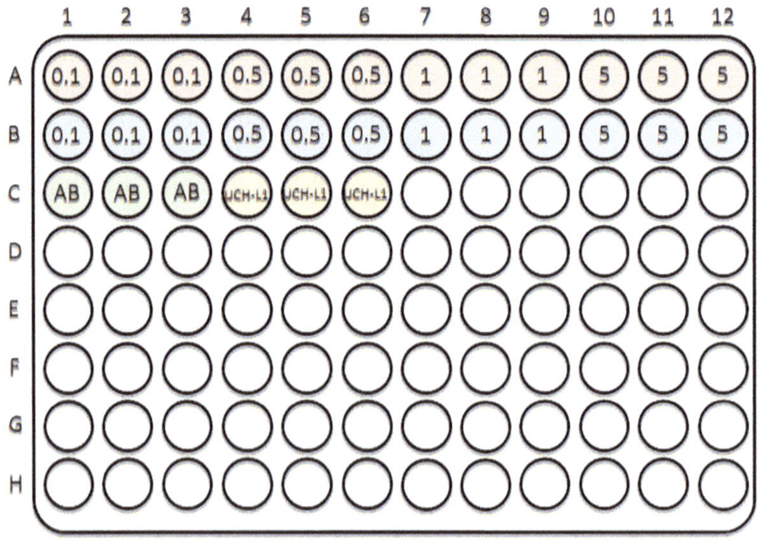

Fig. 1 Sample layout of a 96-well plate to study single-enzyme reaction kinetics for 100 nM UCH-L1 using four different concentrations of Ub-AMC. Reaction wells in row A (wells A1-A12) are designated for the DUB activity experiment (Ub-AMC + UCH-L1), while row B (wells B1-B12) serves as the probe-only background control (Ub-AMC). Numbers inside the reaction wells (rows A and B) indicate Ub-AMC concentration in the total volume of 100 μL per well. Reaction wells C1-C3, and C4-C6 consist of 100 μM of assay buffer and 100 μM of 100 nM UCH-L1 serving as background controls. Each reaction condition is performed in triplicate to allow for statistical analyses

3. Add 20 μL of the respective Ub-AMC secondary stock solutions to all reaction wells in rows A and B (*see* **Note 7**).
4. Add 100 μL of assay buffer to wells C1–C3, and add 10 μL of 1000 nM UCH-L1 and 90 μL of assay buffer to wells C4–C6 for background controls (*see* **Note 11**).
5. Once the 96-well plate is ready, the 6-h course of the experiment starts. Take measurements of the fluorescent signal at λ_{ex}=341 and λ_{em}=441 for AMC using the plate reader every 60 min (*see* **Notes 12** and **13**).

3.2 Measure DUB Activity in Cellular Lysates (See Note 14) (Fig. 2).

3.2.1 Cell Lysate Generation (See **Note 15**)

1. Culture four T-175 flasks of OPM.2 cells at a seeding density of 3 × 10^5 cells/mL 3 days prior to the experiment (*see* **Note 16**).
2. On the experiment day, transfer the contents of each cell culture flask (cells and growth media) to a separate 50 mL centrifuge tube (for a total of 4 tubes), and spin the cells down at 1800 × *g* for 2.5 min at room temperature.
3. Carefully remove the supernatant from each tube without disturbing the cell pellet, and serially resuspend each pellet with 10 mL of 1X PBS, combining the contents from the four flasks into a single 50 mL tube. Centrifuge at 1800 × *g* for 2.5 min (*see* **Note 17**).

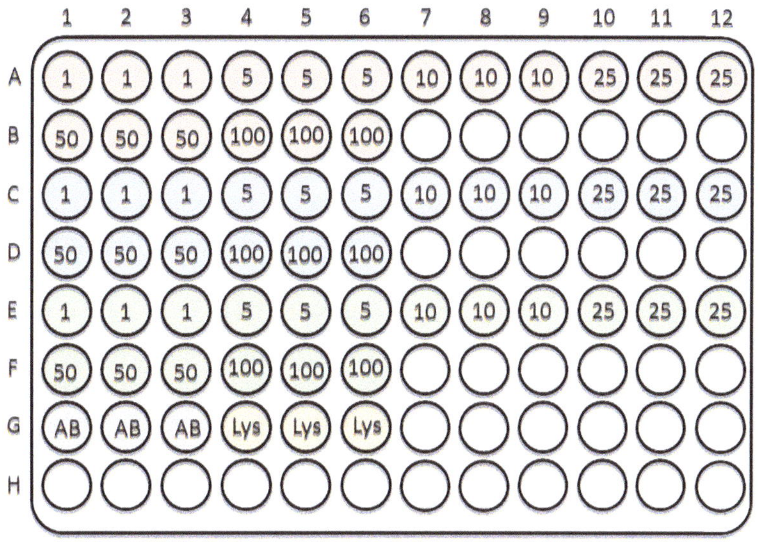

Fig. 2 Sample layout of a 96-well plate to study DUB activity kinetics in cell lysates using a peptide-based probe (Z-LRGG-AMC). Reaction wells A1-B6 are designated for the DUB activity experiment (lysates + DUB probe), while wells C1-D6 are for the inhibited DUB experiment (lysates + DUB probe + 50 μM DUB inhibitor PR-619). Reaction wells E1-F6 serve as the probe-only background control. Numbers inside the reaction wells indicate probe concentration in the total volume of 100 μL per well. Reaction wells G1-G3 contain assay buffer background control, and G4-G6 consist of lysate-only background control. Each reaction condition is performed in triplicate to allow for statistical analyses

4. Carefully remove the PBS without disturbing the cell pellet, and resuspend the pellet in an approximately equivalent volume of M-PER (*see* **Note 18**). This step results in a total volume of 2.4 mL (2400 μL) of cells in M-PER solution.
5. Transfer 800 μL of the cell suspension to four 1.5 mL microcentrifuge tubes (*see* **Note 19**), and vortex the tubes for 10 min at room temperature.
6. Following the vortex, centrifuge the tubes at 14,000 × g for 15 min at 4 °C.
7. Following the centrifuge step, the resulting soft agglomerate observed in the tubes is the cellular membrane and other cellular debris. Carefully remove this agglomerate from each tube, and mix all the remaining protein contents together.
8. Quantify the total protein concentration in the harvested cellular lysates (*see* **Note 20**).

3.2.2 DUB Probe Preparation

1. Prepare a master stock concentration of 1 mg/mL Z-LRGG-AMC by adding 1 mL of DMSO to 1 mg of Z-LRGG-AMC peptide powder (*see* **Note 21**). This mixture will yield a stock concentration of 1443.5 μM.
2. Prepare six secondary stock solutions of Z-LRGG-AMC at concentrations of 5, 25, 50, 125, 250, and 500 μM, by serial dilution from the 1443.5 μM stock solution in assay buffer (*see* **Note 22**).
3. Prepare a secondary stock DUB inhibitor PR-619 solution in DMSO at an experiment-specific concentration of 500 μM (*see* **Note 23**).

3.2.3 Fluorometric Analysis of DUB Activity

1. Based on the OPM.2 lysate concentration, calculate how much of this stock solution is needed for a final reaction volume of 100 μL to yield a final lysate concentration of 4 mg/mL per well (*see* **Note 24**).
2. Add 45 μL of assay buffer to each of the reaction wells A1–B6 and C1–D6 (*see* Fig. 2 for plate layout for this and the subsequent steps) (*see* **Note 25**).
3. Add 25 μL of 16 mg/mL stock OPM.2 lysates to each reaction well mentioned in **step 2**.
4. Add 80 μL of assay buffer to wells E1–F6.
5. Add 10 μL of DMSO to wells A1–B6 and 10 μL of 500 μM stock PR-619 to wells C1–D6 (*see* **Note 26**).
6. Incubate the plate in a 30 °C incubator for 60 min (*see* **Note 27**).

7. Add 20 μL of the respective Z-LRGG-AMC secondary stock solutions to all reaction wells from the prior steps (including DUB activity, inhibited DUB, and background control experiments; *see* **Note 22**).
8. Add 100 μL of assay buffer to wells G1–G3, and add 75 μL of assay buffer and 25 μL of stock lysates solution to wells G4–G6 for background controls (*see* **Note 11**).
9. Begin the 6-h experimental time course. Take measurements of the fluorescent signal at λ_{ex}=341 and λ_{em}=441 for AMC using the plate reader every 60 min (*see* **Notes 12, 13,** and **28**).

3.3 Measure DUB Activity in Live Intact Cells (See Note 29) (Fig. 3)

3.3.1 Cell Solution Preparation (See Note 30)

1. Culture ten T-75 flasks of HeLa cells to achieve ~90% confluent cells after 3 days of subculture.
2. On the day of experiment, carefully remove the culture media, and wash the cells with 7 mL of 1X PBS per flask.
3. Carefully remove the 1X PBS, and add 2 mL of trypsin to each flask to detach cells from the flask's inner surface (*see* **Note 30**). Incubate the flask in the 37 °C, 5% CO_2 incubator for 4 min.
4. Add 8 mL of fresh culture media to each flask bringing the total volume to 10 mL per flask (*see* **Note 31**).

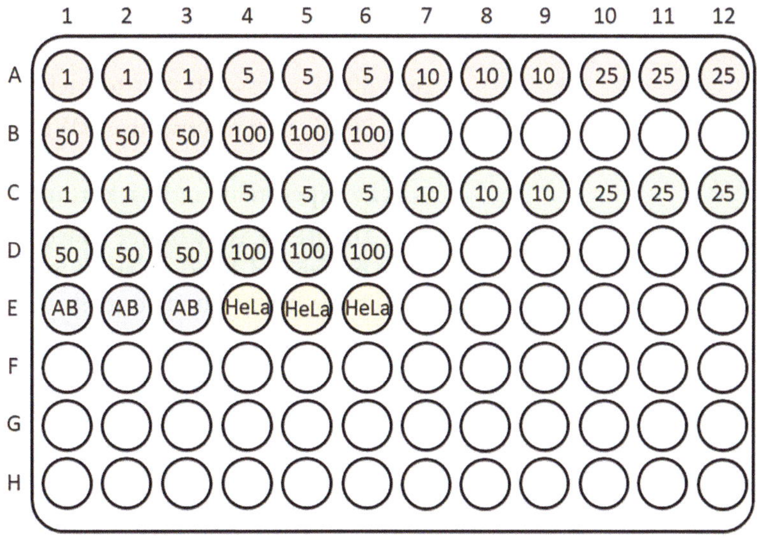

Fig. 3 Sample layout of a 96-well plate to study DUB activity in live intact HeLa cells using a cell-permeable, peptide-based probe (RWRWR-LRGG-AFC). Reaction wells A1-B6 are designated for the DUB activity experiment (HeLa cells + DUB probe), while wells C1-D6 are for the probe-only background control. Numbers inside the reaction wells indicate probe concentration in the total volume of 100 μL per well. Reaction wells E1-E3 contain assay buffer background control, and E4-E6 consist of live cell-only background control. Each reaction condition is performed in triplicate to allow for statistical analyses

5. Transfer the contents of each culture flask to a separate 50 mL centrifuge tube (10 total tubes for 10 flasks), and spin the cells down at 1800 × *g* for 2.5 min.
6. Carefully remove the supernatant without disturbing the cell pellet, and serially resuspend the cell pellets in a total volume of 10 mL 1X PBS. Centrifuge the suspension at 1800 × g for 2.5 min (*see* **Note 17**).
7. Carefully remove the PBS without disturbing the cell pellet, resuspend the pellet in 10 mL of ECB, and calculate the cell density. The desired final density of cells of this solution is 10^5 cells/mL (*see* **Note 32**).

3.3.2 DUB Probe Preparation

1. Reconstitute 1 mg of the peptide probe, RWRWR-LRGG-AFC, in 1 mL of sodium phosphate buffer, and quantify the total peptide concentration (*see* **Note 33**).
2. Prepare six secondary stock solutions of RWRWR-LRGG-AFC at concentrations of 5, 25, 50, 125, 250, and 500 μM, by serial dilution from the master stock solution in assay buffer (*see* **Note 34**).

3.3.3 Fluorometric Analysis of DUB Activity

1. Add 70 μL of assay buffer to each reaction well A1–B6 (*see* Fig. 3 for plate layout for this and the subsequent steps).
2. Add 10 μL of the HeLa cell suspension (10^5 cells/mL) to each of the wells mentioned in **step 1** (A1–B6).
3. Add 80 μL of assay buffer to each of the wells C1–D6.
4. Add 20 μL of the respective RWRWR-LRGG-AFC secondary stock solutions to each of the wells from prior steps (DUB activity experiment and background control) (*see* **Note 34**).
5. Add 100 μL of assay buffer to wells E1–E3, and add 90 μL of assay buffer and 10 μL of stock live HeLa cell solution to wells E4–E6 for background controls (*see* **Note 11**).
6. Begin the 6-h course of the experiment. Take measurements of the fluorescent signal at λ_{ex}=376 and λ_{em}=482 for AFC using the plate reader every 60 min (*see* **Notes 12, 13,** and **35**).

3.4 Fluorophore Standard Curve (See Note 36) (Fig. 4)

1. Prepare a stock concentration of 1 mg/mL AMC by adding 1 mL of DMSO to 1 mg of AMC. This dilution will yield a final concentration of 5708.414 μM (*see* **Note 37**).
2. Perform serial dilution of the stock in assay buffer to generate two secondary AMC stock solutions of 400 and 10 μM. This protocol is designed to measure the fluorescent signal for an AMC concentration range of 0.1–100 μM. The 10 μM secondary stock solution will be used for fluorophore concentrations below 10 μM, while the 400 μM stock will be used for concentrations of 10 μM and above (*see* **Notes 38**).

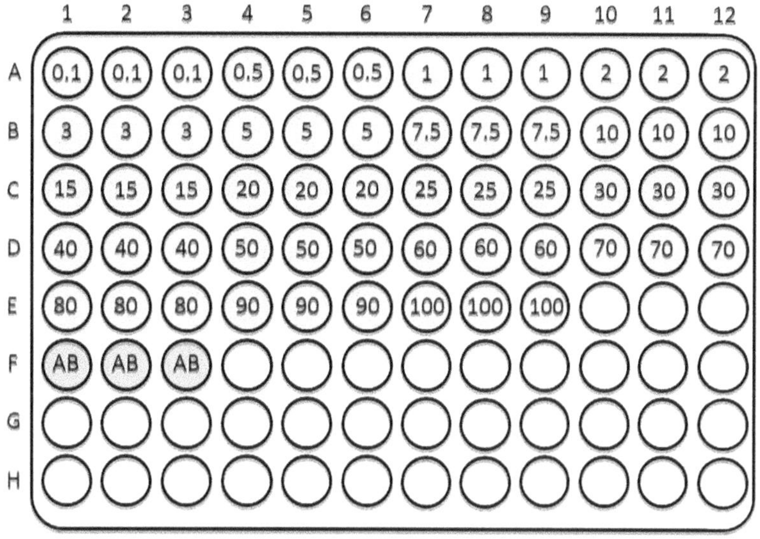

Fig. 4 Sample layout of a 96-well plate, designed to generate an AMC standard curve. The numbers inside the reaction wells in rows A-E (wells A1-E9) indicate the total concentration of AMC in each well in a total reaction volume of 100 μL. Wells F1-F3 labeled "AB" contain 100 μL of assay buffer used as a background control. Each sample is performed in triplicate to allow for statistical analyses

3. Add the calculated volume of assay buffer to each of the wells A1–E9.
4. Add the calculated volumes of AMC from the 400 and 10 μM stock solutions to each of the wells A1–E9.
5. Add 100 μL of assay buffer to wells F1–F3 for the background control.
6. Perform the measurements using a plate reader at λ_{ex}=341 and λ_{em}=441 for AMC using the plate reader every 60 min (*see* **Notes 12, 39,** and **40**).

4 Notes

1. Store the powder and stock solutions reconstituted in DMSO at −20 °C for the following reagents: UCH-L1, AMC, Z-LRGG-AMC, Ub-AMC, and PR-619. UCH-L1 is a model DUB used in the protocols introduced in this chapter. The user can replace this enzyme with any other DUB of interest.
2. Ub-AMC is an example of a DUB probe compatible with single-enzyme and lysate-based studies. Any other probe of similar compatibility can be used interchangeably in the protocols.
3. PR-619 is an example DUB inhibitor used in this chapter usually prepared as a concentrated stock of 500 μM. Any other DUB inhibitor of interest can be used interchangeably.

4. HeLa cells are used as a model adherent cell line for these protocols. The reader can use any other adherent cell line; however, they are encouraged to review the vendor's (e.g., ATCC) proper media formulations and culturing procedures.
5. OPM.2 cells are used as a model suspension cell line for these protocols. The reader can use any other suspension cell line; however, they are encouraged to review the vendor's proper media formulations and culturing procedures.
6. This protocol is designed for the Ub-AMC (DUB probe) at concentrations of 0.1, 0.5, 1, and 5 μM to be incubated with 100 nM UCH-L1 pure enzyme. The layout of the 96-well plate for this experiment is presented in Fig. 1. Probe and pure enzyme concentrations can be changed based on the reader's need. This experiment can be performed using other DUB probes and including the other two used in this chapter, commercially available Z-LRGG-AMC and cell-permeable RWRWR-LRGG-AFC. However, due to the lower sensitivity of the peptide-based probes (Z-LRGG-AMC and RWRWR-LRGG-AFC), the enzymatic reaction cleaving the fluorophore would require a substantially higher concentration of the target DUB. For a single-enzyme reaction, the Ub conjugated probes are recommended. The choice of DUB can be different as well, including other DUB family members like USP2.
7. The five secondary stock solutions are prepared by serial dilution of the 250 μM stock solution such that the 250 μM Ub-AMC solution is used to make the 25 μM secondary stock. Then the 5 μM secondary stock is prepared by diluting a portion of the 25 μM stock in assay buffer. The purpose of preparing the five stock solutions is to optimize the amount of probe used and make the experimental steps of adding reagents to the 96-well plate more convenient, as equal volumes of respective secondary stock solutions (20 μL) will be needed for each reaction well. For this protocol, 20 μL of the 0.5 μM stock is required for reaction wells A1–A3 (DUB activity experiment) and B1–B3 (probe-only control experiment) that are designated for a final Ub-AMC concentration of 0.1 μM. Similarly, 20 μL of the 25 μM secondary stock is required for reaction wells A9–A12 (DUB activity experiment) and B9–B12 (probe-only control experiment) that are designated for a final Ub-AMC concentration of 5 μM. Pipetting the same volume of the DUB probe reduces the potential for user error. The user can adjust the stock concentration values as needed depending on the range of interest.
8. Commercially available purified enzymes are sold as a powder. Once the powder is reconstituted in the proper solvent, it yields a highly concentrated stock solution that needs to be diluted prior to addition to the reaction wells. A starting UCH-L1

concentration of 40 μM can be diluted to a secondary stock concentration of 1000 nM by adding 25 μL of the 40 μM stock to 965 μL of assay buffer and 10 μL of 1M DTT for a total volume of 1000 μL. The presence of 10 mM DTT in the reaction mixture is recommended to facilitate DUB activity.

9. The addition of buffer is required to bring the total volume of each reaction well to 100 μL. These volumes can be subject to change depending on the calculations performed in the prior steps (namely, changes in probe concentration/volume per well and enzyme concentration/volume per well). A minimum volume of 100 μL is required to obtain a sufficient signal using the plate reader.

10. The volume of enzyme used per well can be subject to change as well, depending on the concentration range of interest to the user. Different stock enzyme concentrations would also require different volumes added to each well. Moreover, the same layout of wells (rows A and B, Fig. 1) can be repeated on the 96-well plate for another enzyme if the user aims to study multiple enzymes using single or multiple DUB probes simultaneously.

11. The background control reaction wells, including only assay buffer, only DUB, only lysates, or only intact cells, are designed to determine the background fluorescence intensity of the reaction components. The observed fluorescence intensity using the probe can then be normalized against these background values as needed.

12. Some instances occur where the exact excitation and emission wavelengths for the fluorogenic molecules are not available in the plate reader the user has access to equipment. However, in these experiments and for every fluorophore, there are filter sets with an optimally close set of excitation and emission values available that can be used. For example, in the case of AMC, the exact wavelengths for excitation and emission are 341 nm and 441 nm, respectively. However, filter sets with these exact configurations are not available, and one with λ_{ex}= 355, and λ_{em}= 460 that shows to be well inside the AMC fluorescence spectrum and close enough to the peaks of both excitation and emission, can be used. The same approach of implementing a filter set with configurations closest to the precise *Ex/Em* values can also be applied to other fluorescent molecules.

13. In between measurements, cover the plate using an adhesive plastic cover, and wrap in aluminum foil to avoid any light exposure and place in a 30 °C incubator. Please note that the transfer of the plate to the incubator would not be necessary if the plate reader used is equipped with temperature control, in which case the 96-well plate can be kept inside of it throughout the course of the experiment.

14. This protocol is designed for the peptide-based commercially available DUB probe Z-LRGG-AMC (substrate) at concentrations of 1, 5, 10, 25, 50, and 100 μM using a cell lysate concentration of 4 mg/mL. For this protocol, OPM-2 cells (a model multiple myeloma cell line) were used as they have been shown to exhibit high DUB activity. The inhibited study is designed for 50 μM concentration of DUB inhibitor PR-619; however, other inhibitors at other concentrations can be used. All quantities mentioned above, including DUB probe concentration, lysate concentration, and DUB inhibitor concentration, can be subject to change based on the reader's need and the 96-well plate. The use of a spreadsheet is recommended to keep track of the concentrations and volumes to organize the plate. The 96-well plate layout presented in Fig. 2 will need to be adjusted if other reaction conditions are used. Moreover, the DUB probe Z-LRGG-AMC can be replaced with any other DUB probe. Any other cell line (adherent or non-adherent) can replace the OPM.2 line used in this protocol.
15. This protocol is designed for suspension cell lines and should be adjusted accordingly if an adherent cell line is used. Moreover, different cell lines yield higher or lower amounts of total protein. It is always safer to culture a larger number of flasks to ensure enough lysates for the experiment.
16. For the protocol designed here, with the sample 96-well plate layout presented in Fig. 2, a total of 4, T-175 culture flasks of OPM.2 cells at a seeding density of 3×10^5 cells/mL is recommended to generate sufficient amounts of lysate for the experiment. However, the number of flasks, flask size, and seeding density can all be changed based on user needs. Adjustments to the other steps of the protocol directly related to these variables (e.g., amount of M-PER needed in **step 5, Note 16**) should be made accordingly. It is always recommended to prepare 20% more lysates than are needed to avoid running out during the experiment.
17. This step is performed by serial resuspension of the cells using a total volume of 10 mL 1X PBS. The dilution series is accomplished by adding 10 mL to the first tube, resuspending the cell pellet, transferring the suspension to the next tube, and repeating this for the remaining three tubes.
18. A recommended rule of thumb for deciding the volume is using 200 and 600 μL M-PER per T-75 and T-175 flask of cells used, respectively. This protocol uses M-PER to lyse the cells; however, the user may use other methods of lysate generation depending on preference.
19. Using the recommended small volumes (<1000 μL per tube) for the vortex and centrifuge steps makes removing the solid cellular components easier.

20. A spectrophotometer such as NanoDrop (Thermo Scientific, Madison, WI) can be used to quantify lysate concentration. An alternative approach, also widely used, is the Bradford assay, a Coomassie dye-binding assay performed at room temperature.
21. The peptide-based DUB probe is sold as a small amount of powder (1 mg) making it difficult to accurately transfer small volumes without a loss of powder. It is recommended to reconstitute the entirety of the peptide-based probe in the vessel it was shipped in by adding 1 mL of DMSO to a glass bottle. Aliquoting the probe into separate 1.5 mL microcentrifuge tubes is also recommended to avoid loss of efficacy due to repeated freeze/thaw. Since reconstitution of probe powder in a solvent at 1 mg/mL yields a relatively high concentration of 1443.5 μM, it is recommended to aliquot the resulting solution at 150–200 μL/tube for storage purposes.
22. The six secondary stock solutions are prepared by serial dilution of the 1443.5 μM stock solution such that the 1443.5 μM Z-LRGG-AMC solution is used to make the 500 μM secondary stock. Then the 250 μM secondary stock is prepared by diluting a portion of the 500 μM stock in assay buffer. The purpose of preparing the six stock solutions is to optimize the amount of probe used and make the experimental steps of adding reagents to the 96-well plate more convenient, as equal volumes of respective secondary stock solutions (20 μL) will be needed for each reaction well. For this protocol, 20 μL of the 250 μM stock is required for reaction wells B1–B3 (DUB activity experiment), D1–D3 (inhibited DUB experiment), and E1–E3 (probe-only background control) that are designated for a final Z-LRGG-AMC concentration of 50 μM.
23. The concentrated stock PR-619 of 500 μM is recommended for the protocol described here to achieve the final concentration of 50 μM per reaction well. This recommended stock concentration is used to avoid the need to pipette small volumes (e.g., <10 μL) to reduce potential errors. The user can adjust both the final (per well) and stock concentrations based on need.
24. This protocol and the amounts of lysates required per well are based on the hypothetical stock lysate concentration of 16 mg/mL. However, the volume of lysates needed for each reaction well can and will be subject to change depending on the concentration of stock cellular lysates. A sample containing 25 μL of lysates must be added to each reaction well for a hypothetical stock lysate concentration of 16 mg/mL to yield a final 4 mg/mL concentration. This protocol is designed based on this hypothetical stock lysate concentration. Note that changes in the lysate volume are offset by assay buffer changes to maintain a final volume of 100 μL.

25. The volume of buffer needed per reaction well is decided at the final calculations and protocol design step. For the protocol designed and explained in this chapter, knowing that 20 μL of respective stock probe solutions, 10 μL of DMSO (for DUB activity experiment) or PR-619 (for inhibited DUB experiment), and 25 μL of 16 mg/mL stock OPM.2 lysates (hypothetical, can be subject to change; *see* **Note 20**) is required per reaction well, 45 μL will be needed to bring up the total reaction volume of each well to 100 μL. The volume of the assay has to be adjusted should any of the concentrations mentioned above (probe, inhibitor, or lysates) be subject to change.
26. Since DMSO is used as a solvent for PR-619, DMSO is added to the reaction wells as vehicle control in the DUB activity experiment at a volume equal to that of PR-619 in inhibited DUB experiment.
27. This incubation period of OPM.2 lysates with the DUB inhibitor PR-619 is designed to allow for successful inhibition of DUBs before adding the probe to the reaction mix so that a significant decrease in DUB activity can be observed.
28. Sample results from analysis of DUB activity using Z-LRGG-AMC in OPM.2 lysates, performed by Safa et al. [8], are presented in Fig. 5. Briefly, the fluorescent intensities are measured throughout the experiment, which is then converted

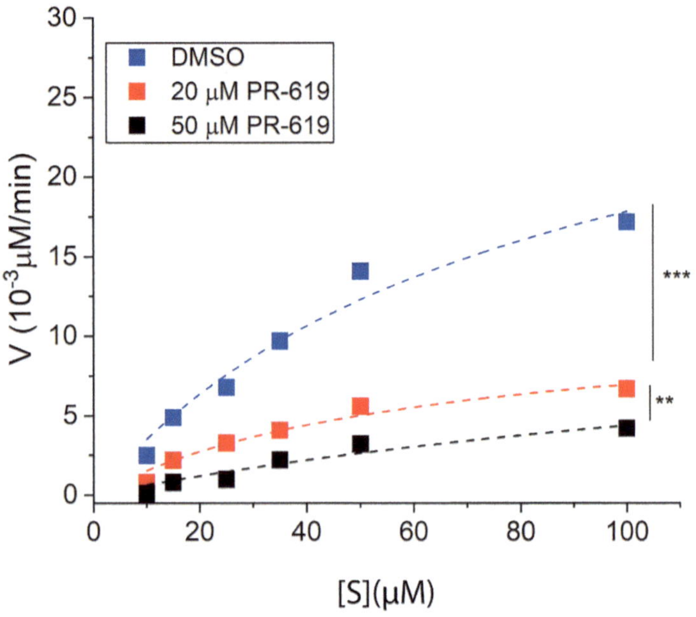

Fig. 5 Kinetic analysis of DUB-mediated hydrolysis of a commercial peptide-based DUB probe (Z-LRGG-AMC) in OPM.2 lysates. Uninhibited (DMSO control) and inhibited (PR-619) DUB activity was measured via fluorometry. (Reproduced from [8] by permission of Elsevier)

to free fluorophore concentration using a standard curve (*see* Subheading 3.4 for methods). The concentration of free fluorophore is proportional to the concentration of DUB probe reacted with enzyme allowing for a direct metric of the enzymatic reaction. The slope of the free fluorophore concentration (*y*-axis) vs. time (*x*) achieved during the time course can be used to approximate the reaction rate (ν) at each concentration of the probe used in the experiment. Figure 5 presents the reaction rate (ν) vs. probe (substrate) concentration ([S]) for uninhibited DUB activity and inhibited DUB activity at two different concentrations of the DUB inhibitor PR-619 (20 and 50 μM). For more information on further analysis steps, (*see* **Note 39**).

29. This protocol is designed for the cell-permeable DUB probe RWRWR-LRGG-AFC (substrate) introduced by Safa et al. [8] at concentrations of 1, 5, 10, 25, 50, 75, and 100 μM, using a live cell density of 10^4 cells/mL. HeLa cells (a model adherent, cervical cancer cell line) were used for this protocol. All of the quantities mentioned above, including probe concentrations and cell density, can be subject to change based on the reader's need and the 96-well plate. Using a spreadsheet is recommended to keep track of the concentrations and volumes to organize the plate. The 96-well plate layout presented in Fig. 3 will need to be adjusted if other reaction conditions are used. Although several cell-permeable probes that implement different approaches in design and functionality have been and are currently being developed [12], no cell-permeable DUB probe is commercially available.

30. This section of the protocol is designed for adherent cell lines and should be adjusted accordingly if a suspension cell line is used. Moreover, different cell lines yield higher or lower cell densities, and the recommended number of culture flasks in this protocol is specific to HeLa cells and the designed experiment and quantities required in this protocol. All the volumes of reagents required in the cell preparation protocol (e.g., PBS, trypsin) assume the use of T-75 cell culture flasks; however, these numbers can be subject to change, should any other flask size be used.

31. At this step, ensure total detachment of the cells from the flask by repeatedly pipetting the suspension and flowing the media across the culture surface in the flask.

32. As mentioned before, the desired final cell density in the stock solution is 10^5 cells/mL. Hence, depending on the initial density obtained (in the 10 mL cell in ECB solution), either dilute (add more ECB) or concentrate (centrifuge and resuspended the cell pellet in a smaller volume of ECB) the solution to achieve the final cell density of 10^5 cells/mL. This cell

density is recommended for the specific cell line (HeLa) and probe concentration range used in the design of this protocol; hence, it can be subject to change based on reader's need.

33. The reconstitution of the custom peptide-based probe at 1 mg/1 mL of sodium phosphate buffer is recommended to get desirable stock concentration (e.g., greater than or at least equal to the highest secondary stock solution concentration of 500 μM; *see* **Note 32**). Sodium phosphate buffer is used as the solvent for this probe because DMSO can inhibit probe uptake in intact cells. The yields of the peptide-based probe can vary since the probe is custom-made, with the final stock concentration being determined by spectrophotometry (*see* [8] for additional information and instructions). The stock peptide concentration can be quantified by measuring the absorbance of the two tryptophan residues (W) on the β-hairpin sequence motif (RWVRVpGRWIRQ) 280 nm. This absorbance can be used to calculate the final stock peptide concentration via Beer's law. Dilution of a separate small volume of the reconstituted peptide-based probe is recommended (just enough to perform the spectrophotometry) in 6M guanidium hydrochloride at a 1:1 ratio. This dilution step is required to open the β-hairpin sequence. Note that the dilution factor should be considered in calculating the final peptide concentration. Moreover, the final value should be adjusted to account for the observed absorbance results from the two residues being measured.
34. The six secondary stock solutions are prepared by serial dilution of the master stock solution, assuming its concentration is greater than 500 μM, the highest secondary stock concentration. For example, a master stock concentration of 600 μM can be used to make the 500 μM secondary stock for the subsequent serial dilutions. The 250 μM secondary stock is prepared by diluting a portion of the 500 μM stock in assay buffer. The purpose of preparing the six stock solutions is to optimize the amount of peptide used and make the experimental steps of adding reagents to the 96-well plate more convenient, as equal volumes of respective secondary stock solutions (20 μL) will be needed for each reaction well. For this protocol, 20 μL of the 5 μM stock is required for reaction wells A1–A3 (DUB activity experiment) and C1–C3 (probe-only background control), which are designated for a final RWRWR-LRGG-AFC concentration of 1 μM.
35. Sample results from analysis of DUB activity using RWRWR-LRGG-AFC in live intact HeLa cells, performed by Safa et al. [8], are presented in Fig. 6. Briefly, the fluorescent intensities are measured throughout the experiment, which is then converted to free fluorophore concentration using a standard

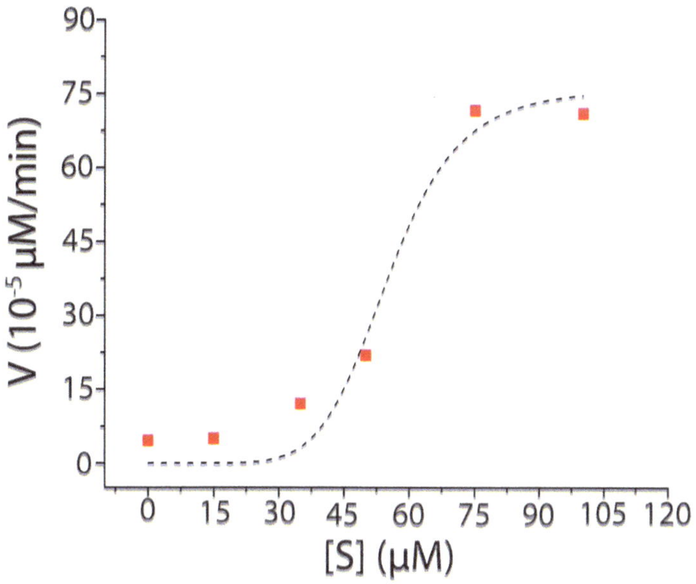

Fig. 6 Kinetic analysis of DUB-mediated hydrolysis of a cell-permeable peptide-based DUB probe (RWRWR-LRGG-AFC) in live intact HeLa cells. Uninhibited DUB activity was measured via fluorometry. (Reproduced from [8] by permission of Elsevier)

curve (*see* Subheading 3.4). The concentration of free fluorophore is proportional to the concentration of DUB probe reacted with enzyme allowing for a direct metric of the enzymatic reaction. The slope of the free fluorophore concentration (*y*-axis) vs. time (*x*) achieved during the time course can be used to approximate the reaction rate (ν) at each concentration of the probe used in the experiment. Figure 6 presents the reaction rate (ν) vs. probe (substrate) concentration ([S]) for uninhibited DUB activity. For more information on further analysis steps, (*see* **Note 39**).

36. A standard curve for any fluorophore used in the structure of the DUB probe is required to convert the measured fluorescent intensity to the concentration of free fluorophore in the solution. This concentration is significant as it is indicative of the enzymatic reaction and can be used in quantitative analyses of DUB activity kinetics. This standard curve is generated by measuring the fluorescent signal received from known concentrations of a free fluorophore. The provided protocol is designed to measure the fluorescent signal received from free (e.g., non-conjugated) AMC at concentrations in the range of 0.1–100 μM. The concentration range and the sample layout of the 96-well plate presented in Fig. 4 can be subject to change based on the user's needs.

37. The stock concentration is calculated from the molecular weight of AMC. This stock concentration can be adjusted according to the user's needs.

38. Using two diluted secondary stock solutions makes things more convenient when working with a wide range of concentration as the user can avoid working with small (less than 0.5 μL) volumes that can lead to pipetting error. For example, to make the solution required for each of the reaction wells A1–A3 (0.1 μM AMC), the user would need to add 0.025 μL of a 400 μM stock solution to 99.975 μL of assay buffer (for a total volume of 100 μL). Performing all necessary calculations on a spreadsheet is recommended to determine how much assay buffer and stock solution are needed for each fluorophore concentration to maintain a constant volume of 100 μL per reaction well. Below is an example calculation for the protocol presented here. For example, for the 25 μM concentration (reaction wells C7–C9):

    1. $(400\ \mu M)(V_1) = (25\ \mu M)(100\ \mu L)$
    2. $V_1 = 6.25\ \mu L$

    Add 6.25 μL of the 400 μM AMC stock solution to 93.75 μL of assay buffer for a total volume of 100 μL in the reaction well to obtain a final concentration of 25 μM that is desired.

39. This step is to assure the time independence of the measurements as the received signal for each concentration should stay constant throughout the course of the experiment. After the experiment, the user should plot the fluorescence intensity vs. fluorophore concentration for each time point. If the data points for all time points align, that means there is no decay in the fluorescent signal and that the standard curve can be used for experimentations. If a time-dependent increase or decrease in signal is observed for any of the concentrations, the experiment should be repeated. This potential variation in observed signal with respect to time usually occurs due to machine or user errors. An example standard curve for AMC is presented in Fig. 7.

40. A variety of approaches are implemented in analyzing fluorometric data from enzyme reactions. In comparing different probes in characterizing a single-enzyme, or comparing different enzymes' activity using a single probe, plots of raw fluorescent intensity vs. time can be informative. Moreover, the effects of different inhibitor levels on enzyme activity can be investigated using the same approach. Based on the user's approach, the raw signal can also be normalized against the background control signal (*see* **Note 11**) either by subtraction or division. The normalization step is only necessary in the instances where

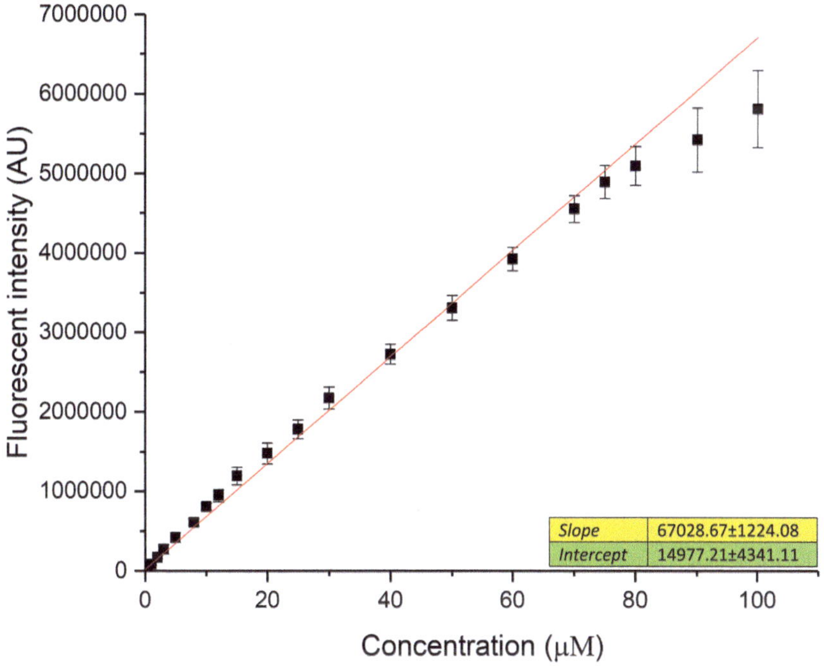

Fig. 7 Sample calibration curve generated for pure AMC at concentrations ranging from 0–100 μM. Linear regression was performed to generate the slope and intercept for the curve. Data is representative of triplicate experimentation

a high background signal is received. The same aforementioned plots can then be made using the normalized fluorescent intensities. Another method is demonstrating enzyme reaction rates (ν) for different probe (substrate) concentrations ([S]). This type of calculation requires converting fluorescent intensity (raw or normalized) to free fluorophore concentration, which is a direct correspondent of enzyme reaction levels. This conversion is done using standard fluorophore curves (*see* Subheading 3.4). Then, the reaction rates are obtained by finding the slope of the concentration vs. time plot for each probe concentration. The ν vs. [S] data can then be fitted to different enzyme reaction kinetics models such as Michaelis-Menten, Hill, and others. This approach can also be informative in studies such as the comparison of different probes' potency in the characterization of different enzymes.

References

1. Kessler BM, Edelmann MJ (2011) PTMs in conversation: activity and function of deubiquitinating enzymes regulated via post-translational modifications. Cell Biochem Biophys 60(1):21–38. https://doi.org/10.1007/s12013-011-9176-6
2. Hideshima T, Qi J, Paranal RM, Tang W, Greenberg E, West N, Colling ME, Estiu G, Mazitschek R, Perry JA, Ohguchi H, Cottini F, Mimura N, Görgün G, Tai Y-T, Richardson PG, Carrasco RD, Wiest O, Schreiber SL, Anderson KC, Bradner JE (2016) Discovery of selective small-molecule HDAC6 inhibitor for overcoming proteasome inhibitor resistance in multiple myeloma. Proc Natl Acad Sci 113(46):13162–13167. https://doi.org/10.1073/pnas.1608067113
3. Edelmann MJ, Nicholson B, Kessler BM (2011) Pharmacological targets in the ubiquitin system offer new ways of treating cancer, neurodegenerative disorders and infectious diseases. Expert Rev Mol Med 13:e35. https://doi.org/10.1017/S1462399411002031
4. Gupta N, Li W, McIntyre TM (2015) Deubiquitinases modulate platelet proteome ubiquitination, aggregation, and thrombosis. Arterioscler Thromb Vasc Biol 35(12):2657–2666. https://doi.org/10.1161/ATVBAHA.115.306054
5. Cohen P, Tcherpakov M (2010) Will the ubiquitin system furnish as many drug targets as protein kinases? Cell 143(5):686–693. https://doi.org/10.1016/j.cell.2010.11.016
6. Harrigan JA, Jacq X, Martin NM, Jackson SP (2018) Deubiquitylating enzymes and drug discovery: emerging opportunities. Nat Rev Drug Discov 17(1):57–78. https://doi.org/10.1038/nrd.2017.152
7. Dang LC, Melandri FD, Stein RL (1998) Kinetic and mechanistic studies on the hydrolysis of ubiquitin C-terminal 7-amido-4-methylcoumarin by deubiquitinating enzymes. Biochemistry 37(7):1868–1879. https://doi.org/10.1021/bi9723360
8. Safa N, Pettigrew JH, Gauthier TJ, Melvin AT (2019) Direct measurement of deubiquitinating enzyme activity in intact cells using a protease-resistant, cell-permeable, peptide-based reporter. Biochem Eng J 151:107320. https://doi.org/10.1016/j.bej.2019.107320
9. Hassiepen U, Eidhoff U, Meder G, Bulber J-F, Hein A, Bodendorf U, Lorthiois E, Martoglio B (2007) A sensitive fluorescence intensity assay for deubiquitinating proteases using ubiquitin-rhodamine110-glycine as substrate. Anal Biochem 371(2):201–207. https://doi.org/10.1016/j.ab.2007.07.034
10. Gui W, Ott CA, Yang K, Chung JS, Shen S, Zhuang Z (2018) Cell-permeable activity-based ubiquitin probes enable intracellular profiling of human deubiquitinases. J Am Chem Soc 140(39):12424–12433. https://doi.org/10.1021/jacs.8b05147
11. Safa N, Anderson JC, Vaithiyanathan M, Pettigrew JH, Pappas GA, Liu D, Gauthier TJ, Melvin AT (2019) CPProtectides: rapid uptake of well-folded β-hairpin peptides with enhanced resistance to intracellular degradation. Pept Sci 111(2):e24092. https://doi.org/10.1002/pep2.24092
12. Conole D, Mondal M, Majmudar JD, Tate EW (2019) Recent developments in cell permeable deubiquitinating enzyme activity-based probes. Front Chem 7:876. https://doi.org/10.3389/fchem.2019.00876

Chapter 4

Using Activity-Based Proteomics for the Quantification of Deubiquitinases in Animal Tissue

Bindu Nanduri, Leslie A. Shack, John Santelices, and Mariola J. Edelmann

Abstract

Ubiquitination is a post-translational modification, that regulates essential cellular functions, and the enzymes that control the removal of this modification, deubiquitinases (DUBs), have been well described for the model organisms. However, the information about DUBs is still largely lacking for the non-model organisms, such as agriculturally relevant animals. To understand the expression of these enzymes in animal tissues, we have used chemical proteomics which can be used to identify biologically active DUBs present in tissues based on their reactivity with the activity-based probes (ABPs). Here we describe a sample preparation protocol for ABP-based purification of DUBs from animal tissue using two approaches to homogenize and lyse the animal tissue compatible with ABP labeling of DUBs, including an ultrasonication-based tissue processing method and bead-beating method. Both of these methods retain the enzymatic activity of DUBs. In addition, we describe a protocol for ABP labeling of DUBs in tissue lysates and the immunoprecipitation of the probe-reactive DUBs that can be used along with mass spectrometric identification of proteins and the detection of these DUBs by Western blotting.

Key words Deubiquitinases, Mass spectrometry, Tissue, Proteomics

1 Introduction

Insights into the changes in post-translational modifications of proteins (PTMs) such as ubiquitination (Ub) can help understand genome regulation at the protein level. Deubiquitinases, DUBs, are involved in health and disease, including such diseases as malignancies, neurodegenerative disorders, and pathogenicity associated with microbial infections [1–5]. The model systems such as human, mouse, or rat have been well described in terms of the expression and activity of DUBs, where >90 DUBs have been identified in humans. However, for many other species of animals, DUBs have not been well evaluated. For example, while a similar number of DUBs is expected in chickens as in humans, only a few of these enzymes have been identified as catalytically active in this bird species, including UBP41 [6], UBP46, UBP52, UBP66 [7],

Julie A. Maupin-Furlow and Mariola J. Edelmann (eds.), *Deubiquitinases: Methods and Protocols*, Methods in Molecular Biology, vol. 2591, https://doi.org/10.1007/978-1-0716-2803-4_4,

UCH6 [8, 9], UCH8 [10], UCH-L5 [11], or USP1 [12–14]. DUBs could be used in agriculture, including poultry-based research, such as biomarkers, prognostic disease markers, or drug target discovery research. However, to understand ubiquitination dynamics in animal disease, we need to understand the expression of these enzymes in tissues. Although there are bioinformatics resources such as ubiquitin and ubiquitin-like conjugation database (UUCD) using hidden Markov models (HMM) [15] that can be used to identify putative DUBs in non-model organisms, these proteins remain as predictions unless experimental approaches are applied for validation of these predicted enzymatic functions. Moreover, DUBs from most livestock species are not yet represented in these databases. Another weakness of bioinformatic-based approaches is that the putative annotation might not be complete [16], and the DUBs also might not be biologically active or expressed in certain tissues and under specific conditions.

Hence, chemical proteomics approaches using activity-based probes (ABPs) can be invaluable to studying the enzymatic activity of proteins in tissues and cell culture. These approaches depend on the design of probes, often inhibitors or other compounds that can form covalent bonds with the enzyme of interest. Such ABPs designed for the DUBs usually consist of a ubiquitin molecule fused to the cysteine-reactive electrophilic group and often to a tag allowing for the purification or detection of any probe-reactive proteins [16, 17]. Examples of the electrophilic groups reactive with the cysteine residue or a warhead are propargylamide (PA) [18, 19], vinyl methyl ester (VME) [20], vinyl methyl sulfone (VS) [17], chloroethylamine [20], bromoethylamine [20], and others.

A chemical proteomics-based approach based on ABPs combined with mass spectrometry-based proteomics can be used to identify and semi-quantify DUBs in the cell lysates. However, this approach can also be altered for tissue homogenization, such as those from agriculturally relevant organisms. The protocols developed for cell lysates must be further amended for animal tissues to ensure that the DUBs retain activity in the homogenates, which is one of the requirements of this chemical proteomics approach. Using the ABPs, there will be some non-DUB proteins whose cysteine residues form the Ub-electrophile conjugates. Hence for the genuine discovery of new DUBs in non-model animal species by using this approach, bioinformatics can still be a helpful tool that should be used in combination with chemical proteomics [21].

Here, we describe a sample preparation protocol for ABP-based purification of DUBs from animal tissue using two approaches to homogenize and lyse the animal tissue compatible with downstream DUB labeling with Ub-VS or similar probes followed by immunoprecipitation and mass spectrometric identification of DUBs in tissue (Fig. 1). First, we will describe the ultrasonication-based sample preparation method that uses adaptive focused acoustics (AFA), which delivers specifically targeted

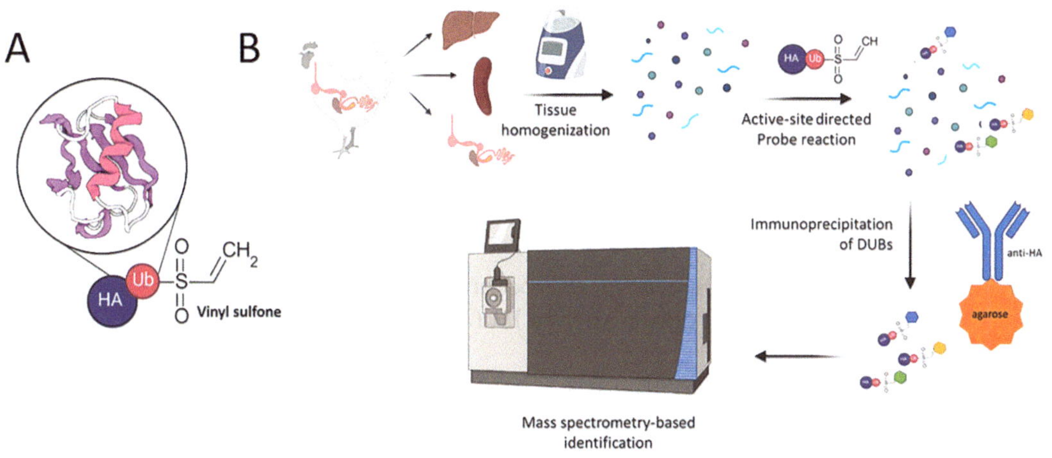

Fig. 1 An approach combining active site-directed probes and quantitative proteomics to identify novel DUBs in chicken tissues. (**a**) Ubiquitin vinyl sulfone HA (Ub-VS-HA) active site-directed probe used for chemical proteomics. The probe contains warhead (VS), ubiquitin molecule, and human influenza hemagglutinin (HA) tag. (**b**) Schematic of the chemical proteomics profiling. Each chicken was dissected, and the spleen, liver, and cecum were collected, homogenized, and lysed, followed by labeling with active site-directed probes (Ub-VS-HA) specific to deubiquitinases (DUBs). The probe-containing DUBs were purified by anti-HA immunoprecipitation. The DUBs were analyzed by mass spectrometry (Orbitrap Fusion). (Reprinted with permission from Nanduri et al. [21])

high-frequency energy into the sample [22, 23]. Second, we describe more traditional bead-beating technology. Both of these methods do not require harsh detergents and have a rapid processing time necessary for the retention of the enzymatic activity of DUBs [21, 23]. In addition, we describe the probe labeling of DUBs in tissue lysates, immunoprecipitation of the probe-reactive DUBs that can be used along with mass spectrometric identification of proteins, and the detection of these DUBs by Western blotting.

2 Materials

2.1 Tissue Sample Preparation by Ultrasonication (Alternative 1)

1. 15 mL tubes.
2. 1.5 mL tubes.
3. NP-40 lysis buffer: 0.5% NP-40, 150 mM NaCl, 20 mM CaCl·2H2O, 50 mM Tris, pH 7.4 supplemented with 1 mM phenylmethanesulfonyl fluoride (PMSF).
4. Ultrafree-MC 0.1 μm units (Millipore, USA).
5. Protein concentration kits such as Bradford-based protein assay kit or DC Protein Assay Kit.
6. Covaris tissue tube (Cat. # 520071, Covaris, Woburn, MA).
7. Covaris glass tube (Cat. #520031, Covaris, Woburn, MA).

8. Covaris cryoPREP Manual Dry Pulverizer (Covaris, Woburn, MA).
9. Covaris S220 instrument (Covaris, Woburn, MA).
10. Cuvette spectrophotometer, such as Evolution 160 UV-Vi (Thermo Scientific, Rockford, IL).
11. Benchtop centrifuge, refrigerated, such as the Sorvall ST40R (Thermo Scientific, Rockford, IL).
12. Benchtop microcentrifuge, refrigerated, such as the Sorvall Legend Micro 21R (Thermo Scientific, Rockford, IL).

2.2 Tissue Sample Preparation by Bead Beating (Alternative 2)

1. 15 mL tubes.
2. 1.5 mL tubes.
3. Matrix A beads (MP Biomedicals, Inc., Irvine, CA).
4. Dry ice.
5. NP-40 lysis buffer: 0.5% NP-40, 150 mM NaCl, 20 mM CaCl·2H2O, 50 mM Tris, pH 7.4 supplemented with 1 mM phenylmethanesulfonyl fluoride (PMSF).
6. Centrifugal filter unit: 10 kDa MWCO (15 mL sample or 0.5 mL sample).
7. Ultrafree-MC 0.1 μm units (Millipore, USA).
8. Protein concentration kit, such as Bradford-based protein assay kit or DC Protein Assay Kit.
9. Cuvette spectrophotometer, such as Evolution 160 UV-Vis (Thermo Scientific, Rockford, IL).
10. Benchtop centrifuge, refrigerated, such as the Sorvall ST40R (Thermo Scientific, Rockford, IL).
11. Benchtop microcentrifuge, refrigerated, such as the Sorvall Legend Micro 21R (Thermo Scientific, Rockford, IL).
12. FastPrep-24 instrument (MP Biomedicals, Inc., Irvine, CA).

2.3 Deubiquitination (DUB) Probe Assay

1. 1.5 mL tubes.
2. Reaction sample buffer: 0.15 M NaCl, 0.005 M CaCl2, 0.25 M sucrose, 0.05 M Tris-HCl, pH 7.4.
3. 1 M dithiothreitol (DTT).
4. NP-40 lysis buffer: 0.5% NP-40, 150 mM NaCl, 20 mM CaCl·2H2O, 50 mM Tris, pH 7.4 supplemented with 1 mM phenylmethanesulfonyl fluoride (PMSF).
5. Ubiquitin vinyl sulfone (VS) probe including HA tag (rh hemagglutinin-ubiquitin-vinyl sulfone; Boston Biochem, Cambridge, MA).
6. 3X Reducing Laemmli's sample buffer: 2.5 mL 0.5 M Tris-HCl (pH 6.8), 1 mL β-mercaptoethanol, 1.5 mL 1% bromophenol blue, 5 mL 10% SDS, 1.5 mL 75% glycerol.

7. Benchtop microcentrifuge, refrigerated, such as the Sorvall Legend Micro 21R (Thermo Scientific, Rockford, IL).
8. Heat block.

2.4 Immunoprecipitation

1. Agarose A beads.
2. 1.5 mL tubes.
3. NP-40 lysis buffer: 0.5% NP-40, 150 mM NaCl, 20 mM $CaCl{\cdot}2H2O$, 50 mM Tris, pH 7.4 supplemented with 1 mM phenylmethanesulfonyl fluoride (PMSF).
4. HA-Agarose A beads, such as EZview anti-HA (Sigma-Aldrich).
5. 100 mM glycine, pH 2.5.
6. Rotator.
7. Parafilm.
8. Benchtop microcentrifuge, refrigerated, such as the Sorvall Legend Micro 21R (Thermo Scientific, Rockford, IL).

2.5 Protein Precipitation

1. Methanol.
2. Chloroform.
3. Milli-Q-H_20.
4. Vacuum centrifuge.
5. 1.5 mL tubes.
6. Benchtop microcentrifuge, such as the Sorvall Legend Micro 21R (Thermo Scientific, Rockford, IL).

2.6 Protein Digestion

1. Milli-Q-H_20.
2. 0.4 M Tris stock (pH 7.8).
3. 6 M urea in Tris: 2 g of urea, 1.25 mL 0.4 M Tris stock, 5 mL Milli-Q-H_20.
4. 0.2 M DTT in Tris: 0.031 g DTT, 750 μL of Milli-Q-H_20, 250 μL 0.4 M Tris stock.
5. 0.2 M iodoacetamide (IAA) in Tris: 0.037 g iodoacetamide, 250 μL 0.4 M Tris stock, 750 μL Milli-Q-H_20.
6. 1 M Tris buffer pH 8.0.
7. Mass Spectrometry Grade trypsin (such as VA9000 from Promega).
8. 0.2 μg/μl trypsin: 25 μL of ice-cold 0.4 M Tris stock, 75 μL ice-cold Milli-Q-H_20, 20 μg sequencing-grade modified trypsin (Promega).
9. Acetic acid.
10. pH paper.

11. Sonicator bath.
12. Vacuum centrifuge.
13. 37 °C incubator or water bath.
14. Benchtop microcentrifuge, refrigerated, such as the Sorvall Legend Micro 21R (Thermo Scientific, Rockford, IL).

2.7 Peptide Purification Using C18 ZipTips

1. ZipTip with 0.2 μL C18 resin (Cat No #ZTC18M096, Millipore).
2. Acetonitrile. HPLC grade.
    1. Trifluoroacetic acid (TFA). HPLC grade.
    2. Fluoroacetic acid (FA). HPLC grade.
    3. 1.5 mL Eppendorf tubes.
    4. Vacuum centrifuge.
3. Benchtop microcentrifuge, refrigerated, such as the Sorvall Legend Micro 21R (Thermo Scientific, Rockford, IL).

2.8 Immunoblotting

1. 10X SDS-PAGE buffer (1 L): 30.3 g Tris base, 144 g glycine, 10 g SDS. Dissolve Tris base and glycine in 700 mL H_2O. Add H_2O to 1 L and add SDS.
2. Protein electrophoresis and Trans-Blot Turbo Transfer System (Mini-PROTEAN Tetra Vertica Cell, Bio-Rad, Hercules, CA) or another similar system.
3. Trans-Blot Turbo 0.2 um PVDF Transfer Membranes.
4. TBST buffer: 150 mM NaCl, 20 mM Tris (pH 7.5), 0.1% Tween-20.
5. Blocking solution: 5% nonfat dry milk in TBST Buffer.
6. Biosafe Coomassie blue stain (e.g., SimplyBlue Biosafe Stain, Novex, Life Technologies).
7. Tris-glycine transfer buffer: 25 mM Tris base, 192 mM glycine, 20% methanol. Store at 4 °C.
8. Antibody solution: 1% non-fat milk powder in TBST buffer.
9. Mouse monoclonal anti-HA antibody.
10. Anti-mouse, horseradish peroxidase (HRP)-linked antibodies.
11. ECL Western Blotting substrate (e.g., Pierce).
12. iBright Imaging System (Thermo Fisher Scientific, Waltham, MA) or X-ray film (Hyperfilm; Amersham Biosciences) with X-ray film developer.
13. Restore Western Blot Stripping Reagent (Thermo Fisher Scientific, Waltham, MA).
14. 1.5 mL Eppendorf tubes.
15. Benchtop microcentrifuge, refrigerated, such as the Sorvall Legend Micro 21R (Thermo Scientific, Rockford, IL).

3 Methods

3.1 Tissue Sample Preparation by Ultrasonication

1. Weigh 200 mg of frozen sections from an animal (*see* **Note 1**), and insert them into the tissue tube. Keep on dry ice.
2. Crush each section by using a cryoPREP Manual Dry Pulverizer. Towards this, raise the cylinder of the Pulverizer, and insert the tissue tube containing the tissue. Release the cylinder, and hit it with the hammer included with the cryoPREP Pulverizer to pulverize the sample.
3. Solubilize the sample in cold NP-40 lysis buffer.
4. Process the sections by a Covaris S220 instrument (Covaris, Woburn, MA) at the following settings: 200 W, 20%, 500 cycles/burst.
5. After the lysis, clarify the lysates by centrifugation at 20,000 × g, 15 min, 4 °C. Repeat two more times.
6. Filter the lysate with Ultrafree-MC 0.1 μm filter units by centrifuging twice at 20,000 × g, 15 min, 4 °C.
7. Transfer the lysate to a fresh centrifuge tube.
8. Measure the protein concentration of the lysates by using a protein concentration kit according to the manufacturer's instructions and using a cuvette spectrophotometer.
9. Prepare aliquots of protein samples containing 10 mg of protein per sample and equal volume. Use NP-40 lysis buffer to adjust the sample volume as needed.
10. Store the sample at −80 °C for extended storage or use fresh for the DUB probe reaction.

3.2 Tissue Sample Preparation by Bead Beating (Alternative)

1. Weigh 200 mg of frozen sections from an animal (*see* **Note 1**).
2. Insert the Matrix A beads into the tube and add the tissue.
3. Homogenize the tissue by using a FastPrep-24 instrument (MP Biomedicals, Inc., Irvine, CA) using the cooling adapter filled with dry ice. Run the instrument four times for 30 s at 6.0 m/s, and let the samples cool on ice for 5 min between each run.
4. Resuspend the homogenized samples in NP-40 lysis buffer. Incubate the samples on ice for 10 min, and centrifuge at 20,000 × g for 10 min at 4 °C.
5. After the lysis, clarify the lysates by centrifugation at 20,000 × g, 15 min, 4 °C three times. Transfer the lysate to a fresh centrifuge tube.
6. Remove the supernatant to a new tube, and filter by using Ultrafree-MC centrifugal filter units.

7. Measure the protein concentration of the lysates by using a protein concentration kit according to the manufacturer's instructions and using a cuvette spectrophotometer.
8. Prepare aliquots of protein samples containing 10 mg of protein per sample and equal volume. Use NP-40 lysis buffer to adjust the sample volume as needed.
9. Store the sample at −80 °C for extended storage or use fresh for the DUB probe assay.

3.3 Deubiquitination (DUB) Probe Assay

1. Add 10 mg of protein from lysed tissue samples (Subheading 3.1 or 3.2.), and add reaction sample buffer containing 0.1 M DTT and incubate for 10 min on ice. Reserve an equivalent of 100 μg of the sample as your "input" material.
2. Add Ub-VS-HA probe (an equivalent of 1.1 μL probe per mg protein) to the rest of the sample, and incubate the mixture for 1 h at 37 °C with gentle shaking or occasional vortexing (*see* **Note 2**).
3. Reserve an equivalent of 100 μg of the sample incubated with the Ub-VS-HA probe as your "reaction" sample. Add the 3X Reducing Laemmli's sample buffer. Boil the sample for 4 min at 95 °C. Centrifuge the samples at 20,000 × g for 5 min, and store them at −80 °C for later use or proceed with the immunoblotting procedure.
4. The remaining reaction sample containing protein and the Ub-VS-HA probe can be stored at −80 °C for later use, or proceed to the immunoprecipitation assay procedure.

3.4 Immunoprecipitation Assay

1. Dilute each reaction sample from subheading 3.3 to a 1–5 mg/mL concentration using NP-40 lysis buffer.
2. Prepare Agarose A beads by washing the beads three times in the NP-40 lysis buffer, followed by spinning the beads down at 1000 × g for 2 min at 4 °C. Resuspend Agarose A beads 1:1 in the NP-40 lysis buffer.
3. Add sample to the beads, and incubate each sample with Agarose A beads for 1 h at 4 °C, with rotation. In this step, any proteins that might bind to the Agarose A beads unspecifically attach to the beads and hence can be removed from the further steps.
4. Spin down the beads at 1000 × g for 2 min at 4 °C, and transfer the supernatant to a new tube.
5. Prepare EZview anti-HA beads by washing the beads according to the manufacturer's recommendations (*see* **Note 3**). Specifically, wash the beads three times in 750 μL of the NP-40 lysis buffer, vortex, and centrifuge in a microcentrifuge for 1 min at 8200 × g at 4 °C. Repeat the step one more time. Resuspend

EZview anti-HA-agarose beads 1:1 in the NP-40 lysis buffer, and set them on ice.

6. Add the supernatant collected at **step 4** to 100 μL of HA-agarose beads, and incubate for 2 h at 4 °C with rotation.
7. Spin down the beads at 8200 × g for 2 min at 4 °C and discard the supernatant.
8. Perform four sequential washes using 800 μL of chilled NP-40 lysis buffer, and after each wash, spin down the sample.
9. Wash the beads with 150 mM NaCl, 50 mM Tris, pH 7.5.
10. Elute the proteins bound to the anti-HA resin by boiling in 0.1% RapiGest SF (Waters Inc.) buffer at 95 °C for 2 min.
11. Spin down the beads at 8200 × g for 3 min, 4 °C, and collect the eluate. Seal the samples with Parafilm to prevent leakage during rotation.

3.5 Protein Precipitation (See Note 4)

1. Use 200 μL of eluted reaction sample (*see* **Note 5**), add 600 μL methanol and 150 μL chloroform, and vortex.
2. Add 450 μL Milli-Q-H_20 and vortex.
3. Centrifuge the sample at 20,000 × g for 2 min at room temperature.
4. Remove the upper layer formed after the centrifugation.
5. Wash the sample by adding 450 μL methanol.
6. Vortex the sample.
7. Centrifuge at 20,000 × g for 2 min at room temperature (*see* **Note 6**), and remove the supernatant.
8. Vacuum centrifuge the samples for approximately 15 min to complete dryness.

3.6 Protein Digestion

1. Solubilize the protein samples from subheading 3.8 in 100 μL 6 M urea for 2 min using a sonicator bath.
2. Reduce the samples by adding 5 μL of 0.2 M DTT for 45 min at room temperature.
3. Alkylate the samples with 20 μL of 0.2 M IAA for 45 min at room temperature in the dark.
4. Add 20 μL of 0.2 M DTT for 45 min at room temperature as a quenching step.
5. Dilute the protein samples in 775 μL water.
6. Adjust the pH to 7.5 using 1 M Tris buffer pH 8.0. Use pH papers to confirm the pH.
7. Add 20 μL of 0.2 μg/μL trypsin (keep trypsin on ice until adding to the sample), vortex, spin down the samples at 20,000 × g for 3 min, and incubate overnight at 37 °C.

8. Add acetic acid dropwise to acidify the sample (pH < 6). Use pH papers to confirm the pH.

3.7 Peptide Purification Using C18 ZipTips

1. Condition the ZipTip with 10 μL 100% Acetonitrile. Repeat twice.
2. Condition the ZipTip with a 10 μL solution containing 50% Acetonitrile and 0.1%-trifluoroacetic acid (TFA). Repeat twice.
3. Condition the ZipTip with 10 μL 0.1% TFA. Repeat twice.
4. Pipet the solution sample up and down the ZipTip ten times. Discard the sample into the original test tube.
5. Wash the ZipTip with 10 μL of 0.1% TFA ten times to remove contaminants.
6. Elute the sample with a 10 μL solution containing 80% Acetonitrile and 0.1%-fluoroacetic acid (FA) by pipetting up and down ten times.
7. Lyophilize the eluates in the vacuum centrifuge to complete dryness.
8. Analyze the entire eluate by mass spectrometry.

3.8 SDS-PAGE and Western Blotting

1. For the Mini-PROTEAN TGX Precast gels, assemble the gel into the Mini-PROTEAN Tetra Vertical Electrophoresis Cell chamber according to manufacturer specifications, and then load the appropriate amount of 1X SDS-PAGE running buffer (700 mL for two gels, 1000 mL for four gels).
2. Add ~20 μg of each protein sample, including supernatants eluted from HA-agarose and whole protein lysates, to each well of the polyacrylamide gel (*see* **Note 7**).
3. Run the samples using an electrophoresis unit at 75 V for 15 min to equilibrate protein samples as they enter the gel. After each sample is appropriately inside the gel, the power can be turned up to 120 V.
4. Once the electrophoresis is complete, sandwich the gel into the Trans-Blot Turbo 0.2 μm PVDF Transfer Membrane, and ensure no air bubbles between the membrane and the gel by rolling the membrane with a roller. Then using the Trans-Blot Turbo Transfer System, transfer the gel to the PVDF membrane using the Turbo setting for 1 Mini TGX. If this setting is not available, the transfer can be performed at 2.5 A with 25 V for 3 min.
5. Discard the gel and carefully transfer the blotted PVDF membrane to a clean washing chamber.
6. Wash the membrane three times with 1X TBST for 5 min to remove residual methanol.

7. Incubate the membrane in blocking solution at room temperature for 1 h.
8. After the 1 h blocking step, prepare the primary antibody [such as mouse anti-hemagglutinin (anti-HA)] in antibody solution, and apply them to the membrane for an overnight incubation on a shaker at 4 °C.
9. The next day, remove the primary antibody solution, and store it for later use at 4 °C.
10. Wash the blot with non-antibody treated antibody solution four times for 5 min each.
11. On the last wash, prepare the secondary antibody (such as polyclonal goat anti-mouse HRP) in antibody solution for 1–2 h.
12. Wash the blot three times in PBS, 0.1% Tween-20 for 5 min each. After the washing is completed, the blot can now be imaged.
13. Incubate the PVDF membrane with 5 mL of both the peroxide and luminol/enhancer solutions included in the ECL Plus Kit for 5 min (*see* **Note 8**).
14. Develop the signal using iBright Imaging System or onto an X-ray film.
15. After imaging has been completed, the blot can be washed three times in PBS, 0.1% Tween-20 for 5 min each, followed by a 15 min treatment with Restore Western Blot Stripping Buffer with another three TBST washes. The blot can now be blocked and treated with a new primary antibody for protein imaging.

4 Notes

1. This procedure might require optimization for other tissues. Ensure that each tissue sample is placed in a sterile bag after dissection and transported on ice to a freezer or submerged into the liquid nitrogen and stored in −80 °C until further use.
2. Other probes can also be used, such as Ub-PA or Ub-VME.
3. To dispense beads, use a wide orifice pipette tip or cut ~1 mm off the tip of a regular pipette tip, to enlarge the opening and allow unrestricted flow of the bead suspension.
4. The eluates can also be precipitated using the TCA method instead of the chloroform/methanol method. For TCA precipitation, add 0.18 sample volumes of 100% TCA for 20 min on ice, and spin for 30 min at 21,000 × g. The precipitated proteins are then washed with 500 μL of −20 °C acetone, and the precipitate is left to air dry.

5. If sample volume is lower than 200 μL, bring the volume up to 200 μL with Milli-Q-$H_2$0.
6. Centrifuge the sample for longer if the sample is still turbid.
7. Be aware of the volume limits for each of the wells to prevent overfilling and inadvertently losing samples.
8. Other similar ECL substrates can also be used according to the manufacturer's instructions.

Acknowledgments

Funds were awarded to MJE via the Department of Defense (W81XWH-20-1-0176).

References

1. Hussain S, Zhang Y, Galardy PJ (2009) DUBs and cancer: the role of deubiquitinating enzymes as oncogenes, non-oncogenes and tumor suppressors. Cell Cycle 8:1688–1697
2. Singhal S, Taylor MC, Baker RT (2008) Deubiquitylating enzymes and disease. BMC Biochem 9(Suppl 1):S3
3. Edelmann MJ, Kessler BM (2008) Ubiquitin and ubiquitin-like specific proteases targeted by infectious pathogens: emerging patterns and molecular principles. Biochim Biophys Acta 1782:809–816
4. Edelmann MJ, Kramer HB, Altun M, Kessler BM (2010) Post-translational modification of the deubiquitinating enzyme otubain 1 modulates active RhoA levels and susceptibility to Yersinia invasion. FEBS J 277:2515–2530
5. Nanduri B, Suvarnapunya AE, Venkatesan M, Edelmann MJ (2013) Deubiquitinating enzymes as promising drug targets for infectious diseases. Curr Pharm Des 19(18): 3234–3247
6. Baek SH, Choi KS, Yoo YJ, Cho JM, Baker RT, Tanaka K, Chung CH (1997) Molecular cloning of a novel ubiquitin-specific protease, UBP41, with isopeptidase activity in chick skeletal muscle. J Biol Chem 272:25560–25565
7. Baek SH, Park KC, Lee JI, Kim KI, Yoo YJ, Tanaka K, Baker RT, Chung CH (1998) A novel family of ubiquitin-specific proteases in chick skeletal muscle with distinct N- and C-terminal extensions. Biochem J 334(Pt 3): 677–684
8. Baek SH, Yoo YJ, Tanaka K, Chung CH (1999) Molecular cloning of chick UCH-6 which shares high similarity with human UCH-L3: its unusual substrate specificity and tissue distribution. Biochem Biophys Res Commun 264:235–240
9. Woo SK, Lee JI, Park IK, Yoo YJ, Cho CM, Kang MS, Ha DB, Tanaka K, Chung CH (1995) Multiple ubiquitin C-terminal hydrolases from chick skeletal muscle. J Biol Chem 270:18766–18773
10. Baek SH, Woo SK, Lee JI, Yoo YJ, Cho CM, Kang MS, Tanaka K, Chung CH (1997) New de-ubiquitinating enzyme, ubiquitin C-terminal hydrolase 8, in chick skeletal muscle. Biochem J 325(Pt 2):325–330
11. Kummari E, Alugubelly N, Hsu CY, Dong B, Nanduri B, Edelmann MJ (2015) Activity-based proteomic profiling of deubiquitinating enzymes in salmonella-infected macrophages leads to identification of putative function of UCH-L5 in inflammasome regulation. PLoS One 10:e0135531
12. Murai J, Yang K, Dejsuphong D, Hirota K, Takeda S, D'Andrea AD (2011) The USP1/UAF1 complex promotes double-strand break repair through homologous recombination. Mol Cell Biol 31:2462–2469
13. Yang K, Moldovan GL, Vinciguerra P, Murai J, Takeda S, D'Andrea AD (2011) Regulation of the Fanconi anemia pathway by a SUMO-like delivery network. Genes Dev 25:1847–1858
14. Zheng H, Wang M, Zhao C, Wu S, Yu P, Lü Y, Wang T, Ai Y (2017) Characterization of the deubiquitination activity and substrate specificity of the chicken ubiquitin-specific protease 1/USP associated factor 1 complex. PLoS One 12:e0186535
15. Gao T, Liu Z, Wang Y, Cheng H, Yang Q, Guo A, Ren J, Xue Y (2012) UUCD: a family-based database of ubiquitin and

ubiquitin-like conjugation. Nucleic Acids Res 41:D445–D451

16. Kessler BM (2006) Putting proteomics on target: activity-based profiling of ubiquitin and ubiquitin-like processing enzymes. Expert Rev Proteomics 3:213–221
17. Borodovsky A, Kessler BM, Casagrande R, Overkleeft HS, Wilkinson KD, Ploegh HL (2001) A novel active site-directed probe specific for deubiquitylating enzymes reveals proteasome association of USP14. EMBO J 20: 5187–5196
18. Ekkebus R, van Kasteren SI, Kulathu Y, Scholten A, Berlin I, Geurink PP, de Jong A, Goerdayal S, Neefjes J, Heck AJ, Komander D, Ovaa H (2013) On terminal alkynes that can react with active-site cysteine nucleophiles in proteases. J Am Chem Soc 135:2867–2870
19. Sommer S, Weikart ND, Linne U, Mootz HD (2013) Covalent inhibition of SUMO and ubiquitin-specific cysteine proteases by an in situ thiol-alkyne addition. Bioorg Med Chem 21:2511–2517
20. Borodovsky A, Ovaa H, Kolli N, Gan-Erdene T, Wilkinson KD, Ploegh HL, Kessler BM (2002) Chemistry-based functional proteomics reveals novel members of the deubiquitinating enzyme family. Chem Biol 9:1149–1159
21. Nanduri B, Gresham CR, Jones G, Bailey RH, Edelmann MJ (2021) Identification of active deubiquitinases in the chicken tissues. Proteomics 22:e2100122
22. Pei Y, Hancock PJ, Zhang H, Bartz R, Cherrin C, Innocent N, Pomerantz CJ, Seitzer J, Koser ML, Abrams MT, Xu Y, Kuklin NA, Burke PA, Sachs AB, Sepp-Lorenzino L, Barnett SF (2010) Quantitative evaluation of siRNA delivery in vivo. RNA 16:2553–2563
23. Nanduri B, Shack LA, Rai AN, Epperson WB, Baumgartner W, Schmidt TB, Edelmann MJ (2016) Use of focused ultrasonication in activity-based profiling of deubiquitinating enzymes in tissue. Anal Biochem 515:9–13

Chapter 5

Design and Synthesis of Ubiquitin-Based Chemical Tools with Unnatural Amino Acids for Selective Detection of Deubiquitinases

Wioletta Rut, Mikolaj Zmudzinski, and Marcin Drag

Abstract

Several chemical approaches have been applied to develop Ub-based substrates and probes selective toward one or a narrow subset of deubiquitinases (DUBs). Since DUBs are highly specific toward ubiquitin and exhibit low activity toward shorter peptides, it is challenging to design truly selective chemical tools to investigate one DUB in biological samples. Incorporating amino acids other than canonical LRG at the P4–P2 positions in the Ub improves DUB activity and selectivity toward Ub derivatives. Here, we describe the protocol for identifying selective peptide sequences using a hybrid combinatorial substrate library (HyCoSuL) approach that can be introduced in the C-terminal motif of Ub. Furthermore, we describe the synthesis protocol of Ub-based probes and substrates containing unnatural amino acids and the application of Ub-based probes to detect DUBs in cell lysates.

Key words DUBs, Ubiquitin, Combinatorial library, Substrate specificity, Unnatural amino acids

1 Introduction

Ubiquitin (Ub) is a small regulatory protein that covalently labels proteins in a process known as ubiquitination. Ubiquitination remains one of the most prevalent posttranslational modifications in cells, and it is tightly controlled by enzymatic machinery consisting of ubiquitin-activating enzymes (E1), ubiquitin-conjugating enzymes (E2), ubiquitin ligases (E3), and deubiquitinases (DUBs) [1, 2]. While the E1-E2-E3 enzymatic cascade mediates in the ubiquitin attachment to a substrate protein, DUBs are proteases that reverse this process as they hydrolyze amide bonds between Ub C-terminal carboxylic group and modified amine groups. Most of the time, ubiquitin is attached to lysine ε-amine groups via isopeptide bond [3], but ubiquitination can occur also in a linear fashion, mediated by linear ubiquitin chain assembly complex (LUBAC) [4].

Julie A. Maupin-Furlow and Mariola J. Edelmann (eds.), *Deubiquitinases: Methods and Protocols*,
Methods in Molecular Biology, vol. 2591, https://doi.org/10.1007/978-1-0716-2803-4_5,

To date, nearly 100 different DUBs have been discovered in humans [5]. Additionally, enzymes possessing deubiquitinating activity are widespread and are encoded even by viruses, indicating the importance of Ub signaling pathways [6]. DUBs have been recognized as a relevant target in therapies of various health conditions. Therefore, developing novel, potent, and selective inhibitors remains a fundamental challenge for researchers [7–10].

One of the most versatile tools used in protease investigation is activity-based probes (ABPs) – molecules that enable visualization of the active form of proteases of interest. A "classic" ABP design includes (1) reactive chemical moiety that binds to an active site of the enzyme, commonly referred to as the warhead; (2) reporter group that provides a trackable signal (e.g., fluorophore, biotin, radioisotopes); and (3) linker, usually a short peptide sequence providing selective recognition by the enzyme of interest [11]. However, most of ABPs designed for deubiquitinases requires the usage of a ubiquitin molecule as a linker. Protein-protein interactions between ubiquitin and ubiquitin binding domains (UBDs) on the DUB surface trigger conformational changes within the enzyme, providing an optimal active site arrangement [12]; thus, short peptide linkers might be insufficient for efficient recognition and binding by the enzyme. Currently, standard chemical reagents in DUB research are based on functionalized Ub molecules or short poly-Ub chains. Poly-Ub-based probes are tools that enable distinguishing between DUBs that process ubiquitin chains of different topologies [13–16].

Usage of a single ubiquitin molecule as a universal recognition pattern generates a major issue with the selectivity of Ub-based chemical tools. To tackle this problem up to date, various chemical approaches have been employed to design and synthesize Ub-based ABPs, substrates, and inhibitors. In independent studies by Borodovsky et al. [17] and Love et al. [18], the teams synthesized sets of hemagglutinin (HA)-tagged probes with various electrophilic warheads. Dependently on the C-terminal electrophilic group, labeling profiles in cell lysates experiments were different, indicating the importance of the choice of the warhead for the probe's further applicability. In two studies by Gjonaj et al. [19, 20], selectivity toward USP7 and USP16 was achieved by introducing mutations in the Ub primary structure. Sites for the mutations were rationally predicted using computational analysis. In comparison with the wild-type Ub, mutations disrupted interactions between the Ub mutants and other enzymes, on the other hand preserving their recognizability by the investigated deubiquitinases.

Recently we proposed a novel approach in the generation of selective Ub-based tools [21]. The method is based on the finding that introducing mutations within the conservative LRGG

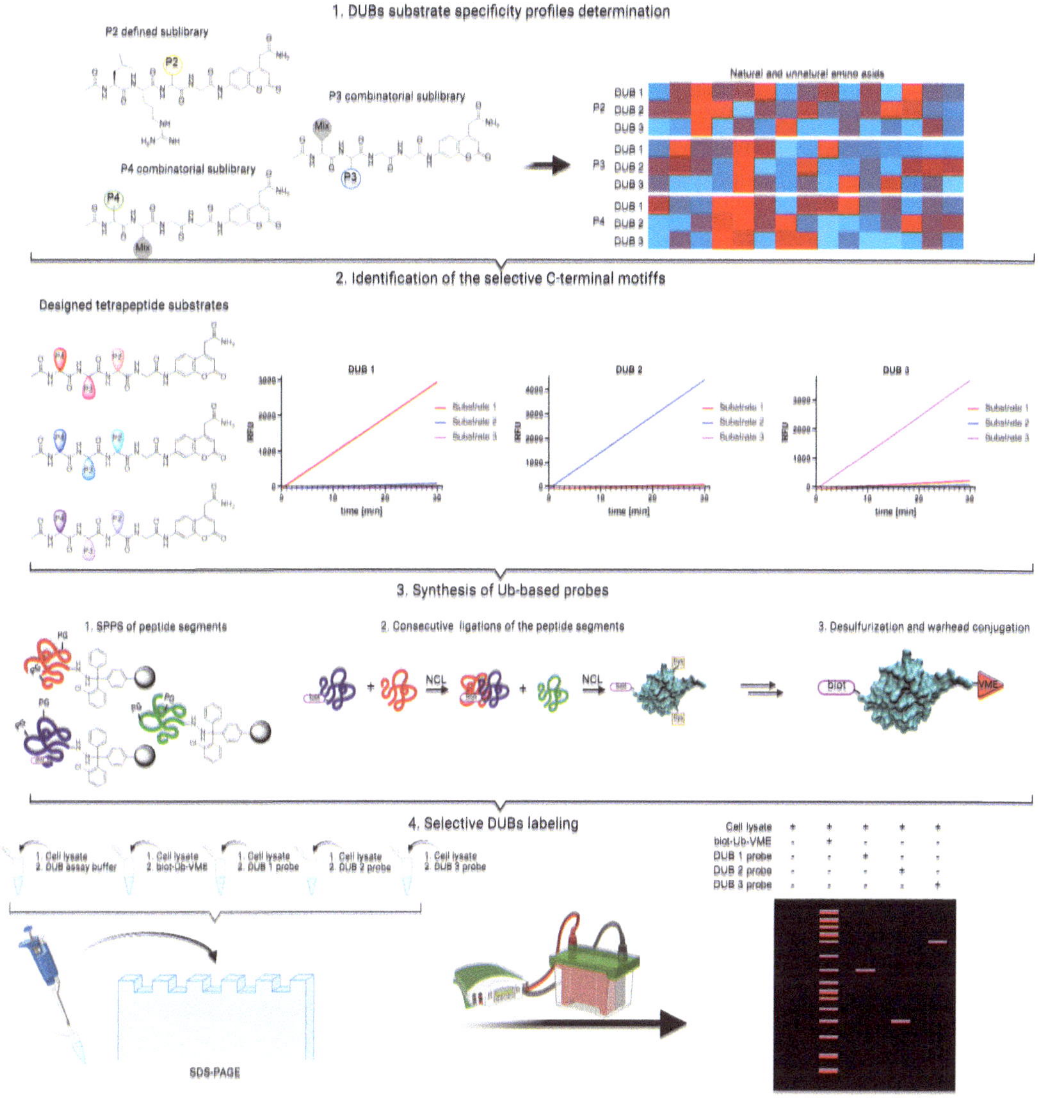

Fig. 1 A general framework for designing, synthesizing, and applying Ub-based probes. (Created with BioRender.com)

C-terminal epitope of Ub can lead to selective recognition by particular DUBs [22]. Herein we present a protocol for generating selective Ub-based tools for deubiquitinases. Our strategy includes three main steps: (1) substrate specificity determination using hybrid combinatorial substrate library (HyCoSuL) and defined library; (2) design, synthesis, and selectivity evaluation of short peptide substrates; and (3) synthesis of selective Ub derivatives with specific sequences on the C-terminus (Fig. 1). By using more than 100 non-proteinogenic amino acids, HyCoSuL screening provides detailed information about the binding pocket architecture of the enzyme [23–27]. Obtained specificity profiles are compared and analyzed to identify peptide sequences that might provide selectivity toward the DUB of interest. Next, selected amino acids are incorporated into the structures of short fluorogenic substrates. The substrates are evaluated in terms of their selectivity toward the enzyme. Then, selective sequences are incorporated into the structure of ubiquitin at its C-terminal epitope within the LRGG motif. Ubiquitin derivatives are synthesized in a stepwise manner using Fmoc/tBu solid-phase peptide synthesis (SPPS) and native chemical ligation (NCL) techniques. The current protocol presents a general procedure for the development of new Ub-based tools. Total chemical synthesis of modified Ub enables incorporation of various C-terminal electrophiles and labels other than biotin, e.g., fluorophores with required spectral properties or chemical moieties, enabling further functionalization. The presented chemical approach might also be expanded and applied in developing ubiquitin-like modifier derivatives targeted by enzymes displaying deSUMOylating, deISGylating, or other related activities.

2 Materials

2.1 DUB Profiling by HyCoSuL and Defined Library

1. Hybrid combinatorial substrate library (HyCoSuL) and defined P2 library (Fig. 2):

   P4 and P3 sublibraries: Ac-P4-Mix-Gly-Gly-ACC (ACC, 7-amino-4-carbamoylmethylcoumarin) and Ac-Mix-P3-Gly-Gly-ACC (where P4 and P3 represent 19 natural and 119 unnatural amino acids; Mix represents an equimolar mixture of natural amino acids excluding cysteine and methionine and including norleucine as a structural analog of methionine).

   P2 sublibrary: Ac-Leu-Arg-P2-Gly-ACC, where P2 represents 128 natural and unnatural amino acids. The sublibraries are dissolved in biochemical grade DMSO at the concentration of 10 or 20 mM.
2. Recombinant deubiquitinating enzymes.
3. DUB buffer: 20 mM Tris, 150 mM NaCl, 5 mM DTT, pH 8.0, or 50 mM HEPES, 0.5 mM EDTA, 5 mM DTT, pH 7.5 (*see* **Note 1**).
4. Spectrofluorometer to monitor substrate hydrolysis (e.g., Molecular Devices SpectraMax Gemini XPS).
5. Round-bottom 96-well plates (e.g., Corning 96-well plates, opaque bottom).
6. A set of single-channel pipettes with different capacities (0.5–10 μL, 2–20 μL, 10–100 μL, 100–1000 μL) and one multichannel pipette for delivering 8 × 100 μL.
7. 15 mL Falcon tubes for DUB incubation.
8. Reagent reservoir (50 mL size).

Sublibrary	Structure of sublibrary	Number of amino acids at a fixed position
P2	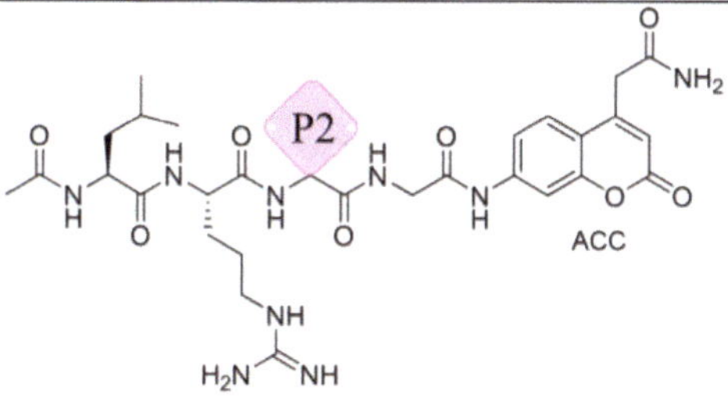	128
P3	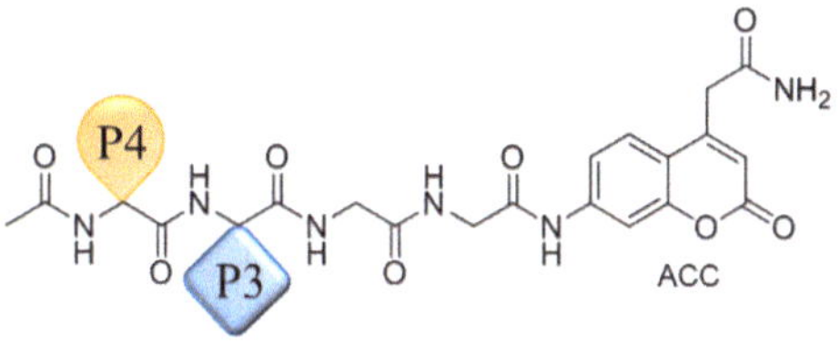	138
P4	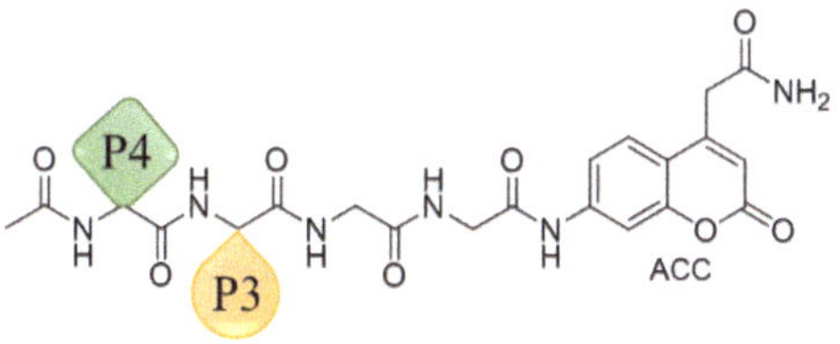	138

 an equimolar mixture of natural amino acids excluding cysteine and methionine and including norleucine

 natural or unnatural amino acid

Fig. 2 Substrate library for DUB profiling

9. 37 °C incubator.
10. Microcentrifuge.
11. Vortex mixer.

2.2 Kinetic Analysis of Tetrapeptide Fluorogenic Substrates

1. Individual tetrapeptide substrate dissolved in peptide-grade DMSO to the final concentration of 10 or 20 mM.
2. Other materials as described in Subheading 2.1 (**steps 2–11**).

2.3 Synthesis of Ub-Based Substrate and Activity-Based Probe

1. 2-Chlorotrityl chloride resin (particle size 100–200 mesh, loading 0.97 mmol/g).
2. Fmoc-protected amino acids, purity ≥97%.
3. Dipeptide building blocks: Fmoc-Asp(OtBu)-(Dmb)Gly-OH and Fmoc-Ser(tBu)-Thr(ψMe,Mepro)-OH, purity >98%.

4. 9-Fluorenylmethyl carbazate, Fmoc-NH-NH_2, purity ≥98%.
5. Tris(2-carboxyethyl)phosphine hydrochloride, TCEP, purity ≥98%.
6. 2,2′-Azobis[2-(imidazolin-2-yl)propane], VA-044, purity ≥98%.
7. 4-Mercaptophenylacetic acid, MPAA, purity ≥97%.
8. Guanidine hydrochloride, Gn·HCl, purity ≥99%.
9. Sodium phosphate monobasic monohydrate, $NaH_2PO_4 \cdot H_2O$, purity ≥99%.
10. 4.0 M HCl in dioxane.
11. 2-Methyl-2-propanethiol, tBuSH, purity ≥99%.
12. Sodium nitrite, $NaNO_2$, purity ≥99%.
13. N,N′-diisopropylethylamine, DIPEA, purity ≥99%.
14. 2-(1-H-7-azabenzotriazol-1-yl)-1,1,3,3-tetramethyluranium hexafluorophosphate, HATU, peptide grade.
15. 2,4,6-Trimethylpyridine, 2,4,6-collidine, peptide grade.
16. Piperidine, PIP, purity >99%.
17. N,N ′-dimethylformamide, DMF, peptide grade.
18. Dichloromethane, DCM, pure p.a.
19. Methanol, MeOH, pure p.a.
20. Acetonitrile, ACN, HPLC gradient grade.
21. Diethyl ether, Et_2O, pure p.a.
22. Phosphorus pentoxide, P_2O_5, purity 98%.
23. Cleavage mixture (95%:2.5%:2.5%, %v/v/v): Trifluoroacetic acid, TFA, purity 99%, Triisopropylsilane, TIPS, purity 99%, and H_2O.
24. Ligation buffer: 6 M Gn·HCl, 0.2 M NaH_2PO_4, pH 3.0–3.1.
25. Desulfurization buffer: 6 M Gn·HCl, 100 mM Na_2HPO_4, 375 mM TCEP, 37.5 mM VA-044, 150 mM tBuSH, pH 6.9.
26. TFA·H_2N-Gly-VME synthesized according to the literature protocol [28].
27. TFA·H_2N-Gly-ACC synthesized on Rink Amide resin using standard solid-phase synthesis method [29].
28. Vacuum line with a trap.
29. Forty mL solid-phase peptide synthesis vessel (e.g., Chemglass).
30. Microcentrifuge.
31. Shaker.
32. Magnetic stirrer.

33. High-performance liquid chromatography system (HPLC).
34. Analytical (Jupiter 10 μm particle size C4 300 Å column (250 × 4.6 mm)) and semi-preparative columns (Wide Pore C8 Discovery column 10 μm particle size (250 × 21.2 mm) and Jupiter 10 μm particle size C4 300 Å column (250 × 10 mm)).
35. Lyophilizer.

2.4 DUB Labeling in Cell Lysates

1. Selected cell lines (e.g., HeLa or HEK-293T).
2. Culture medium (e.g., Dulbecco's Modified Eagle Medium (DMEM)).
3. Fetal bovine serum, FBS.
4. L-Glutamine.
5. Antibiotics: penicillin, streptomycin.
6. Phosphate buffered saline, PBS (155 mM NaCl, 1 mM KH_2PO_4, 3 mM $Na_2HPO_4{\cdot}7H_2O$, pH 7.4).
7. Lysis buffer: 20 mM Tris, 150 mM NaCl, and 5 mM DTT, pH 8.0.
8. DUB activity-based probes dissolved in biochemical grade DMSO at a 5 or 10 mM concentration.
9. 3× SDS/DTT loading buffer: 150 mM Tris-HCl, 300 mM DTT, 6% SDS, 0.3% bromophenol blue, 30% glycerol, pH 6.8.
10. 4–12% Bis-Tris Plus gels.
11. Nitrocellulose membrane (0.2 μm, Bio-Rad).
12. Tris-buffered saline with 0.1% (v/v) Tween 20 (TBS-T: 20 mM Tris, 150 mM NaCl, 0.1% (v/v) Tween 20, pH 7.6).
13. 2% Bovine Serum Albumin (BSA) in TBS-T.
14. Fluorescent streptavidin conjugate (e.g., Alexa Fluor 647 conjugate).
15. Primary and secondary antibodies.
16. CO_2 incubator for cell culture (humidified 5% CO_2 atmosphere at 37 °C).
17. Sonicator.
18. Centrifuge.
19. Protein gel electrophoresis and transfer system with MES or MOPS running buffer and transfer buffer (Invitrogen).
20. Imaging system with software (e.g., Azure Biosystems Sapphire Biomolecular Imager and Azure Spot Analysis Software).

3 Methods

3.1 DUB Profiling by HyCoSuL and Defined Library

1. To determine library screening conditions, prepare several DUB solutions in assay buffer at different enzyme concentrations ranging from 0.5 to 20 μM (e.g., 0.5, 2.5, 5, 10, and 20 μM). DUBs recognize the C-terminal motif of ubiquitin-LRGG sequence; thus, select Ac-Leu-Mix-Gly-Gly-ACC, Ac-Mix-Arg-Gly-Gly-ACC, and Ac-Leu-Arg-Gly-Gly-ACC from P4, P3, and P2 sublibraries, respectively.
2. Perform the initial library screening. Preincubate DUBs at aforementioned concentrations in the assay buffer for 10–30 min at 37 °C (*see* **Note 2**). Test HyCoSuL at the final concentration of 200 μM and the defined P2 sublibrary at 50 or 100 μM. If substrates from the combinatorial library are dissolved in DMSO at a final concentration of 20 mM, vortex and add 1 μL of the selected substrate from each sublibrary into a well of a 96-well plate, and then add 99 μL of a DUB diluted in assay buffer.
3. Monitor substrate hydrolysis on a spectrofluorometer at appropriate wavelengths (for ACC substrate λ_{ex} = 355 nm, λ_{em} = 460 nm) for 20–30 min (set the interval between fluorescence readings every 15–30 s).
4. Appropriate DUB concentration is when the rate of substrate hydrolysis of Ac-Leu-Mix-Gly-Gly-ACC, Ac-Mix-Arg-Gly-Gly-ACC, and Ac-Leu-Arg-Gly-Gly-ACC expressed as RFU/s (relative fluorescence unit per second) is at 5–50 RFU/s. These assay conditions allow detecting very poor DUB substrates (RFU/s $<$ 0.5). Substrates displaying RFU/s lower than 5% of the best DUB substrate cannot be considered DUB substrates. The P4 and P3 sublibraries should be screened at the same DUB concentration, while the P2 sublibrary as a defined library can be screened at a lower DUB concentration.
5. To determine DUB substrate specificity at the P4 position, vortex and add 1 μL of each substrate from the P4 sublibrary into wells of 96-well plate. After 10–30 min of DUB preincubation at 37 °C, add 99 μL of the enzyme into each well with the substrate using an 8-channel pipette, and run the measurement. Due to the pipetting time and different rates of substrate hydrolysis by DUBs, substrates with high RFU/s values from the first columns may be significantly hydrolyzed before the measurement is started. Thus it is recommended to use only half of the 96-well plate (6 out of 12 columns) for one measurement to reduce the time of the assay preparation. Perform three separate measurements to screen the P4 sublibrary containing 138 substrates. Repeat each experiment at least two times (*see* **Note 3**).

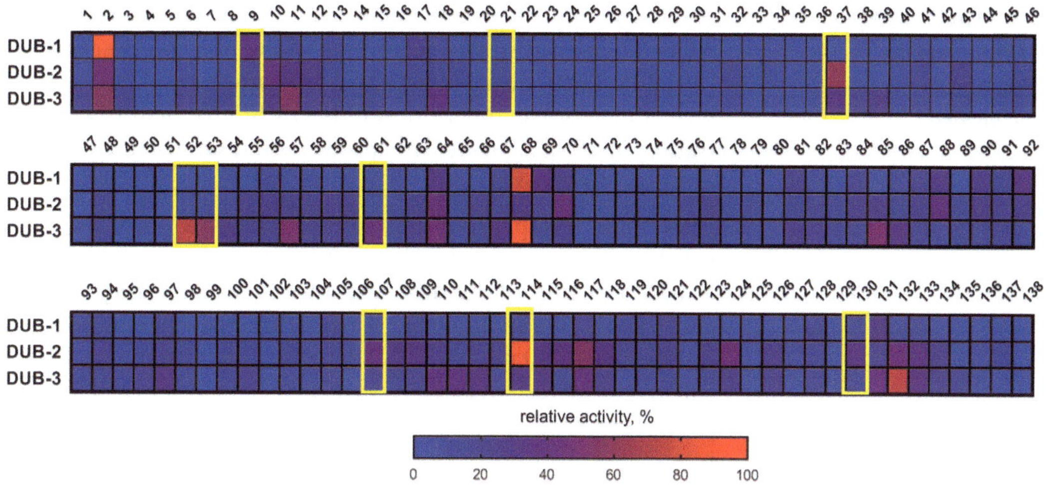

Fig. 3 DUBs substrate specificity at the P3 position is presented as a heat map. The yellow box indicates the amino acids that should be selected as building blocks in designing selective substrates for tested DUBs

6. Monitor substrate hydrolysis on a spectrofluorometer at appropriate wavelengths (for ACC substrate $\lambda_{ex} = 355$ nm, $\lambda_{em} = 460$ nm) for 40–60 min (set the interval between fluorescence readings every 15–30 s). Stop the measurement when the signal from hydrolysis of very poor substrates is strong enough to be detected by the spectrofluorometer.
7. To obtain the DUB substrate specificity profile at the P4 position, analyze the data from the sublibrary screening assay. For each substrate, extract only the linear part of the curve. Select the same time interval from hydrolysis curves for all substrates. However, sometimes each substrate hydrolysis curve should be analyzed separately due to the different hydrolysis rates. In the case of very poor substrates (RFU/s < 0.5), the time range for a linear curve is longer than for substrates with high RFU/s values. On the other hand, the time range for substrates with the highest RFU/s values can be shorter than for very poor substrates. Set the highest value of RFU/s for the best substrate as 100%, and adjust the RFU/s values of other substrates accordingly. If the standard deviation for any substrate is higher than 10% of the mean value, repeat the measurement for these substrates (*see* **Note 4**).
8. To determine the DUB substrate specificity profile at P3 and P2 positions, repeat **steps 5–7**.
9. To design selective substrates for each DUB, analyze substrate specificity profiles, and select amino acids that are recognized by one tested enzyme and are not recognized or are very weakly recognized by other enzymes (*see* Fig. 3).

3.2 Kinetic Analysis of Tetrapeptide Fluorogenic Substrates

1. After selecting the amino acids at P4–P2 positions, synthesize ACC-labeled substrates according to the method described in [29, 30]. Then, purify tetrapeptide substrates by HPLC, lyophilize, and dissolve in peptide-grade DMSO to the final concentration of 10 or 20 mM. Substrates should be stored at −20 °C until use (*see* **Note 5**).
2. Determine tetrapeptide substrate selectivity toward DUBs. If substrates are dissolved in DMSO to the final concentration of 1 mM, add 1 μL of the substrate into the well on the plate, and add 99 μL of the enzyme. Preincubate the enzyme in assay buffer for 10–30 min at 37 °C before the measurement. If ≤12 substrates are tested, three repetitions can be performed on one plate during one experiment. Use Ac-Leu-Arg-Gly-Gly-ACC reference substrate as a positive control. DUBs concentration depends on their activity toward tetrapeptide substrates. To select appropriate DUB concentration, test its activity toward 10 μM of Ac-Leu-Arg-Gly-Gly-ACC. If DUB activity toward the control substrate is too low even at a high enzyme concentration, use a higher substrate concentration of 15–30 μM (*see* **Note 6**).
3. Analyze the results of the kinetic assays. For each substrate determine the hydrolysis rate (RFU/s) of substrates by tested DUBs, and calculate the mean values with standard deviations. As hits, select substrates, which are recognized by one enzyme but are not recognized (or weakly recognized) by other enzymes.
4. Determine the catalytic efficiency of the enzyme toward selected substrates. Since DUBs do not efficiently hydrolyze tetrapeptide substrates, high substrate concentrations are required to measure kinetic parameters. Due to the precipitations at high concentrations for most DUB substrates, only the specificity constant (k_{cat}/K_M) can be determined.
5. To determine kinetic parameters for ACC-labeled substrates, measuring how much a certain amount of ACC fluorophore produces fluorescence signal (RFU) is required. To estimate this value, add 1 μL of 1 mM of Ac-LRGG-ACC (final substrate concentration 10 μM) into the well of 96-well plate, and add 99 μL of the DUB preincubated in the assay buffer at 37 °C for 10–30 min. Monitor substrate hydrolysis on a spectrofluorometer for 2–3 h at $\lambda_{ex} = 355$ nm and $\lambda_{em} = 460$ nm wavelengths (set the interval between fluorescence readings every 30–40 s). The time of measurement depends on the DUB activity toward a substrate. After the Ac-LRGG-ACC substrate is hydrolyzed, divide the maximum RFU by 10 to obtain the RFU value produced by 1 μM of ACC fluorophore.

6. To measure k_{cat}/K_M prepare serial dilution of the selected substrate in the assay buffer on a plate (eight different concentrations, 20 μL of substrate in assay buffer in each well), add 80 μL of the enzyme into each well on the plate, and run the experiment. Monitor substrate hydrolysis using a spectrofluorometer. Read the fluorescence every 10–15 s at $\lambda_{ex} = 355$ nm and $\lambda_{em} = 460$ nm wavelengths for 20–60 min. Calculate RFU/s for each well using the linear part of the plots.
7. Draw the plot: the *y*-axis is RFU/s or μM of ACC/s, and the *x*-axis is substrate concentration. Use GraphPad Prism software to analyze the data using linear regression, and calculate the slope value. When $[S_0] << K_M$, the plot of *vi* (the initial velocities) versus $[S_0]$ yields a straight line with slope representing V_{max}/K_M, k_{cat}/K_M = slope/E (E – total enzyme concentration).

3.3 Synthesis of Ub-Based Substrate and Activity-Based Probe

Ubiquitin with the modified C-terminal epitope constitutes a scaffold for Ub-based activity-based probes and substrates. To synthesize these Ub-based tools with good yield and purity, ubiquitin derivative should be divided into three peptide segments: Ub [1-27]-NH-NH_2 or Biot-6-Ahx-Ub[1-27]-NH-NH_2, Ub [28-45]-(A28C)-NH-NH_2, and Ub[46-75]-(A46C)-NH-NH_2. Each peptide segment should be synthesized separately using Fmoc-based SPPS of peptide hydrazides on 2-chlorotrityl chloride resin. Naturally occurring Ala residues on the N-termini of Ub [28-45]-(A28C)-NH-NH_2 and Ub[46-75]-(A46C)-NH-NH_2 are exchanged to Cys residues to enable subsequent native chemical ligation (NCL) reactions. The synthesis of Ub[46-75]-(A46C)-NH-NH_2 requires the usage of dipeptide building blocks at $Asp^{52}Gly^{53}$ (Fmoc-Asp(OtBu)-(Dmb)Gly-OH) and $Ser^{65}Thr^{66}$ (Fmoc-Ser(tBu)-Thr(ΨMe,Mepro)-OH), as reported elsewhere [31]. Purify each peptide hydrazide segment by preparative HPLC and lyophilize the product. Herein, we present a general protocol for Ub derivative synthesis based on the example of Biot-6-Ahx-Ub-VME (*see* Fig. 4).

1. To synthesize Ub[28-45]-(A28C)-NH-NH_2 segment, start the synthesis with the preparation of Fmoc-hydrazide resin. Swell 500 mg of 2-CTC resin (0.97 mmol/g) for 20 min in fresh DCM in a solid-phase peptide synthesis vessel with gentle stirring (once per 4 min).
2. Remove DCM by vacuum filtration and wash the resin three times with DCM.
3. Dissolve 4 eq of Fmoc-NH-NH_2 in 5 mL of DMF:DCM (8:1, v/v) in a 15 mL Falcon tube, and add 10 eq of DIPEA. After 1 min of preactivation, transfer mixture to the reaction vessel with the resin.

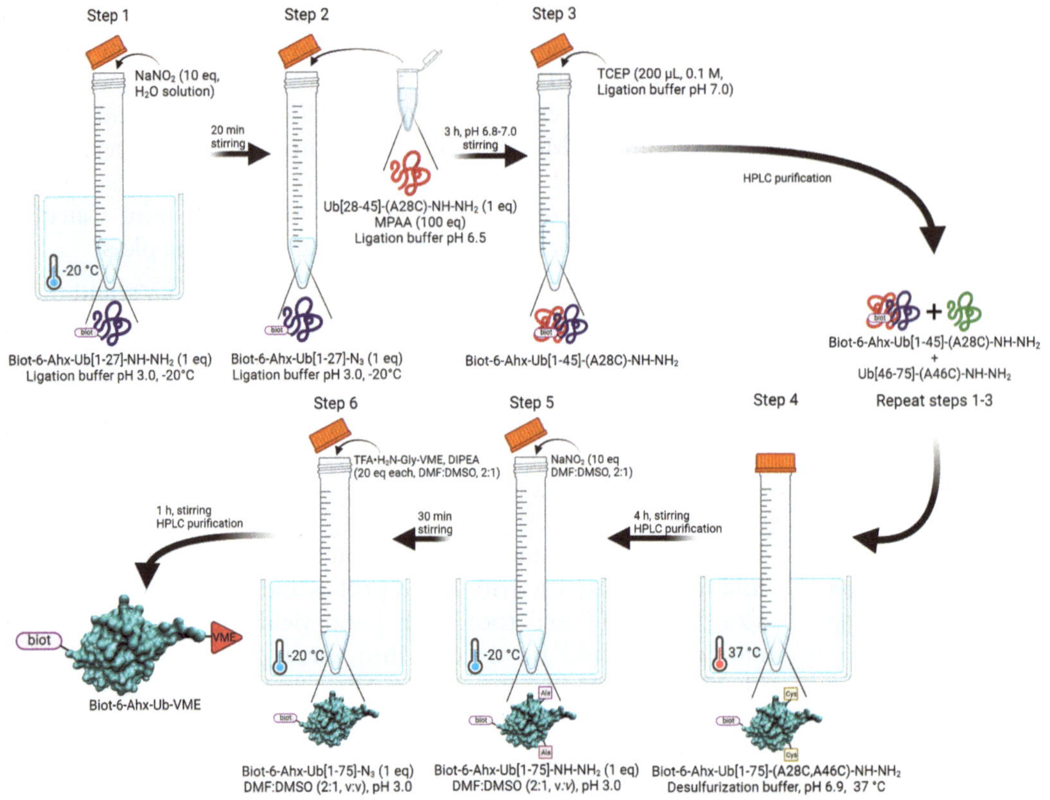

Fig. 4 A general outline of Ub-based probe synthesis. (Created with BioRender.com)

4. Perform the reaction at room temperature with slow agitation. After 1 h, add 5 eq of DIPEA and continue Fmoc-NH-NH_2 coupling for 3 h.
5. Remove the reaction mixture by vacuum filtration, and wash the resin three times with DMF.
6. Perform end-capping of residual active 2-CTC resin. Add 4.2 mL of a capping mixture (DMF/MeOH/DIPEA, 3.5/0.5/0.2, v/v/v) to the reaction vessel, and incubate for 30 min with slow agitation.
7. Remove the reaction mixture by vacuum filtration, and wash the resin three times with DMF.
8. Remove Fmoc groups by incubating the resin in 5 mL of 20% piperidine in DMF for 5 and 25 min. Each time agitate the resin gently, and remove the solution by vacuum filtration. After the first 5 min, wash the resin once with 5 mL of DMF, and add 5 mL of 20% piperidine in DMF. After the second deprotection (25 min), wash the resin seven times with 5 mL of DMF.

9. To ensure that amine groups are deprotected, perform a ninhydrin test. Transfer a few resin beads to a glass test tube, and add 500 μL of ninhydrin ethanol solution. Incubate the tube at 95 °C for 5 min. If beads turn dark blue (positive test), start the next step of the synthesis. If beads are yellow (negative test), repeat **step 8** and perform ninhydrin test once again (*see* **Note 7**).
10. In a 15 mL Falcon tube, dissolve 2.5 eq of the first amino acid and 2.5 eq of HATU in DMF. Add 2.5 eq of 2,4,6-collidine. Activate the functional groups for 1 min, and transfer the mixture to the reaction vessel. Agitate gently for 24 h.
11. After coupling, remove the reaction mixture by vacuum filtration, and wash the resin three times with DMF. Perform ninhydrin test (**step 9**). If the test is negative, proceed to **step 12**. If positive, repeat **step 10** with half reagents to increase coupling efficiency.
12. To attach another amino acids, repeat **steps 8–11**, but agitate gently for 2.5 h. Sometimes double coupling or elongation of coupling time is needed. If the ninhydrin test is positive after the first coupling of amino acid, perform a second coupling with half reagents to increase coupling efficiency.
13. After the last amino acid coupling, remove the Fmoc group (repeat **step 8**). Next, wash the resin three times with DCM and three times with MeOH, and dry over P_2O_5 overnight.
14. To remove peptide from the resin, add a cleavage cocktail (8 mL, TFA/H_2O/TIPS, 95/2.5/2.5, %v/v/v) to the dry resin. After 2 h, collect the peptide in a 50 mL Falcon tube, and precipitate in 30 mL of cold Et_2O at −20 °C for 30 min.
15. Centrifuge precipitate at 3000 × g for 5 min at 4 °C, and remove supernatant. Add 15 mL of cold Et_2O, shake, precipitate at −20 °C for 20 min, and centrifuge again. Remove supernatant and dry crude peptide on air.
16. Purify crude peptide hydrazide by preparative HPLC and lyophilize.
17. To synthesize Ub[46-75]-(A46C)-NH-NH_2 segment, repeat **steps 1–16**.
18. To synthesize Biot-6-Ahx-Ub[1-27]-NH-NH_2, the last two positions to be attached are a linker, Fmoc-6-Ahx-OH, and a tag – biotin. Repeat **steps 8–11** (coupling time 2.5 h) to attach the linker.
19. Next, attach a biotin tag. Using 15 mL Falcon tubes, dissolve 2.5 eq of biotin and 2.5 eq of HATU in DMF/DMSO (1:1, v/v). Add 2.5 eq of 2,4,6-collidine. Activate the functional groups for 1 min, and transfer the mixture to the reaction vessel. Agitate gently for 2.5 h.

20. Remove reaction mixture by vacuum filtration, wash the resin three times with DMF, and perform the ninhydrin test (**step 9**). If the test result is negative, wash the resin three times with DCM and three times with MeOH, and dry over P_2O_5 overnight. Then repeat **steps 14–16**.
21. To assemble peptide hydrazide segments, perform native chemical ligation reactions according to a published protocol [32]. In a 15 mL Falcon tube, dissolve Biot-6-Ahx-Ub [1-27]-NH-NH_2 (27.4 mg, 7.13 μmol) in 2 mL of ligation buffer, and cool the mixture in an ice-salt bath to −20 °C for 20 min with stirring.
22. Prepare a fresh $NaNO_2$ water stock solution by dissolving 61.5 mg in 500 μL of water. Add 10 eq of $NaNO_2$ by adding 40 μL of the stock solution to the reaction mixture. Generate the acyl azide for 20 min at −15 °C with stirring.
23. Dissolve Ub[28-45]-(A28C)-NH-NH_2 (1 eq, 19.3 mg) in 1.5 mL of ligation buffer, add 100 eq of MPAA, and adjust pH to 6.5 using 6 M NaOH.
24. Add solution from **step 23** to the Falcon tube with the generated acyl azide Biot-6-Ahx-Ub[1-27]-N_3, remove the ice-salt bath, and, when the reaction mixture reaches room temperature, slowly adjust the pH to 6.8 (*see* **Note 8**).
25. Monitor the progress of the reaction by LC-MS. Prepare the sample for the LC-MS analysis by adding 10 μL of the reaction mixture to a vial with 650 μL of water and 40 μL of TCEP (200 μL, 0.1 M in ligation buffer, pH 7.0), shake the vial using vortex, and incubate for 15 min at room temperature. Observe the consumption of the substrates and product formation using UV-Vis detector ($\lambda = 220$ nm) and mass detector (e.g., Waters ACQUITY QDa). The reaction is usually completed within 3 h.
26. After completing the reaction, add a reducing agent (TCEP, 200 μL, 0.1 M in ligation buffer, pH 7.0).
27. Purify the ligation product (Biot-6-Ahx-Ub[1-45]-(A28C)-NH-NH_2) by semipreparative HPLC, freeze the collected product, and lyophilize it overnight using lyophilizer.
28. To ligate Biot-6-Ahx-Ub[1-45]-(A28C)-NH-NH_2 with Ub [46-75]-(A46C)-NH-NH_2, repeat **steps 21–27** with adjusted amounts of reagents.
29. Purify Biot-6-Ahx-Ub[1-75]-(A28C, A46C)-NH-NH_2 and lyophilize.

30. To convert cysteine residues back to alanine, perform a radical desulfurization reaction. Dissolve Biot-6-Ahx-Ub [1-75]-(A28C,A46C)-NH-NH_2 (20 mg, 1.92 μmol) in 1.8 mL of desulfurization buffer (*see* **Notes 9–10**).
31. Run the desulfurization for 4 h at 37 °C with stirring.
32. Monitor the progress of the reaction by LC-MS. Prepare the sample for the LC-MS analysis by adding 10 μL of the reaction mixture to a vial with 690 μL of water, and shake the vial using vortex. Observe the consumption of the substrates and product formation using UV-Vis detector ($\lambda = 220$ nm) and mass detector (e.g., Waters ACQUITY QDa). The reaction is usually completed within 4 h.
33. Purify Biot-6-Ahx-Ub[1-75]-NH-NH_2 by semipreparative HPLC and lyophilize.
34. To obtain Ub-based probe, dissolve Biot-6-Ahx-Ub [1-75]-NH-NH_2 (6.7 mg, 0.65 μmol) in DMF:DMSO (0.9 mL, 2:1, v/v).
35. Cool the mixture in an ice-salt bath to −20 °C for 20 min with stirring.
36. Add 10 eq of $NaNO_2$ in DMF:DMSO, and adjust the pH of the reaction mixture to 3.0–4.0 with 4 M HCl in dioxane solution (*see* **Notes 11–12**).
37. After 30 min monitor the acyl azide generation progress using LC-MS. Prepare the sample for the LC-MS analysis by adding 10 μL of the reaction mixture to a vial with 690 μL of H_2O: ACN (3:1, v/v), and shake the vial using vortex. Observe the consumption of the substrates and product formation using UV-Vis detector ($\lambda = 220$ nm) and mass detector (e.g., Waters ACQUITY QDa). Reaction is completed usually within 30 min. When the acyl azide is generated, add 20 eq of TFA·H_2N-Gly-VME dissolved in 100 μL of DMF:DMSO (2: 1, v/v) and 20 eq of DIPEA. Adjust the pH to 8.0–9.0 with DIPEA (*see* **Note 12**).
38. Perform the reaction at −20 °C for 1 h with stirring. Monitor the progress of the reaction by LC-MS. Repeat the procedure for sample preparation from **step 37**. The reaction is usually completed within 1 h.
39. When the reaction is completed, dilute the reaction mixture with DMSO to avoid product precipitation. Purify Biot-6-Ahx-Ub-VME by analytical HPLC, lyophilize, and characterize by a high-resolution mass spectrometry (HRMS).
40. To synthesize Ub-based substrate, repeat **steps 1–17** and **21–39**, but use Ub[1-27]-NH-NH_2 segment instead of Biot-6-Ahx-Ub[1-27]-NH-NH_2 and TFA·H_2N-Gly-ACC instead of TFA·H_2N-Gly-VME.

3.4 DUB Labeling in Cell Lysates

1. Culture selected cell lines in appropriated medium (e.g., for HeLa and HEK-293T cells, use DMEM) supplemented with 10% FBS, 2 mM L-glutamine, and antibiotics (100 U/mL penicillin, 100 μg/mL streptomycin) in a humidified 5% CO_2 atmosphere at 37 °C.
2. Harvest approximately 1,200,000 cells and wash three times with PBS.
3. To obtain cell lysate, suspend the cell pellet in 300–400 μL of lysis buffer and sonicate.
4. Centrifuge cell lysate for 10 min and collect the supernatant. Cell lysate can be stored at this stage for later use by freezing the supernatant and storage in liquid nitrogen.
5. Prepare samples for SDS-PAGE analysis. Incubate 20 μL of cell lysate with 80 μL of Ub-based probes at different concentrations in lysis buffer for 30 min at 37 °C. The amount of cell lysate and probe concentrations depend on DUB activity (*see* **Notes 13–14**).
6. Add 50 μL of 3× SDS/DTT loading buffer and boil the samples for 5 min at 95 °C.
7. Add 30 μL of samples into wells of 4–12% Bis-Tris Plus 12-well gel, and perform electrophoresis at 165 V for 45 min. Next, transfer the proteins to a nitrocellulose membrane for 60 min at 10 V.
8. Block the membrane with 2% BSA in TBS-T for 60 min at room temperature.
9. Wash the membrane with TBS-T for 5 min. Repeat this step three times.
10. Incubate the membrane with a fluorescent streptavidin Alexa Fluor 647 conjugate (1:10000) in TBS-T with 1% BSA for 1 h.
11. Wash the membrane with TBS-T for 5 min with stirring. Repeat this step three times.
12. Detect biotinylated Ub-based probe using imaging system with software (e.g., Azure Biosystems Sapphire Biomolecular Imager and Azure Spot Analysis Software).
13. To confirm that Ub-based probe detects selected DUB, use primary antibody. Incubate the membrane from **step 11** with selected primary antibody overnight at 4 °C with gentle agitation.
14. Wash the membrane with TBS-T for 20 min. Repeat this step three times.
15. Incubate the membrane with secondary antibody conjugated with a fluorescent tag for 1 h at room temperature with gentle agitation (*see* **Note 15**).

16. Wash the membrane with TBS-T for 5 min. Repeat this step three times.
17. Scan the membrane using an imaging system with software (e.g., Azure Biosystems Sapphire Biomolecular Imager and Azure Spot Analysis Software).

4 Notes

1. DUB activity in substrate kinetic assays depends on the composition of assay buffer. Sometimes DUBs for optimal activity require buffer with a low concentration of salt or buffer supplemented with sodium citrate (kosmotropic salt) [33]. An optimal buffer for each DUB should be used to achieve good enzyme activity in the kinetic assay.
2. Different DUBs require different preincubation times at 37 °C before kinetic assay. The enzyme should be preincubated at 37 °C in assay buffer for 5, 10, 15, 30, 45, and 60 min, and its activity toward Ac-LRGG-ACC should be tested to determine the preincubation time for each DUB.
3. To ensure that the enzyme exhibits the same activity during the screening of one sublibrary (three separate measurements), add the same substrate into each plate as a control (Ac-Leu-Mix-Gly-Gly-ACC, Ac-Mix-Arg-Gly-Gly-ACC, and Ac-Leu-Arg-Gly-Gly-ACC for P4, P3, and P2 sublibrary, respectively).
4. It is important to add the best substrate (the highest RFU/s value = 100%) from each of the sublibraries when the measurement is repeated only for selected substrates, for which the standard deviation was higher than 10% of the mean value.
5. For long-term storage, substrates should be kept at −80 °C.
6. To compare DUB activity toward tetrapeptide substrates, the concentration of the substrates should be the same in all DUB assays.
7. Ninhydrin test after amine deprotection gives decisive results only for primary amines.
8. To avoid local hydrolysis of thioester generated in situ at high pH, add NaOH dropwise using a pipette. Put a drop of NaOH solution on the wall of the reaction vessel, and then use the vortex to mix it with the reaction mixture.
9. During the desulfurization reaction, tBuSH is used. The reagent has a strong odor, and while handling it, use protective equipment and work in a well-ventilated room under the fume hood.

10. Add tBuSH to the reaction mixture when all the other reagents are added. In this manner, you reduce the exposure time to the substance.
11. Prepare 10 mg/ml solution of $NaNO_2$ in DMF:DMSO 2:1. Add a suitable volume of the solution to the reaction mixture.
12. To measure the pH of the organic mixture, you can use pH-indicator paper. Pipette approx. 5 μL of water on a paper, and then add 1 μL of the reaction mixture to check the pH.
13. To determine the optimal amount of cell lysate to detect DUBs, prepare samples with different amounts of cell lysate from 5 to 50 μL, and add the same concentration of Ub-based probe (probe containing a natural sequence of Ub) in assay buffer to the final volume of 100 μL to each sample. Incubate the samples at 37 °C for 30 min. Then, repeat **steps 6–12** from Subheading 3.4.
14. The concentration of Ub-based probe required to detect selected DUB depends on DUB activity toward the designed probe and the amount of active enzyme in the selected cell line. To determine the optimal concentration of the probe for DUB detection in cell lysate, incubate each designed probe at different concentrations (ranging from 5 nM to 500 nM) with cell lysates for 30 min at 37 °C. Then, 3× SDS/DTT should be added to each sample. Boil the samples for 5 min at 95 °C, and run on a gel. Electrophoresis, protein transfer to a nitrocellulose membrane, and probe visualization should be carried out in the same manner as described in **steps 7–12** from Subheading 3.4.
15. If streptavidin conjugated with Alexa Fluor 647 is used to detect biotinylated Ub-based probe, the secondary antibody should be a different fluorophore conjugate with a spectrum that does not overlap with the spectrum of Alexa Fluor 647 (e.g., Alexa Fluor 488, Alexa Fluor 800).

Acknowledgments

This work was supported by the National Science Center grant 2015/17/N/ST5/03072 (Preludium 9) in Poland (W.R.) and the “TEAM/2017-4/32” project, which is carried out within the TEAM program of the Foundation for Polish Science, co-financed by the European Union under the European Regional Development Fund (M.D.).

References

1. Hershko A, Ciechanover A (1998) The ubiquitin system. Annu Rev Biochem 67:425–479
2. Mevissen TET, Komander D (2017) Mechanisms of Deubiquitinase specificity and regulation. Annu Rev Biochem 86:159–192
3. Komander D, Rape M (2012) The ubiquitin code. Annu Rev Biochem 81:203–229
4. Dittmar G, Winklhofer KF (2019) Linear ubiquitin chains: cellular functions and strategies for detection and quantification. Front Chem 7:915
5. Clague MJ, Urbe S, Komander D (2019) Breaking the chains: deubiquitylating enzyme specificity begets function. Nat Rev Mol Cell Biol 20(6):338–352
6. Gu Z, Shi W (2016) Manipulation of viral infection by deubiquitinating enzymes: new players in host-virus interactions. Future Microbiol 11:1435–1446
7. Harrigan JA et al (2018) Deubiquitylating enzymes and drug discovery: emerging opportunities. Nat Rev Drug Discov 17(1):57–78
8. Schauer NJ et al (2020) Advances in discovering Deubiquitinating enzyme (DUB) inhibitors. J Med Chem 63(6):2731–2750
9. Farshi P et al (2015) Deubiquitinases (DUBs) and DUB inhibitors: a patent review. Expert Opin Ther Pat 25(10):1191–1208
10. Fraile JM et al (2012) Deubiquitinases in cancer: new functions and therapeutic options. Oncogene 31(19):2373–2388
11. Fonovic M, Bogyo M (2008) Activity-based probes as a tool for functional proteomic analysis of proteases. Expert Rev Proteomics 5(5): 721–730
12. Komander D, Clague MJ, Urbe S (2009) Breaking the chains: structure and function of the deubiquitinases. Nat Rev Mol Cell Biol 10(8):550–563
13. Iphofer A et al (2012) Profiling ubiquitin linkage specificities of deubiquitinating enzymes with branched ubiquitin isopeptide probes. Chembiochem 13(10):1416–1420
14. McGouran JF et al (2013) Deubiquitinating enzyme specificity for ubiquitin chain topology profiled by di-ubiquitin activity probes. Chem Biol 20(12):1447–1455
15. Flierman D et al (2016) Non-hydrolyzable Diubiquitin probes reveal linkage-specific reactivity of Deubiquitylating enzymes mediated by S2 pockets. Cell Chem Biol 23(4):472–482
16. Li G et al (2014) Activity-based diubiquitin probes for elucidating the linkage specificity of deubiquitinating enzymes. Chem Commun (Camb) 50(2):216–218
17. Borodovsky A et al (2002) Chemistry-based functional proteomics reveals novel members of the deubiquitinating enzyme family. Chem Biol 9(10):1149–1159
18. Love KR et al (2009) Ubiquitin C-terminal electrophiles are activity-based probes for identification and mechanistic study of ubiquitin conjugating machinery. ACS Chem Biol 4(4): 275–287
19. Gjonaj L et al (2019) Development of a DUB-selective fluorogenic substrate. Chem Sci 10(44):10290–10296
20. Gjonaj L et al (2019) USP7: combining tools towards selectivity. Chem Commun (Camb) 55(35):5075–5078
21. Rut W et al (2020) Engineered unnatural ubiquitin for optimal detection of deubiquitinating enzymes. Chem Sci 11(23):6058–6069
22. Drag M et al (2008) Positional-scanning fluorigenic substrate libraries reveal unexpected specificity determinants of DUBs (deubiquitinating enzymes). Biochem J 415(3):367–375
23. Poreba M et al (2014) Unnatural amino acids increase sensitivity and provide for the design of highly selective caspase substrates. Cell Death Differ 21(9):1482–1492
24. Rut W et al (2017) Extended substrate specificity and first potent irreversible inhibitor/activity-based probe design for Zika virus NS2B-NS3 protease. Antivir Res 139:88–94
25. Rut W et al (2018) Selective substrates and activity-based probes for imaging of the human constitutive 20S proteasome in cells and blood samples. J Med Chem 61(12): 5222–5234
26. Poreba M et al (2016) Counter selection substrate library strategy for developing specific protease substrates and probes. Cell Chem Biol 23(8):1023–1035
27. Poreba M et al (2018) Selective imaging of cathepsin L in breast cancer by fluorescent activity-based probes. Chem Sci 9(8): 2113–2129
28. Vedrenne E et al (2005) Dramatic effect of boron-based Lewis acids in cross-metathesis reactions. Synlett 4:670–672
29. Maly DJ et al (2002) Expedient solid-phase synthesis of fluorogenic protease substrates using the 7-amino-4-carbamoylmethylcoumarin (ACC) fluorophore. J Org Chem 67(3):910–915

30. Poreba M et al (2014) Positional scanning substrate combinatorial library (PS-SCL) approach to define Caspase substrate specificity. Caspases, Paracaspases, and Metacaspases: Methods and Protocols 1133:41–59
31. El Oualid F et al (2010) Chemical synthesis of ubiquitin, ubiquitin-based probes, and Diubiquitin. Angewandte Chemie-International Edition 49(52):10149–10153
32. Zheng JS et al (2013) Chemical synthesis of proteins using peptide hydrazides as thioester surrogates. Nat Protoc 8(12):2483–2495
33. Drag M et al (2008) Positional-scanning fluorigenic substrate libraries reveal unexpected specificity determinants of DUBs (deubiquitinating enzymes). Biochem J 415:367–375

Chapter 6

High-Throughput Assay for Characterizing Rpn11 Deubiquitinase Activity

Gang Xie, Ken C. Dong, Evan J. Worden, and Andreas Martin

Abstract

Rpn11 is an essential metalloprotease responsible for the en bloc removal of ubiquitin chains from protein substrates that are targeted for degradation by the 26S proteasome. A unique feature of Rpn11 is that its deubiquitinase (DUB) activity is greatly stimulated by the mechanical translocation of the substrate into the proteasomal AAA+ (ATPase Associated with diverse cellular Activities) motor, which delivers the scissile isopeptide bond between a substrate lysine and the proximal moiety of an attached ubiquitin chain to the DUB catalytic active site. As a consequence, Rpn11 cleaves at the base of ubiquitin chains and lacks selectivity towards specific ubiquitin-chain linkage types, which is in contrast to other DUBs, including the related AMSH that selectively cleaves Lys63-linked chains. Prevention of Rpn11's deubiquitinase activity leads to inhibition of proteasomal degradation by stalling substrate translocation. With the proteasome as an approved anticancer target, Rpn11 is therefore an attractive point of attack for the development of new inhibitors, which requires robust biochemical assays to measure DUB activity. Here we describe a method for the purification of the Rpn8/Rpn11 heterodimer and ubiquitin-GC-TAMRA, a model substrate that can be used to characterize the DUB activity of Rpn11 in isolation without the need of purifying 26S proteasomes. This assay thus enables a high-throughput screening platform for Rpn11-targeted small-molecule discovery.

Key words Deubiquitinase (DUB), Ubiquitin, High-throughput screening (HTS), 26S proteasome

1 Introduction

The 26S proteasome is the major compartmental protease in eukaryotic cells, responsible for the degradation of proteins through the ubiquitin-proteasome system (UPS) [1–3]. It contains the barrel-shaped 20S core particle (CP), which sequesters the proteolytic sites in an internal chamber, and the 19S regulatory particle (RP) that caps the entrances of the CP and regulates the ubiquitin-dependent delivery of protein substrates into the proteolytic core (Fig. 1a). The RP contains ubiquitin receptors, a hetero-hexameric AAA+ (ATPase Associated with diverse cellular Activities) ATPase motor, and the deubiquitinase (DUB) Rpn11

Julie A. Maupin-Furlow and Mariola J. Edelmann (eds.), *Deubiquitinases: Methods and Protocols*,
Methods in Molecular Biology, vol. 2591, https://doi.org/10.1007/978-1-0716-2803-4_6,

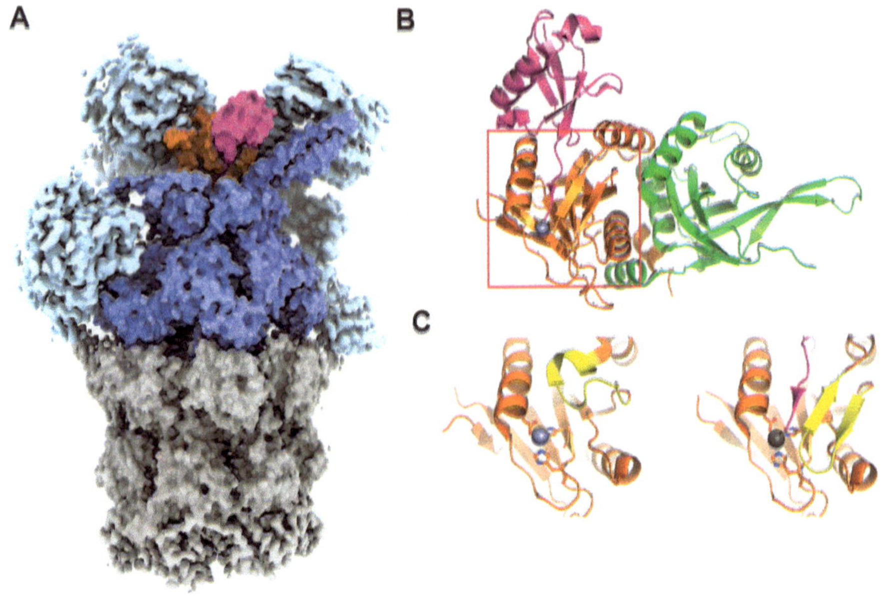

Fig. 1 Rpn11 is the essential deubiquitinase of the 26S proteasome. (**a**) Overview of the 26S proteasome, with the 20S core particle colored in gray and the 19S regulatory particle in blue. The hexameric ATPase motor of the 19S RP is shown in dark blue, Rpn11 in orange, and a Rpn11-bound ubiquitin in pink (EMDB ID: 9045). (**b**) The structure of the Rpn8/Rpn11 heterodimer bound to wild-type ubiquitin, with Rpn8 shown in green, Rpn11 in orange, and ubiquitin in pink (PDB ID: 5u4p). (**c**) Insert-1 conformational change. Left: Close-up of the Rpn11 region boxed in panel B in the ubiquitin-free state with the Insert-1 region (yellow) forming an inhibitory loop over the active site and blocking the catalytic Zn^{2+} (grey ball) (PDB ID: 4o8×). Right: In the ubiquitin-bound structure, the Insert-1 region forms a beta-hairpin and a small beta sheet with the C-terminus of ubiquitin (PDB ID: 5u4p). The Zn^{2+}-coordinating Glu and His residues are shown as sticks

positioned in key locations to aid in ATP- and ubiquitin-dependent protein degradation. For a protein to be recognized as a substrate, it requires both a polyubiquitin signal for binding to a proteasomal receptor and an unstructured initiation region to insert into and be engaged by the ATPase motor [4–6]. The small 8.6 kDa ubiquitin is post-translationally attached to misfolded, damaged, or obsolete proteins by a cascade of E1, E2, and E3 enzymes that form an isopeptide bond between ubiquitin's C-terminus and a lysine residue in the substrate [7]. Additional ubiquitin moieties can then be added using one of the seven lysines (K6, K11, K27, K29, K33, K48, and K63) or the N-terminal methionine (M1 or linear) in ubiquitin to homotypic, heterotypic, or branched chains. Depending on linkage type, length, and branching, these chains can be used as signals for diverse biological pathways, such as degradation (K11 and K48-linked chains), trafficking and autophagy (K63-linked chains), and immune response (M1-linked chains) [8–11]. Substrates targeted for degradation bind the 26S proteasome through one or several of the intrinsic ubiquitin receptors: Rpn1 [12], Rpn10 [13], and Rpn13 [14], allowing for the engagement of the unstructured initiation region by the ring-shaped AAA+

ATPase motor [5]. Rpn11 is positioned above the entrance to the central channel of the ATPase [15], such that ATP-hydrolysis-dependent substrate translocation delivers the isopeptide bond of a substrate-attached ubiquitin to the DUB active site for efficient cleavage and recycling of ubiquitin [16, 17].

Rpn11 was identified to be the essential deubiquitinase of the 26S proteasome and a member of the Jab1/MPN (JAMM) protease family, containing the MPN (Mpr1/Pad1 N-terminal) protein fold and an EXn-HXHXX10D motif that coordinates a catalytic Zn^{2+} ion for isopeptide-bond hydrolysis [18–20]. Rpn11 is therefore resistant to probes such as ubiquitin-vinyl sulfone that target cysteine-dependent deubiquitinases. Mutations to the Zn^{2+}-coordinating motif are lethal in yeast [19], and the addition of the zinc chelator o-phenanthroline (OPA) inhibits deubiquitinase activity [17, 19].

Rpn11 forms an obligate heterodimer with another proteasomal subunit, Rpn8, both within the context of the 26S proteasome holoenzyme and in isolation (Fig. 1b) [21, 22]. This heterodimer can be purified and is capable of cleaving ubiquitin dimers of all linkage types with the exception of M1-linked, linear ubiquitin. However, its ubiquitin affinity (K_M > 100 μM) and rate of isopeptide cleavage (k_{cat} ~ 0.3–0.7 min^{-1}) are relatively low [21]. This is in stark contrast to the related JAMM protease AMSH that has a preference for K63-linked chains through specific contacts with the proximal ubiquitin and cleaves ubiquitin chains ~50–100-fold faster than Rpn11 [23, 24]. To prevent rampant deubiquitinase activity of Rpn11 prior to substrate engagement by the proteasomal motor, the active site is sequestered by the Insert-1 loop that physically blocks the catalytic Zn^{2+} [21, 22]. For Rpn11 to be fully active, this inhibitory loop transitions to a beta-hairpin conformation that forms a small sheet with the C-terminal beta strand of ubiquitin and stabilizes the isopeptide bond in the active site for cleavage (Fig. 1c) [17]. Importantly, mechanical substrate translocation by the ATPase motor delivers ubiquitin to Rpn11, facilitates the rate-limiting loop-to-hairpin transition of the Insert-1 region, and thus accelerates ubiquitin cleavage, with these translocation-coupled deubiquitination rates approaching the cleavage rates of AMSH [16, 17].

Due to its key role in ubiquitin-dependent degradation by the 26S proteasome, Rpn11 has emerged as a potential target for small-molecule drug discovery. The 26S proteasome is an important clinical target for the treatment of multiple myeloma, for instance, by bortezomib and carfilzomib that reversibly or irreversibly inhibit the proteolytic activity of the 20S CP [25–29]. Unfortunately, these cancers eventually develop resistance to the CP inhibitors, necessitating alternative approaches [30]. Bortezomib-resistant multiple myeloma cell lines overexpress Rpn11, and the addition of the Rpn11 inhibitor OPA leads to apoptosis [31], which opens a possible new avenue for targeted inhibition of proteasomal degradation. The Deshaies group used 26S proteasome holoenzymes in high-throughput screens and identified capzimin as a small-molecule inhibitor with specificity for Rpn11 over other related metalloproteases [32–36]. Building off this discovery, novel ubiquitin-capzimin-based probes have also been used to search for other metalloprotease DUBs [37].

The development of novel DUB inhibitors requires biochemical assays that can reliably monitor the activity of Rpn11. Here, we describe a robust enzymatic assay based on a ubiquitin with a 5-carboxytetramethylrhodamine (TAMRA) modification attached via maleimide to a C-terminal Gly-Cys extension (Ub-GC-TAMRA) that is suitable for characterizing Rpn11 activity and the discovery of small-molecule inhibitors.

2 Materials

2.1 General Reagents for the Expression of Rpn8/Rpn11 and Ub-GC-TAMRA

1. Chemically competent *E. coli* BL21 Star (DE3) cells.
2. Expression plasmids for Rpn8/Rpn11 and ubiquitin-GC (*see* Table 1) are available from Addgene.
3. 1000X ampicillin: 300 mg/mL ampicillin sodium in 50% EtOH, stored at –20 °C.
4. 1000X kanamycin: 50 mg/mL kanamycin monosulfate in water, stored at –20 °C.
5. 1000X chloramphenicol: 25 mg/mL chloramphenicol in 100% EtOH, stored at –20 °C.
6. Luria-Bertani (LB) broth-antibiotic plates: 1% (w/v) tryptone, 0.5% (w/v) yeast extract, 0.5% (w/v) NaCl, 1.5% (w/v) agar. Autoclave, cool to 60 °C or lower in a water bath before adding appropriate antibiotics to 1X from 1000X stocks.
7. Double Yeast extract Tryptone (dYT) liquid media: 1.6% (w/v) tryptone, 1% (w/v) yeast extract, 0.5% (w/v) NaCl. Autoclave.
8. Terrific broth (TB): 47 g of Novagen terrific broth (Millipore catalog number 71754–5) in 1 L of water +4 mL of glycerol. Autoclave.

Table 1
Expression plasmids used in this protocol

Plasmid #	Description	Vector backbone	Antibiotic marker
pEW26	Rpn8(1–179)-PreScission-strep tag, His_6-precission-Rpn11(2–239)	pETDuet-1	Ampicillin
pEW145	pET28a MWG-ubiquitin-G77-C78	pET-28a(+)	Kanamycin/Chloramphenicol

9. 1 M $ZnCl_2$.
10. 2.5 L Ultra Yield baffled shaking flask (Thompson).
11. DuPont RC-3B 1 L bucket centrifuge using a H6000A rotor or equivalent.
12. 1 M IPTG: 1 M isopropyl-beta-D-thiogalactoside in water (Gold Bio).
13. 100× PMSF: 0.2613 g phenylmethylsulfonyl fluoride in 15 mL ethanol (*see* **Note 1**).
14. 1000X aprotinin: 1 mg/mL aprotinin in 25% methanol/75% H_2O (*see* **Note 1**).
15. 1000X pepstatin = 1 mg/mL pepstatin A in 90% methanol/10% acetic acid (*see* **Note 1**).
16. 1000X leupeptin = 1 mg/mL leupeptin hemisulfate in H_2O (*see* **Note 1**).
17. Benzonase nuclease (Millipore, Cat. # 70664-3).
18. Ultrasonic homogenizer with horn attachment. We use the Q sonica sonicator Q500 model.
19. 125 mL metal beaker (Polar Ware).
20. Sorvall Super T21 refrigerated centrifuge.
21. Econo-Column™ chromatography column (Bio-Rad, 7371517).
22. 50 mL conical centrifuge tubes.
23. Quick Start ™ Bradford Protein Assay reagent (Bio-Rad).
24. 30 kDa cutoff Ultra-15 centrifugal filter concentrator (Amicon).
25. Thermo Scientific ST40R refrigerated centrifuge.
26. Spin-X 0.22 0.5 mL cellulose acetate centrifuge filter (Corning).
27. AKTA™ FPLC, Explorer, or Pure (Cytiva, formerly GE) or equivalent purification device.
28. 2× SDS loading buffer: 100 mM Tris-HCL, pH 6.8, 4% (w/v) sodium dodecyl sulfate (SDS), 0.2% (w/v) bromophenol blue, 20% (w/v) glycerol. Add dithiothreitol (DTT) just before using (final concentration of 0.1 M).
29. 0.5 M TCEP: Add 5.73 g tris(2-carboxyethyl)phosphine-HCl to 30 mL of water. Adjust pH to 7.0 with 5 M NaOH, and then adjust final volume to 40 mL with water. Aliquot and store at −20 °C.
30. 4–20% Mini-PROTEAN® TGX™ Precast Protein Gels and Mini-PROTEAN Tetra Vertical Electrophoresis Cell (Bio-Rad).

31. Coomassie blue stain: 0.25% (w/v) Coomassie blue R-250, 0.25% (w/v) Coomassie blue G-250, 45% methanol, 45% water, 10% glacial acetic acid. Filter before using.
32. Coomassie destain: 20% methanol, 10% glacial acetic acid in water.
33. UV spectrophotometer.
34. Quartz cuvette, 20 μL sample volume (Hellma).
35. Liquid nitrogen (for freezing proteins).

2.2 Additional Reagents and Equipment for Expression and Purification of Rpn8/Rpn11 Heterodimer

1. His-Pur Ni-NTA resin (Thermo Fisher).
2. NiA buffer: 60 mM HEPES, pH 8.0, 100 mM NaCl, 100 mM KCl, 10 mM $MgCl_2$, 20 mM imidazole, 10% glycerol.
3. NiB buffer: NiA buffer +250 mM imidazole. Add dry imidazole to the NiA buffer, and then readjust pH to 8.0.
4. HiLoad 16/60 Superdex 200 pg column (Cytiva, formerly GE Life Sciences).
5. GF buffer: 60 mM HEPES, pH 8.0,100 mM NaCl, 100 mM KCl, 10% glycerol, 1 mM TCEP.

2.3 Additional Reagents for Purification of Ubiquitin and Labeling Ub-GC-TAMRA

1. Chemically competent *E. coli* BL21 Star (DE3) cells.
2. 5 mL HiTrap SP FF column (Cytiva, formerly GE).
3. Ubiquitin lysis buffer: 50 mM Tris HCl, pH 7.6, 1 mM DTT, 0.02% NP-40, protease inhibitors (1 μg/mL aprotinin, 1 μg/mL pepstatin, 1 μg/mL leupeptin, 174 μg/mL PMSF), 2 mg/ml lysozyme, and 50 units of Benzonase.
4. 3.5 K MWCO Spectra/Por dialysis tubing (Thermo Fisher).
5. Ubiquitin dialysis buffer: 50 mM ammonium acetate, pH 4.5, 1 mM EDTA, 1 mM DTT. Adjust the buffer pH with acetic acid.
6. SP Buffer A: 50 mM ammonium acetate pH 4.5, 1 mM EDTA, 1 mM DTT.
7. SP Buffer B: 50 mM ammonium acetate, pH 4.5, 500 mM NaCl, 1 mM EDTA, 1 mM DTT.
8. 3 kDa cutoff Ultra-15 centrifugal filter concentrator (Amicon, Cat. # UFC901024).
9. Glacial acetic acid.
10. 0.2 micron bottle top filter.
11. Ubiquitin storage buffer: 20 mM Tris, pH 7.5, 150 mM NaCl, 1 mM DTT.
12. Activated thiol Sepharose 4B (Cytiva, formerly GE) (Product number 17064001).
13. Thiol 4B binding buffer: 100 mM Tris, pH 7.5, 300 mM NaCl, 1 mM EDTA.

14. Thiol 4B elution buffer: 100 mM Tris, pH 7.5, 300 mM NaCl, 1 mM EDTA, 35 mM DTT.
15. Labeling buffer: 50 mM HEPES, pH 7.0, 150 mM NaCl, 1 mM TCEP.
16. Size-exclusion buffer: 60 mM HEPES, pH 7.0, 150 mM NaCl, 1 mM TCEP.
17. Tetramethylrhodamine-5-(and-6) C2 maleimide (AnaSpec, Cat. # AS-81441-5).
18. HiLoad Superdex 75 16/60 pg column (Cytiva, formerly GE).

2.4 Materials for Ub-GC-TAMRA Degradation Using a Plate Reader

1. Assay buffer: 60 mM HEPES, pH 8.0, 100 mM NaCl, 100 mM KCl, 10% glycerol.
2. Corning 384-well solid black polystyrene microplates (Thermo Fisher, Cat # CLS3820).
3. A plate reader capable of measuring fluorescence polarization of TAMRA with excitation at 530 nm and emission at 590 nm. In this example we use either a BioTek Gen5™ plate reader or a BMG Labtech CLARIOstar ® Plus.
4. 8-quinolinethiol (8QT) (Sigma, Cat # 359785).
5. O-phenanthroline (OPA) (Sigma, Cat # 131377).
6. 8-channel or 12-channel pipette (Rainin Pipet-Lite).
7. 50 μM Usp2 (from Addgene #36894). Usp2 is purified as previously described [38].
8. 32 μM Yuh1 (from Addgene #18895). pET3a-Yuh1 was a gift from Cecile Pickart (Addgene plasmid # 18895; http://n2t.net/addgene:18895; RRID:Addgene_18895). The protein is purified as described by Raasi and Pickart [39].

3 Methods

3.1 Expression and Purification of Rpn11/Rpn8 Heterodimer

Rpn11 forms an obligate dimer with Rpn8, and we therefore inserted *S. cerevisiae* Rpn8 (UniProt ID: P43588, amino acids 1–179) with a C-terminal PreScission-cleavable Strep tag and *S. cerevisiae* Rpn11 (UniProt ID: Q08723, amino acids 2–239) with a C-terminal PreScission-cleavable His_6 tag into a pETDuet vector (EMD Millipore) for co-expression in *E. coli*. Both Rpn8 and Rpn11 have C-terminal helices that were removed from the expression constructs. As Rpn8/Rpn11 form a stable heterodimer during purification, we run a Ni-NTA column (the Strep tag is not used in this case) and then a size-exclusion Superdex 200 16/60 column (Fig. 2a). The yields after size-exclusion chromatography are very high, in the range of 5 mg/L dYT culture, which we that we typically concentrate to about 1 mM (~50 mg/mL), flash freeze, and store at −80 °C.

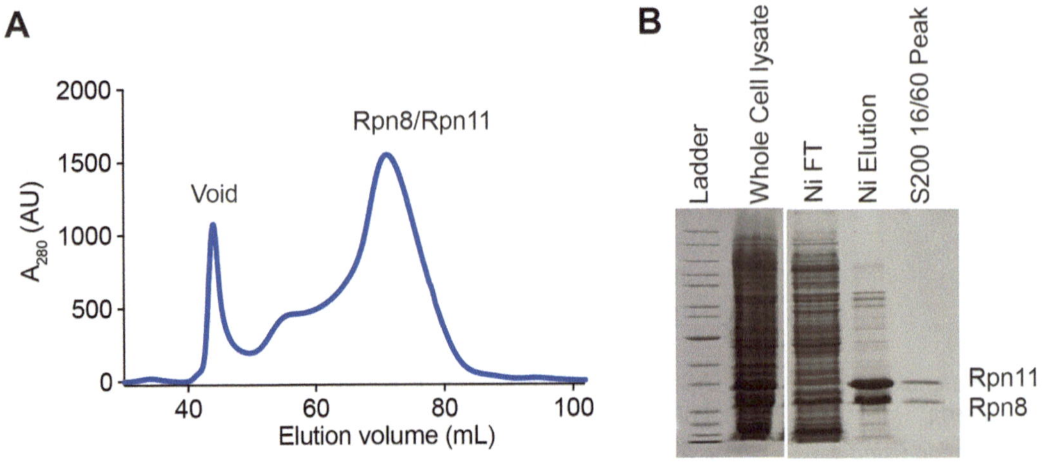

Fig. 2 Purification of Rpn8/Rpn11 heterodimer. (**a**) Elution profile (A_{280}) for the size-exclusion chromatography of the Rpn8/Rpn11 dimer on a Superdex S200 16/60. (**b**) Representative Coomassie-stained SDS-PAGE gradient gel (4–20% in Tris-Glycine) showing individual steps of Rpn8/Rpn11 purification as described in the procedure. The left lane contains a PageRuler™ unstained ladder from Thermo Fisher

3.1.1 Expression of Rpn11/Rpn8

1. Prepare chemically competent *E. coli* BL21 Star (DE3) cells, and transform them with pEW26 (*see* Table 1 for more information on the plasmid). Plate on LB + ampicillin. Grow at 37 ° C overnight.
2. Select a single colony from the plate to inoculate 50 mL of dYT medium containing ampicillin (300 μg/mL). Grow overnight at 30 °C, shaking at 180 rpm.
3. Pellet the overnight culture in a 50 mL conical tube by centrifuging at 3710 rcf for 4 min at 4 °C, pour off the supernatant, and resuspend the cell pellet in 50 mL of fresh dYT. Repeat this wash step three times to completely remove any β-lactamase released from bacteria into the media (*see* **Note 2**).
4. Resuspend the pellet in 25 mL of dYT, and divide evenly between three 2.5 L baffled flasks, each containing 1 L of TB + ampicillin (300 μg/mL) and 125 μM $ZnCl_2$ (*see* **Note 3**).
5. Grow cells at 37 °C in a shaker at 180 rpm for 2–3 h until OD_{600} reaches 1.0. Turn down the temperature to 18 °C, wait until OD_{600} reaches 1.5, induce protein expression by adding 1 mL of 1 M IPTG (1 mM final concentration) to each flask, and continue shaking at 18 °C for protein expression overnight (roughly 18 h).
6. Take an OD_{600} reading of the cultures (the final OD_{600} is usually 4.5). Spin down the culture in three 1 L centrifuge bottles for 15 min at 3500 rcf at room temperature, pour off the supernatant, and resuspend the cell pellets in cold NiA buffer + protease inhibitors (2 mg/mL lysozyme, 1 μg/mL aprotinin, 1 μg/mL pepstatin, 1 μg/mL leupeptin, 174 μg/mL PMSF), lysozyme (2 mg/mL), and 50 units of Benzonase (*see* **Note 4**).

7. The resuspended pellets can then be transferred to 50 mL conical tubes and stored at −80 °C or carried on directly to the purification.

3.1.2 Purification of the Rpn8/Rpn11 Heterodimer

1. If using frozen pellets, thaw them in a water bath at room temperature. All remaining purification steps should be performed at 4 °C. Combine the lysates into one 125 mL metal beaker, and sonicate the lysate on ice for 3 min with an ultrasonic homogenizer at 75% amplitude, using a cycle of 15 s on and 60 s off (*see* **Note 5**).
2. Centrifuge the lysate for 30 min at 26,000 rcf in a pre-chilled rotor at 4 °C.
3. While the lysate is clarifying, wash fresh Ni-NTA beads in NiA buffer three times in a 50 mL centrifuge tube. Spin at 3710 rcf rpm for 5 min at 4 °C between washes to pellet beads. When the cell lysate is done spinning, pour off the supernatant into the 50 mL conical tube containing washed Ni-NTA beads.
4. Batch bind the lysate with the Ni-NTA beads for 45 min on a rotator at 8 rpm and 4 °C (*see* **Note 6**). This can be done in 50 mL conical tubes that are rated for 12500 × g spins.
5. Spin down the Ni-NTA beads at 3710 rcf for 4 min at 4 °C.
6. Pour off the supernatant into a clean beaker on ice (*see* **Note 7**).
7. Batch wash the Ni-NTA beads in 45 mL NiA buffer for three times (*see* **Note 6**).
8. After the third wash, pour beads into a cleaned gravity drip column.
9. Continue washing the beads in the column with ~150 mL of NiA buffer until the elution shows no longer a blue color with Bradford reagent (*see* **Note 8**).
10. Elute the Ni-NTA-bound protein from the column with NiB buffer in 1 mL fractions. Monitor the elution by Bradford reagent.
11. Pool the fractions that contain protein, and concentrate them to ~2 mL in a 30 K MWCO Amicon spin concentrator at 4380 rcf and 4 °C.
12. Filter the protein through a 0.2 μm Costar spin-X spin filter.
13. Load and run the protein on a pre-equilibrated Superdex 200 size-exclusion chromatography column using GF buffer (*see* Fig. 2a for an example size-exclusion trace and Fig. 2b for a typical SDS gel with fractions from the Ni-NTA elution and size-exclusion chromatography).
14. Collect the peak indicated in Fig. 2, typically at elution volumes from 65 mL to 80 mL when run at 0.8 mL/min at 4 °C. Concentrate the protein by successive spins in a 30 kDa cutoff spin concentrator. Take 1 μL of concentrated protein for a gel sample.

15. Quantify the concentration by measuring the A_{280}. The extinction coefficient calculated from the sequence is 48,670 M^{-1} cm^{-1} (*see* **Note 9**).
16. Mix gel samples with 2X SDS loading buffer, and confirm the purity of the final complex by SDS-PAGE. There are two bands corresponding to Rpn11 (28.8 kDa) and Rpn8 (22.2 kDa); *see* Fig. 2b.
17. Divide the protein sample into 50 μL aliquots, flash freeze in liquid nitrogen, and store at −80 °C. Samples can be stored for at least 2 years.

3.2 Expression, Purification, and Labeling of Ubiquitin-GC-TAMRA

Ubiquitin-GC-TAMRA is a construct of wild-type *S. cerevisiae* Ubiquitin (UniProt: P0CG63, amino acids 1–76) with a Met-Trp-Gly extension at the N-terminus to allow for easy protein quantification using tryptophan absorbance and a Gly-Cys extension on the C-terminus for attachment of a fluorescent dye using maleimide chemistry (Fig. 3a). This construct has been cloned into a pET28a (Invitrogen) plasmid without any tags. We purify ubiquitin using acetic acid precipitation and cation exchange chromatography and typically get 50–75 mg/L of culture. As the C-terminal Gly-Cys extension may be cleaved during expression in *E. coli*, we added a Thiol Sepharose step to enrich proteins that retain the cysteine. Using this method, we can generate ~10 mg of Ub-TAMRA per 3 L of *E. coli* culture and can concentrate it to ~1.2 mM.

3.2.1 Expression of Ubiquitin-GC

1. Transform *E. coli* BL21 star (DE3) with pEW145, the plasmid containing the Ub-GC construct, and plate the cells onto LB Plates containing kanamycin and chloramphenicol.
2. Pick a single colony, and start a 50 mL overnight culture grown in dYT media at 30 °C on a shaker rotating at 180 rpm (*see* **Note 10**).
3. Make 3 L buffered TB media. For 1 L customized TB media: first mix 880 mL H_20, 20 mL 50% glycerol, 24 g yeast extract, and 12 g tryptone, and autoclave the mixture. Then just before inoculation, add 100 mL 0.72 M K_2HPO_4, 0.17 M KH_2PO_4 (sterilized with 0.22 micron bottle filter), 1 mL 1000X kanamycin, and 1 mL 1000X chloramphenicol.
4. Pellet the overnight culture in a conical tube by centrifuging at 3710 rcf for 4 min at 4 °C, pour off the supernatant, and resuspend the pellet of cells in 50 mL of fresh dYT.
5. Resuspend the pellet in 25 mL of dYT, and divide evenly between three 2.5 L baffled flasks, each containing 1 L of TB + kanamycin (50 μg/mL) and chloramphenicol (25 μg/mL).

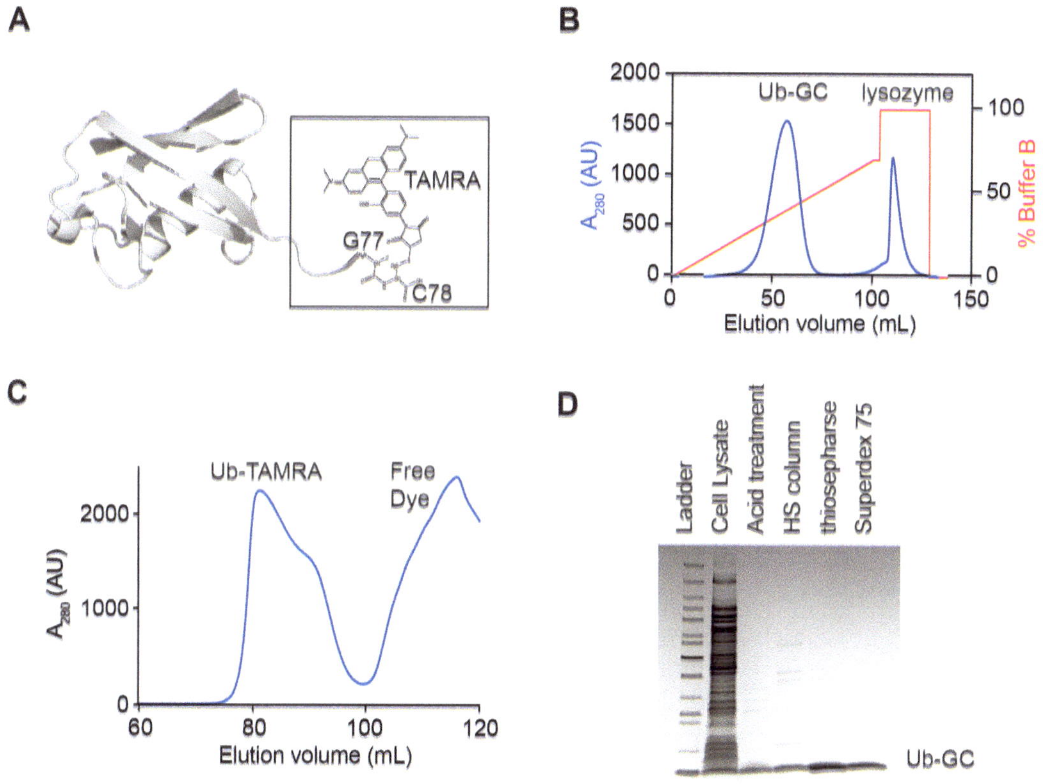

Fig. 3 Purification of Ub-GC-TAMRA. (**a**) Design of the Ub-TAMRA substrate, containing a C-terminal Gly-Cys extension that is labeled with TAMRA-maleimide. (**b**) Example profile for the elution of Ub-GC from the HiTrap SP FF column monitored by A_{280} (blue line) and with the percentage of the high-salt buffer B shown in red. (**c**) Example elution profile (A_{280}) for the size-exclusion chromatography of Ub-GC-TAMRA on a Superdex 75 16/60 column. The second peak is free TAMRA-maleimide dye. (**d**) Coomassie-stained 4–20% Tris-Glycine SDS-PAGE gel showing fractions from individual purification steps of Ub-GC-TAMRA. The left lane contains a PageRuler™ unstained ladder from Thermo Fisher, and Ub-GC-TAMRA runs near the 10 kDa marker

6. Grow cells at 37 °C in a shaker at 180 rpm for 2–3 h until OD_{600} reaches ~1.0. Turn down the temperature to 16 °C, wait until OD_{600} reaches 1.5, induce protein expression by adding 0.5 mL of 1 M IPTG (0.5 mM final concentration) to each flask, and continue shaking for protein expression overnight (roughly 18 h).
7. Pellet the cells by spinning at 3500 rcf for 30 min at 18 °C.
8. Resuspend the pellets in ubiquitin lysis buffer at a ratio of 10 mL lysis buffer per 1 g cells.
9. The resuspended pellets can then be transferred to 50 mL conical tubes and stored at −80 °C or carried on directly to the purification.

3.2.2 Purification of Ubiquitin-GC

1. If using frozen pellets, thaw them in a water bath. All remaining purification steps should be performed at 4 °C. Combine the lysate into one 125 mL metal beaker, and sonicate the lysate on ice for 2 min with an ultrasonic homogenizer at 70% amplitude, using a cycle of 10 s on and 60 s off.
2. Incubate the cell lysate at room temperature for 20–25 min without shaking.
3. Sonicate it again on ice for 2 min at 70% amplitude, using a cycle of 10 s on and 60 s off.
4. Centrifuge the lysate for 30 min at 7650 rcf in a pre-chilled rotor at 4 °C (*see* **Note 11**).
5. Prepare for acetic acid treatment: set up an ice-water bath on a stirrer, and immerse a 125 mL Erlenmeyer flask that is held by a clamp stand and contains a stirring bar. We typically set this up in a chemical hood to contain the glacial acetic acid in the following steps.
6. Decant the supernatant into the flask and stir (*see* **Note 12**).
7. Slowly add glacial acetic acid drop by drop until pH reaches 4.5 (monitor using pH paper). The solution will get cloudy as a result of this process.
8. Stir vigorously covered for at least 20 min on ice in the hood.
9. Centrifuge acid-treated lysate for 30 min at 7650 rcf in a pre-chilled rotor at 4 °C.
10. Dialyze the supernatant against 4 L dialysis buffer at 4 °C for at least 2 h using 3.5 K MWCO dialysis tubing (*see* **Note 13**).
11. Repeat **step 9** at least once with a fresh dialysis buffer. This also can be done overnight. Filter the supernatant through a 0.22 μm vacuum bottle filter.
12. Load all of the dialyzed lysate onto a 5 mL HiTrap SP FF column pre-equilibrated with SP Buffer A. This step can be performed using the 150 mL Superloop at 2–3 mL/min or a peristaltic pump (*see* **Note 14**).
13. Collect flow-through for analysis. Wash the column with 25 mL of SP Buffer A.
14. Run the HiTrap SP FF gradient with SP Buffer A and SP Buffer B, using a gradient elution of 0 to 70% B over 20 column volumes at 3–4 mL/min.
15. Ubiquitin should elute at about 33% SP Buffer B (~165 mM NaCl). The first peak is ubiquitin, and the second peak is lysozyme (*see* Fig. 3b).
16. Collect peak fractions and buffer exchange into ubiquitin storage buffer (20 mM Tris, pH 7.5, 150 mM NaCl, 1 mM DTT) by repeated concentration with a 3 k Amicon concentrator and dilution.

17. Use an extinction coefficient of 0.16 mL mg^{-1} cm^{-1}. Divide by 8.5 to get molar concentration.
18. Buffer exchange the Ub-GC into Thiol Sepharose 4B binding buffer by repeated concentration and dilution in a 3 k Amicon concentrator.
19. Weigh out 5 g of dry Thiol Sepharose 4B resin (*see* **Note 15**).
20. Rehydrate the resin by pouring it into ~70 mL of cold, filtered, de-gassed Milli-Q water. Stir the resin until there are no more chunks.
21. Pour the rehydrated resin into a 3 cm diameter gravity flow column, and wash it with 1 L of the filtered, de-gassed Milli-Q water (*see* **Note 16**).
22. Equilibrate the Thiol Sepharose 4B resin with binding buffer by running ~3 column volumes over the resin in the gravity flow column.
23. Transfer the Thiol Sepharose 4B into a 50 mL conical tube.
24. Add the buffer-exchanged Ub-GC to the equilibrated Thiol Sepharose 4B resin in the 50 mL conical tube (*see* **Note 17**).
25. Incubate the resin with the ubiquitin overnight at 4 °C on a rotator at 8 rpm to maintain mixing.
26. Apply the resin slurry to the same gravity flow column that was used to wash the resin after rehydration.
27. Wash the resin with ~125 mL of binding buffer, and watch the absorbance by A_{280} readings on the spectrophotometer.
28. After the A_{280} has stabilized at 0, elute the Ub-GC with DTT-containing elution buffer in 5 mL fractions, and follow absorbance at 280 nm.
29. Buffer exchange the elution into ubiquitin storage buffer by repeated concentration and dilution using a 3 K Amicon concentrator.
30. Determine the final concentration of Ub-GC. Flash freeze the protein and store it at −80 °C (*see* **Note 18**).

3.2.3 Labeling Ubiquitin-GC with TAMRA-Maleimide

1. Thaw the purified Ub-GC, and buffer exchange it into the labeling buffer by repeated concentration and dilution in a 3 K Amicon concentrator to get rid of any DTT (*see* **Note 19**).
2. Determine the concentration of the buffer-exchanged Ub-GC by measuring A_{280} on the spectrophotometer. The extinction coefficient of Ub-GC is 6990 M^{-1} cm^{-1}.
3. Set up the reaction of Ub-GC and TAMRA-maleimide with a mole ratio of 1:5.
4. Incubate the reaction mixture at room temperature for 2 h.
5. Quench the reaction with 50 mM DTT for 30 min. This is important to prevent the TAMRA-maleimide from labeling the subsequent size-exclusion column.

6. Filter the reaction with 0.22 μm centrifuge filters.
7. Load and run the protein on a pre-equilibrated Superdex 75 16/60 size-exclusion column using size-exclusion buffer at a flow rate of 0.6 mL/min (*see* Fig. 3c for an example size-exclusion trace).
8. Collect the peak fractions indicated in Fig. 3c. The first peak is TAMRA-labeled Ub-GC and the second peak is free dye. Concentrate the Ub-GC-TAMRA by successive spins in a 3 kDa cutoff spin concentrator. We typically aim for 1 mM Ub-GC-TAMRA as a final concentration. A gel for with samples from each step of the purification is shown in Fig. 3d.
9. The Ub-GC-TAMRA is quantified by its absorbance at 556 nm, using an extinction coefficient for TAMRA of 89,000 M^{-1} cm^{-1}.

3.3 Measuring Rpn11 Deubiquitinase Activity Using Ubiquitin-GC-TAMRA

DUB activity has typically been measured using ubiquitin-amino methyl coumarin (Ub-AMC), in which the fluorescence of the C-terminal AMC is quenched when bound to ubiquitin and increases upon AMC liberation. While this is a useful tool for most deubiquitinases, Ub-GC-TAMRA can be produced more economically and in large quantities (Fig. 3c). In addition, unlike Ub-AMC that is commercially available at typical stock concentrations of 100–250 μM, Ub-GC-TAMRA can easily be concentrated to mM concentrations. Rpn11 DUB activity is measured by monitoring the fluorescence polarization (FP) of the TAMRA dye. As the dye is liberated from ubiquitin after cleavage, its fluorescence polarization decreases due to faster tumbling. As a control for complete cleavage, we use another well-characterized and highly active DUB, such as Yuh1 [39] or Usp2 [5]. The FP values for the intact Ub-GC-TAMRA substrate and the completely Yuh1/Usp2-cleaved products represent the upper and lower limit of FP change in order to calculate a steady-state rate of DUB cleavage reactions. Varying the ubiquitin-GC-TAMRA concentrations allows the determination of Michaelis-Menten parameters for Rpn11. In this example, we use the BMG Labtech CLARIOstar ® Plus as the plate reader for FP measurements.

3.3.1 Determining DUB Activity Using Ub-GC-TAMRA

1. Make a 20 μL 2× stock solution of 5.8 μM wild-type Rpn8/Rpn11 (2.9 μM final concentration) in the assay buffer.
2. Make a 2× stock solution of 1.2 μM Yuh1 in assay buffer.
3. Make a 2× substrate mixture containing 500 μM Ub-GC-TAMRA (250 μM final concentration) in assay buffer.
4. Add 10 μL of the 2× substrate mixture to 10 μL of the 2× Yuh1, and incubate for 20 min (note: this reaction is very quick, especially compared to Rpn8/Rpn11). Yuh1 should lead to the complete hydrolysis of the Ub-GC-TAMRA to form a liberated Gly-Cys-TAMRA peptide product.

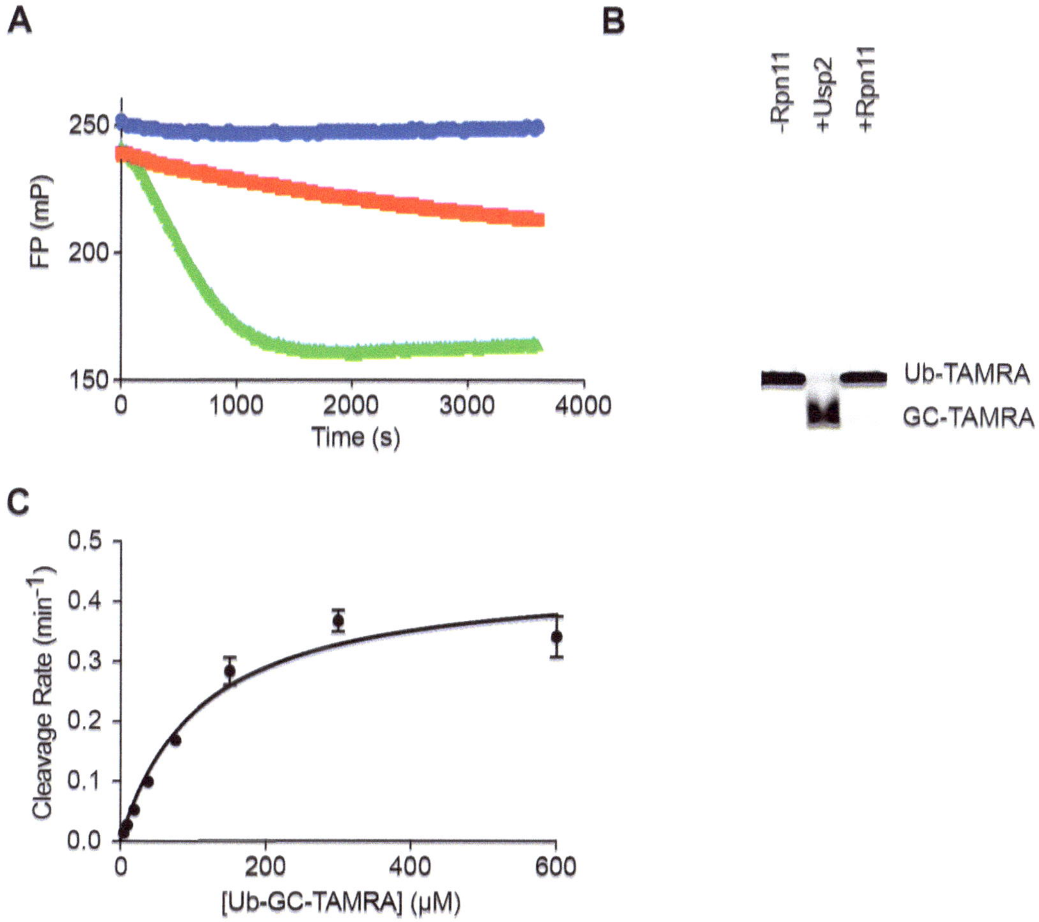

Fig. 4 Using Ub-GC-TAMRA as a probe for Rpn8/Rpn11 DUB activity measurements. (**a**) Cleavage of 150 μM Ub-GC-TAMRA by 2.9 μM Rpn8/Rpn11 (red line) and 0.6 μM Yuh1 (green line) monitored by fluorescence polarization (FP). Substrate in the absence of a DUB is shown in blue. (**b**) SDS-PAGE gel, fluorescence-scanned with a TAMRA filter on a Bio-Rad ChemiDoc XRS, showing Ub-GC-TAMRA cleavage by Usp2 and Rpn8/Rpn11 to yield GC-TAMRA product. (**c**) Michaelis-Menten plot for Ub-GC-TAMRA cleavage by Rpn8/Rpn11. Error bars represent the standard deviation of three technical replicates

5. Add 10 μL of the 2× substrate mixture to 10 μL of assay buffer.
6. To a 384-well plate, add 10 μL of the peptide product to one well and 10 μL of the uncleaved Ub-GC-TAMRA substrate to another well.
7. Preheat the plate reader to 30 °C. Depending on the device, we usually set the gain such that the FP value of the Ub-GC-TAMRA is at 250 mP. The cleaved TAMRA peptide has a FP of about 160 mP (*see* Fig. 4a).
8. Set up the plate reader to read five wells, the positive and negative control described above and three reaction wells, every 30–60 s. To the three reaction wells in the plate, add 5 μL of the 2× wild-type Rpn8/Rpn11 and 5 μL of the 2×

substrate mixture, and start reading on the plate reader. We read the plate for an hour. An example trace is shown in Fig. 4a.

9. When the measurement is finished, take the samples out, add an equal volume of 2× SDS-PAGE loading buffer, and run a 4–20% Tris-Glycine gel (Bio-Rad). An example of the products from the reaction is shown in Fig. 4b. The Yuh1 reaction has gone to completion showing the products, while the negative control shows the TAMRA attached to ubiquitin, and the Rpn8/Rpn11 reaction shows intermediate levels of product formation.

10. To calculate a rate, the negative and positive controls are used to determine the upper and lower boundaries for a 10 μM concentration. Based on the slope for the Rpn8/Rpn11 reaction, the DUB rate is calculated using the following equation:

Equation 1

$$\text{rate} = -\text{slope} \times [\text{substrate}]/(FP\ \text{sub} - FP\ \text{peptide}) \times 60 \div [\text{Enzyme}]$$

Example calculation:

1. The slope from Fig. 4a was calculated to be −0.0062 mP/s using Prism 9 or Excel.
2. To convert this into substrate per minute, we first calculate the linear relationship between the protein concentration and the fluorescence polarization by taking the substrate concentration of 150 μM and dividing it by the difference in FP between the negative control (239 mP) and the Yuh1 control (165 mP) to get 2.65 μM substrate/mP.
3. This relationship is multiplied by the slope and by 60 to convert into minutes and get 0.754 μM substrate per minute.
4. Dividing by the enzyme concentration, in this case 2.9 μM, gives a rate of 0.260 μM substrate per minute per μM Rpn11, equivalent to a turnover number of 0.26 min^{-1}.

3.3.2 Determination of Michaelis-Menten Kinetics Using Ub-GC-TAMRA

1. We first generate a 2× stock solution of 5.8 μM WT Rpn8/Rpn11 (2.9 μM final concentration) in assay buffer +0.5 mM TCEP.
2. We prepare a serial twofold dilution of a 2× stock of Ub-GC-TAMRA from 1200 μM to 9.375 μM in the assay buffer. From 50 μL of a 1200 μM Ub-GC-TAMRA solution, transfer 25 μL to 25 μL assay buffer, mix thoroughly, and repeat (*see* dilutions in Table 2).
3. Transfer 10 μL of the serial dilution to an eight-well PCR strip containing 10 μL of 1.2 μM Yuh1, and incubate at 30 °C for 30 min. This will be the peptide control.
4. We measure in sets of four (first dilutions 1–4 and then dilutions 5–6) to allow the plate reader to take enough data points. We also adjust the readjust gain for the lower dilutions.

Table 2
Substrate dilutions

Dilution	[Working concentration] (μM)	[Final concentration] (μM)
1	1200	600
2	600	300
3	300	150
4	150	75
5	75	37.5
6	37.5	18.75
7	18.75	9.375
8	9.375	4.6875

5. Using an eight-channel pipette, transfer 5 μL of the serial dilution to 5 μL of AB in a 384-well plate. In the adjacent wells, add 10 μL of the Yuh1 control.
6. Place in a preheated plate reader at 30 °C.
7. Adjust the gain as described above to attain a mP value of roughly 250 mP for the negative control and about 160 mP for the Yuh1 peptide control.
8. Add 5 μL of the 2× Rpn8/Rpn11 mixture to 5 μL of the 2× substrate mixture in triplicate, and read on a plate reader every 10 s for 1 h.
9. Calculate the rates for each concentration as shown in the previous section and fit to a Michaelis-Menten plot (*see* Fig. 4c). In this example, we determined a k_{cat} of 0.45 min^{-1} and a K_M of 108 μM. This matches closely the published values for ubiquitin cleavage by Rpn11 [21].

3.4 Using Ub-GC-TAMRA to Monitor Inhibition of Rpn8/Rpn11 by Small Molecules

Rpn11 has been proposed as a target for the small-molecule treatment of cancers that are refractory to bortezomib; however, published screens have relied on using purified 26S proteasomes that are not so well suited for extensive high-throughput screening [32–34]. We therefore use the isolated Rpn8/Rpn11 dimer and its cleavage of Ub-GC-TAMRA as an assay that is easily amenable for a 384-well format. Here we show the determination of IC_{50} values for two previously reported inhibitors, 8QT and OPA, which well agree with previously published results. In this example, we use a BioTek Gen5™ plate reader, showing that this assay is amenable to most commercial plate readers (*see* **Note 20**).

1. Prepare a 4× stock solution of Rpn11/Rpn8 (2.4 μM) and a 2× stock solution of Ub-GC-TAMRA (0.4 μM) in assay buffer (60 mM HEPES, pH 8.0, 100 mM NaCl, 100 mM KCl, and 10% glycerol).

2. Add 10 μL of the Rpn11/Rpn8 stock (final concentration 0.6 μM) to each well of a black low-volume 384-well plate.
3. Prepare a twofold serial dilution of the test compound (inhibitor) (*see* **Notes 21** and **22**). At least 1% DMSO is well tolerated by Rpn8/Rpn11 at room temperature for at least 20 h.
4. Add 10 μL of the compound solutions to each well that contains Rpn8/Rpn11, and mix.
5. Incubate the mixture at room temperature for 15 min.
6. Initiate the enzymatic reaction by adding 20 μL of Ub-GC-TAMRA (final concentration 200 nM) to each well, and mix well.
7. Transfer the plate to a 384-well plate reader (e.g., a Gen5 plate reader), and measure the fluorescence polarization of TAMRA with excitation at 530 nm and emission at 590 nm using FP 530/590 filters. With the gain set to 70/59 (side PMT/top PMT), the substrate alone has a FP value of about 240 mP (Fig. 5).
8. Measure the fluorescence polarization every 10 s at 30 °C for 1 h (*see* Fig. 5). If desired, the reaction temperature and enzyme concentration can be adjusted so that the reaction can be read out for up to 10 h.
9. FP values are plotted versus time, and the slope of the linear portion is determined (*see* **Notes 23** and **24**). For Fig. 5a, we calculated a slope of 0.0029 AU/s for the no-inhibitor control, while the slope for the inhibited Rpn11 was 0.0127 AU/s, equivalent to 23% of the control activity.

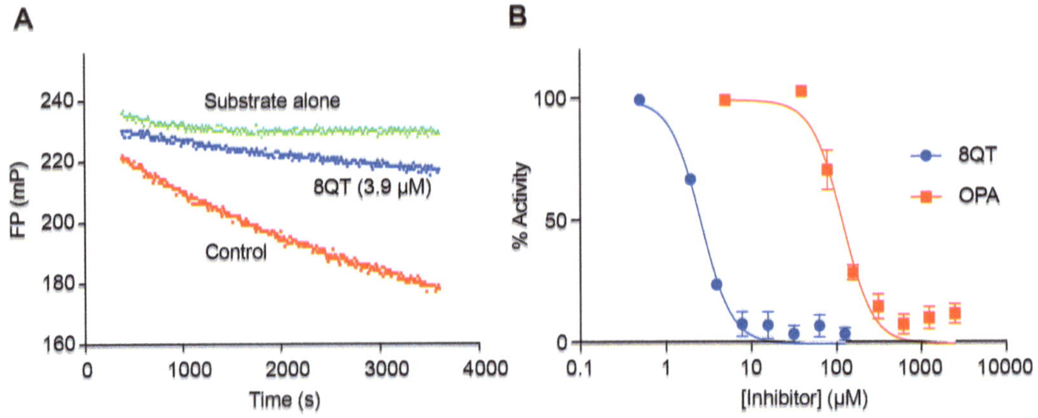

Fig. 5 Small-molecule inhibition of Rpn11 deubiquitination activity measured by Ub-GC-TAMRA cleavage assay. (**a**) Raw data for Ub-GC-TAMRA cleavage by Rpn11 in the absence of inhibitor (red) and the presence of 3.9 μM 8QT (blue) as measured by fluorescence polarization on a Gen5 plate reader. The FP time course for Ub-GC-TAMRA in the absence of Rpn11 is shown in green. (**b**) IC_{50} determination for Rpn11 inhibition by 8QT (2.6 ± 0.2 μM) and OPA (118 ± 16 μM). Error bars represent the standard deviation of three technical replicates

10. The dose-response curve was obtained by plotting a graph of Rpn11 activity (% of control) versus Log concentration of tested compound using GraphPad Prism, and the curve was fitted to determine IC_{50}.

    Equation: % activity = $1/(1 + [\text{inhibitor}]/IC_{50})$).

11. For this example, we determined an IC_{50} of 118 ± 16 μM for OPA and 2.6 ± 0.2 μM for 8QT (Fig. 5b), with the latter being in excellent agreement with the published value of 2.5 μM [34].

4 Notes

1. Protease inhibitors are hazardous and should only be handled wearing gloves and a lab coat. The inhibitor stock solutions are stored at –20 °C.
2. The wash steps repeated three times are important to remove any secreted β-lactamase in the media, which would diminish selection for pEW26-containing cells in the main culture.
3. It is important to add Zn^{2+} (final concentration 125 μM) to the media, because the Zn^{2+} is essential for the activity of Rpn11.
4. The volume (mL) of NiA to be added to the pellets should equal $OD_{600} \times 5$ mL. It is usually 25 mL for each L of cells.
5. The proper sonication conditions will vary depending on the instrument used and should be optimized to maximize lysis without heating the sample.
6. "Batch" means this can be done in 50 mL conical tubes instead of in a gravity drip column.
7. Do not get rid of your supernatant before you see elution from the resin. If things go wrong, you can still salvage your purification.
8. The purification can be monitored using Bradford reagent. Aliquot 100 μL of Bradford reagent into the wells of a 96-well plate. Then take 10 μL of column eluate, and mix thoroughly with the Bradford reagent. The presence of protein will be indicated by a color change from brown to blue.
9. Measurements are typically taken in a 20 μL cuvette using a UV-Vis spectrophotometer.
10. It is important that the cells do not reach stationary phase before being used to inoculate the main culture for protein expression, as plasmid stability decreases when cells stop growing and more β-lactamase is released from lysed cells.
11. The supernatant should be opaque.
12. Make sure the stir bar can spin quickly and freely. It will be important to stir vigorously after adding acetic acid.

13. DTT is added to keep the cysteine of Ub-GC reduced.
14. Precipitates may form during the dialysis. To remove the precipitates, centrifuge the dialyzed protein solution.
15. The binding capacity of the Thiol Sepharose 4B resin is about 3 mg/mL. 5 g of the dry resin should yield about 20 mL of rehydrated resin, with a binding capacity of ~60 mg.
16. The manufacturer's manual recommended 200 mL of water per gram of resin. This step is only required when starting a column from solid powder.
17. The resin was about 20 mL + 30 mL buffer-exchanged Ub-GC, which is close to the 1:1 ratio suggested by the manufacturer.
18. Mass spectrometry analysis showed that no trace of contaminating cleaved ubiquitin is present in the sample.
19. The Ub-GC was stored in ubiquitin storage buffer with 1 mM DTT, which has to be removed prior to maleimide labeling by repeated concentration and dilution. Five to six sets of dilution and concentration are enough to reduce the DTT concentration to a negligible level of 5–10 μM.
20. Any plate reader capable of measuring the FP or TAMRA can be used.
21. The range of the dilution series should be adjusted based on the potency of the inhibitors. The final concentration of 8QT ranged from 125 μM to 2.0 μM, and the concentration of OPA ranged from 2500 μM to 39 μM.
22. 8QT is quite unstable; it is stable only in solid powder form at −80 °C and lasts about 3 weeks when stored in DMSO at −80 °C. It has a violet color in DMSO, and this color correlates with activity. As the color fades, activity decreases.
23. The Rpn8/Rpn11 concentration should be adjusted to ensure a linear FP decay over the time frame tested (e.g., 1 h).
24. For the negative control (in the absence of inhibitors), the slope of the linear regression is typically around −50 mp/h, and the rate of substrate cleavage by Rpn11 equals the absolute value of the calculated slope.

Acknowledgments

We thank the members of the Martin lab for helpful discussions. This research was funded by the Bakar Faculty Fellowship Program (A.M.), the Howard Hughes Medical Institute (K.C.D., and A. M.), and the US National Institutes of Health (R01-GM094497 to A.M.).

Author Contributions G.X, E.J.W, K.C.D., and A.M. designed experiments, G.X. performed biochemical measurements, and K.C.D. assisted with the preparation of materials. All authors contributed to the manuscript preparation.

References

1. Bard JAM et al (2018) Structure and function of the 26S proteasome. Annu Rev Biochem 87: 697–724
2. Greene ER, Dong KC, Martin A (2020) Understanding the 26S proteasome molecular machine from a structural and conformational dynamics perspective. Curr Opin Struct Biol 61:33–41
3. Finley D (2009) Recognition and processing of ubiquitin-protein conjugates by the proteasome. Annu Rev Biochem 78(1):477–513
4. Prakash S, Tian L, Ratliff KS, Lehotzky RE, Matouschek A (2004) An unstructured initiation site is required for efficient proteasome-mediated degradation. Nat Struct Mol Biol 11(9):830–837
5. Bard JAM, Bashore C, Dong KC, Martin A (2019) The 26S proteasome utilizes a kinetic gateway to prioritize substrate degradation. Cell 177(2):286–298. e215
6. Olszewski MM, Williams C, Dong KC, Martin A (2019) The Cdc48 unfoldase prepares well-folded protein substrates for degradation by the 26S proteasome. Commun Biol 2(1):29
7. Komander D, Rape M (2012) The ubiquitin code. Annu Rev Biochem 81(1):203–229
8. Hershko A, Ciechanover A (1998) The ubiquitin system. Annu Rev Biochem 67:425–479
9. Tan JM et al (2008) Lysine 63-linked ubiquitination promotes the formation and autophagic clearance of protein inclusions associated with neurodegenerative diseases. Hum Mol Genet 17(3):431–439
10. Haas TL et al (2009) Recruitment of the linear ubiquitin chain assembly complex stabilizes the TNF-R1 signaling complex and is required for TNF-mediated gene induction. Mol Cell 36(5):831–844
11. Tokunaga F et al (2009) Involvement of linear polyubiquitylation of NEMO in NF-kappaB activation. Nat Cell Biol 11(2):123–132
12. Shi Y et al (2016) Rpn1 provides adjacent receptor sites for substrate binding and deubiquitination by the proteasome. Science 351(6275):aad9421
13. Elsasser S, Chandler-Militello D, Müller B, Hanna J, Finley D (2004) Rad23 and Rpn10 serve as alternative ubiquitin receptors for the proteasome. J Biol Chem 279(26): 26817–26822
14. Husnjak K et al (2008) Proteasome subunit Rpn13 is a novel ubiquitin receptor. Nature 453(7194):481–488
15. Lander GC et al (2012) Complete subunit architecture of the proteasome regulatory particle. Nature 482(7384):186–191
16. de la Pena AH, Goodall EA, Gates SN, Lander GC, Martin A (2018) Substrate-engaged 26S proteasome structures reveal mechanisms for ATP-hydrolysis-driven translocation. Science 362(6418):eaav0725
17. Worden EJ, Dong KC, Martin A (2017) An AAA motor-driven mechanical switch in Rpn11 controls deubiquitination at the 26S proteasome. Mol Cell 67(5):799–811.e798
18. Yao T, Cohen RE (2002) A cryptic protease couples deubiquitination and degradation by the proteasome. Nature 419(6905):403–407
19. Verma R et al (2002) Role of Rpn11 metalloprotease in deubiquitination and degradation by the 26S proteasome. Science 298(5593): 611–615
20. Maytal-Kivity V, Reis N, Hofmann K, Glickman MH (2002) MPN+, a putative catalytic motif found in a subset of MPN domain proteins from eukaryotes and prokaryotes, is critical for Rpn11 function. BMC Biochem 3:28
21. Worden EJ, Padovani C, Martin A (2014) Structure of the Rpn11-Rpn8 dimer reveals mechanisms of substrate deubiquitination during proteasomal degradation. Nat Struct Mol Biol 21(3):220–227
22. Pathare GR et al (2014) Crystal structure of the proteasomal deubiquitylation module Rpn8-Rpn11. Proc Natl Acad Sci U S A 111(8):2984–2989
23. Sato Y et al (2008) Structural basis for specific cleavage of Lys 63-linked polyubiquitin chains. Nature 455(7211):358–362
24. Shrestha RK et al (2014) Insights into the mechanism of deubiquitination by JAMM deubiquitinases from cocrystal structures of the enzyme with the substrate and product. Biochemistry 53(19):3199–3217

25. Richardson PG et al (2005) Bortezomib or high-dose dexamethasone for relapsed multiple myeloma. N Engl J Med 352(24):2487–2498
26. San Miguel JF et al (2008) Bortezomib plus melphalan and prednisone for initial treatment of multiple myeloma. N Engl J Med 359(9): 906–917
27. Steele JM (2013) Carfilzomib: a new proteasome inhibitor for relapsed or refractory multiple myeloma. J Oncol Pharm Pract 19(4): 348–354
28. Gandolfi S et al (2017) The proteasome and proteasome inhibitors in multiple myeloma. Cancer Metastasis Rev 36(4):561–584
29. Fricker LD (2020) Proteasome inhibitor drugs. Annu Rev Pharmacol Toxicol 60:457–476
30. Barrio S et al (2019) Spectrum and functional validation of PSMB5 mutations in multiple myeloma. Leukemia 33(2):447–456
31. Song Y et al (2017) Blockade of deubiquitylating enzyme Rpn11 triggers apoptosis in multiple myeloma cells and overcomes bortezomib resistance. Oncogene 36(40):5631–5638
32. Li J et al (2017) Capzimin is a potent and specific inhibitor of proteasome isopeptidase Rpn11. Nat Chem Biol 13(5):486–493
33. Li J et al (2018) Epidithiodiketopiperazines inhibit protein degradation by targeting proteasome deubiquitinase Rpn11. Cell Chem Biol 25(11):1350–1358. e1359
34. Perez C et al (2017) Discovery of an inhibitor of the proteasome subunit Rpn11. J Med Chem 60(4):1343–1361
35. Perez C et al (2017) Correction to discovery of an inhibitor of the proteasome subunit Rpn11. J Med Chem 60(7):3217
36. Lauinger L et al (2017) Thiolutin is a zinc chelator that inhibits the Rpn11 and other JAMM metalloproteases. Nat Chem Biol 13(7):709–714
37. Hameed DS et al (2019) Development of ubiquitin-based probe for metalloprotease deubiquitinases. Angew Chem 58(41): 14477–14482
38. Bard JAM, Martin A (2018) Recombinant expression, unnatural amino acid incorporation, and site-specific labeling of 26S proteasomal subcomplexes. Methods Mol Biol 1844: 219–236
39. Raasi S, Pickart CM (2005) Ubiquitin chain synthesis. Methods Mol Biol 301:47–55

Chapter 7

Activity-Based Protein Profiling (ABPP) for Cellular Deubiquitinase (DUB) and Inhibitor Profiling at Deep and High-Throughput Levels

Hannah B. L. Jones, Raphael Heilig, Benedikt M. Kessler, and Adán Pinto-Fernández

Abstract

This chapter provides detailed methodology and materials required to profile deubiquitinases (DUBs) in a cellular matrix using specific activity-based probes, with immunoblotting and mass spectrometry proteomics-based readouts. Different types of activity-based protein profiling (ABPP) for studying the potency and selectivity of DUB inhibitors are outlined here, including the standard ABPP, the deep DUBome ABPP, and the ABPP-HT (high-throughput compatible).

Key words Activitomics, Activity-based probes, Chemical biology, Deubiquitinase enzymes, Drug discovery, Pharmacology, Proteomics, Ubiquitin

1 Introduction

Activity-based probes (ABPs) can be used to assess the activity, inhibition, and therefore regulation of the function of a number of different enzyme families. ABPs for the deubiquitinase (DUB) family are of particular interest as these druggable proteases remove ubiquitin from specific protein targets, making them attractive drug targets for precisely tuning cellular pathways [1]. The DUB family consists of both cysteine proteases and metalloproteases, with cysteine proteases making up the majority of these enzymes [2, 3]. While ABPs have been developed to target either cysteine protease or metalloprotease DUBs, the focus of this chapter is the application of ABPs that broadly react with cysteine protease DUBs. Cysteine protease DUB ABPs are typically ubiquitin-based to ensure binding specificity. The C-terminal glycine of ubiquitin (Gly_{76}) is replaced with a reactive warhead, typically an electron-withdrawing group, that binds to the cysteine residue within the

Julie A. Maupin-Furlow and Mariola J. Edelmann (eds.), *Deubiquitinases: Methods and Protocols*,
Methods in Molecular Biology, vol. 2591, https://doi.org/10.1007/978-1-0716-2803-4_7,

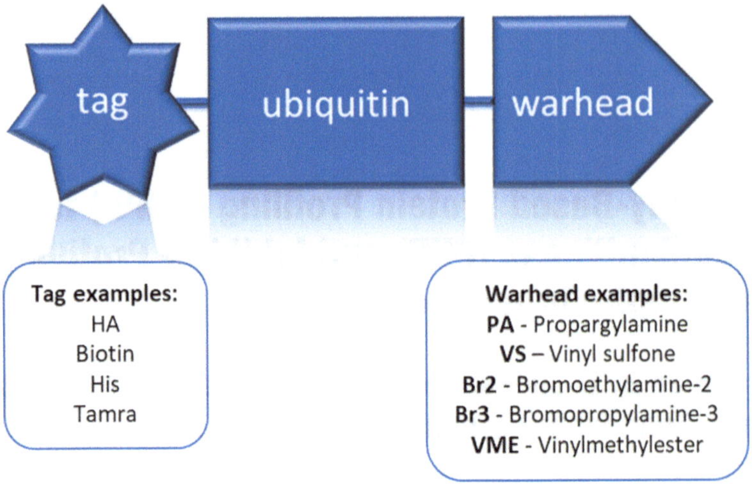

Fig. 1 Typical composition of a DUB ABP

DUB active site. The addition of a tag to this probe is required where visualization or immunoprecipitation is desired (Fig. 1) [4].

The DUB ABPP assay has been particularly useful for discovering DUB inhibitors [5–8]. By incubating cellular lysates with an inhibitor molecule and subsequently with an ABP, inhibitor potency can be assessed by competition with the ABP from binding to its target DUBs. This approach assesses direct target engagement of inhibitor molecule candidates to the active site of cellular DUBs. Detection of inhibitor engagement can be achieved for a specific DUB by Western blot, where the binding of a ubiquitin-based probe will add ~10 kDa to the molecular weight of a DUB, whereas the binding of a small-molecule inhibitor will result in a negligible increase in a DUB mass. In this case, an estimation of the inhibitor concentration dependence can be carried out with densitometric quantification. When compared to a control, densitometric quantification enables the extraction of IC_{50} values for small-molecule inhibitor candidates. Where cross-reactivity of an inhibitor with other DUBs is of interest, immunoblotting against the probe tag can provide information on the selectivity of the small-molecule inhibitor. Due to intrinsic limitations of the Western blotting technique, activity-based protein profiling is recommended as this approach combines affinity purification of the probe with mass spectrometry (MS) analysis, to fully profile an inhibitor against a higher number of cellular DUBs. When DUBs react with an inhibitor, they no longer bind to the probe and are not immunoprecipitated as efficiently as compared to a control immunoprecipitation. Differences in DUB immunoprecipitation efficiency with or without an inhibitor can be quantified by bottom-up proteomics with trypsin digestion and label-free quantitation (LFQ) liquid chromatography combined with tandem mass spectrometry (LC-MS/MS) analysis, informing on the potency and selectivity of the inhibitor.

Although there are multiple available assays for monitoring the inhibition and activity of DUBs, these assays tend to rely on recombinantly expressed DUBs and are not carried out in a cellular environment. The combination of DUB ABPs with immunoblotting or immunoprecipitation and LC-MS/MS (ABPP) readouts is advantageous as these assays can be carried out in a cellular environment, reflecting accurate DUB target engagement of inhibitors.

1.1 Considerations for ABP Labeling

Different Ub-based ABP probes have been described. The probe architecture, in particular the type of chemical warhead and its positioning within the probe molecule, impacts the reactivity of the probe toward different subsets of DUBs. For instance, a linear di-Ub probe has been designed to be selective toward a linear Ub-specific DUB called OTULIN [9], and different C-terminal chemical moieties enabling Michael additions and nucleophilic displacements have been explored [6]. An alkyne-based ABP that reacts with the catalytic cysteine of DUBs, with enhanced reactivity, has been reported [10] and has been broadly recognized by the field as the gold standard due to its superior reactivity and selectivity toward the active site of DUB cysteines [11]. Based on these recent insights, the warhead (propargylamine) is applied in the methodology outlined in this chapter.

ABPs are routinely used to label cellular lysates but have also been applied to intact cell cultures. The latter has been achieved with ubiquitin-based ABPs via electroporation, pore-forming toxins, and the attachment of cell-penetrating peptides [12]. Although DUB ABP labeling of intact cells is the most physiologically relevant method of assessing DUB activity and inhibition, these techniques may negatively affect cellular homeostasis. Small-molecule permeable DUB ABPs may overcome this issue and have been synthesized, thanks to the utilization of a DUB inhibitor as a specificity motif [13]. However, the reactivity of these types of probes is limited, as they only bind to a reduced number of DUBs, compared to the HA-Ub-PA probe applied here. As a consequence of the current limitations of cell-permeable probes, the methodology presented here focuses on applying ABPs to cellular lysates. Cellular extracts from tissue samples, cells in culture, and other sources of cellular material can be used for ABP labeling.

Consideration of probe concentration and incubation times is important to ensure the probe is saturating all active DUBs. If probe available for binding is limited, then the concentration of free probe may be higher where an inhibitor has blocked the active site of a DUB, or a number of DUBs. This increase in free probe may lead to an increase in probe binding of other DUBs, presenting as an apparent increase in activity of these DUBs. This complex kinetic mechanism may also occur where probe labeling time is insufficient.

As cysteine DUB ABP labeling involves strong covalent binding of a warhead to the active site cysteine of the enzyme, the displacement of the inhibitor by the probe itself should also be considered, especially in the case of a non-covalent inhibitor or a reversible-covalent inhibitor with a fast off-rate. While probe inhibitor displacement can inform on the strength of inhibitor binding in these cases, an inhibitor may appear to not bind the DUB due to this displacement. This issue can be overcome by assessing inhibitor displacement with a probe incubation time course. Where the inhibitor can be displaced by probe incubation, a probe-binding concentration/time dependence can be carried out to minimize probe displacement while ensuring optimal labeling of the target DUB.

1.2 Considerations of Inhibitor Treatment

Inhibitor treatment can be applied to cell lysates, intact cells, or in vivo. These applications offer different information about the usability of an inhibitor as a drug (Fig. 2). For direct comparisons of inhibitor potency and for developing inhibitors that bind strongly to a DUB, treating cell lysates with the inhibitor is most beneficial. If the cellular permeability and stability of an inhibitor are of interest, incubating an inhibitor with intact cultured cells is more informative than incubating the same inhibitor with lysate. However, note that when intact cells are treated with a small-molecule inhibitor for a few hours, reduced DUB activity may occur by alteration of other pathways affecting the DUB, not as a direct result of inhibitor binding to the DUB. These complications are less likely to occur in cell extracts. When an inhibitor is both cell-permeable and potent, a DUB ABP can also be applied to lysates from in vivo inhibitor treatment to examine inhibitor target engagement levels. The focus of these studies may be a set of target organs. Carrying out an inhibitor concentration and time dependence study is important in all cases.

1.3 Considerations for Inhibitor Selectivity Analysis: Deep DUBome and High Throughput

The methodology outlined here builds on previously described protocols for profiling DUBs with activity-based probes [14–17]. In addition to the standard ABPP protocol, we have also outlined a deep DUBome ABPP protocol for enhanced sensitivity to be applied where a DUB of interest is not detected or in-depth cross-reactivity data is required [2]. We have also outlined ABPP-HT (high-throughput-compatible activity-based protein profiling) for applications such as drug discovery programs testing a high number of compounds [18] (Fig. 2).

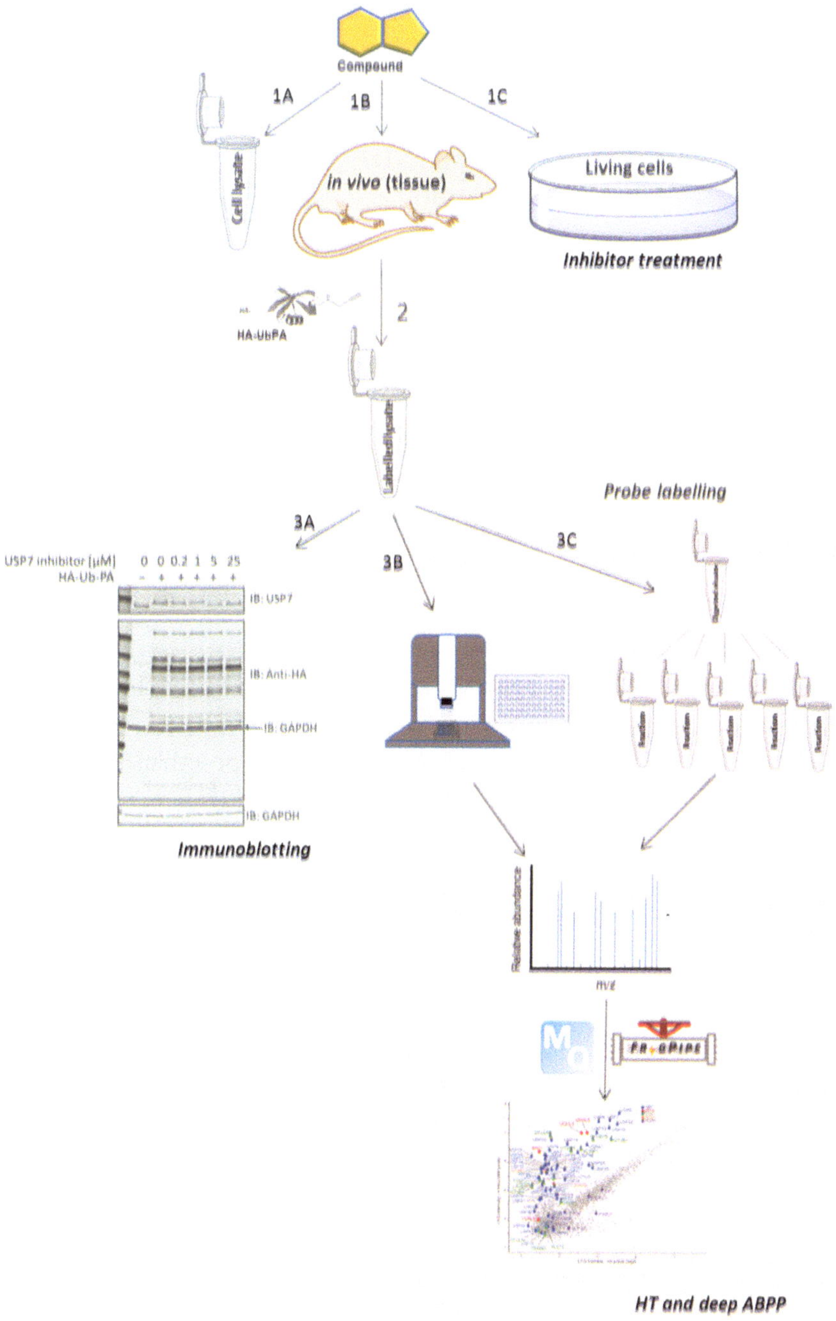

Fig. 2 Optional workflows for assessing a DUB inhibitor via ABP assays (**1**) Inhibitor treatment of either cell lysates to compare inhibitor potency (A), in vivo tissue to assess inhibitor target engagement and stability (B), or intact cells in culture to assess inhibitor potency, permeability, and stability (C). (**2**) HA-Ub-PA probe treatment with cell/tissue lysates. (**3A**) Western blot readout to look for inhibitor potency by immunoblotting for DUBs of interest and an anti-HA blot to rapidly assess inhibitor selectivity. (**3B**) High-throughput inhibitor potency and selectivity screening across a representative panel of DUBs (~20–30) using an Agilent Bravo liquid handling platform combined with a timsTOF LC-MS/MS. (**3C**) Sample fractionation to assess the selectivity of strong inhibitor candidates across a comprehensive cysteine active DUBome (>70 DUBs)

2 Materials

2.1 HA-Ub-PA Probe Synthesis

1. pTYB2 vector [4] encoding for HA, Ubiquitin (Ubiquitin 76delG), intein, and chitin-binding domain.
2. Competent BL21(DE3) *E. coli*.
3. Luria broth (LB) media with 100 μg/mL ampicillin.
4. Luria broth (LB) media with 1.5% agar and 100 μg/mL ampicillin.
5. Sterile SOC media: 2% Tryptone (W/V), 0.5% yeast extract (W/V), 10 mM NaCl.
6. 100 mm Petri dishes.
7. 1 M IPTG stock (isopropyl β-d-1-thiogalactopyranoside).
8. 60% sterile glycerol stock.
9. Buffer A: 50 mM HEPES, 150 mM NaCl, 0.5 mM TCEP or DTT, pH 7.4 (adjusted with NaOH).
10. Chitin bead slurry.
11. Clamp stand.
12. Liquid chromatography column (1.5 × 10 cm).
13. Buffer B: 50 mM HEPES, 150 mM NaCl, 0.5 mM TCEP or DTT, 100 mM Mesna (2-mercaptoethanesulfonic acid sodium salt), pH 7.4 (adjusted with NaOH).
14. 5× Laemmli buffer: 50% glycerol (V/V), 10% SDS, 0.5% bromophenol blue, 500 mM DTT, 250 mM Tris-HCl, pH 6.8.
15. Acrylamide gel (Tris/Gly compatible).
16. Gel tank.
17. Tris-glycine gel running buffer: 25 mM Trizma base, 192 mM glycine, 0.1% SDS (W/V).
18. Pre-stained protein molecular weight marker.
19. Instant blue Coomassie stain (or alternative stain/destain).
20. Western blot membrane (PVDF or nitrocellulose, 0.45 μm pore size).
21. Thick blot filter paper.
22. Western blot apparatus – tank, cassette, ice block, and foam pads.
23. Towbin buffer: 20 mM Trizma base, 200 mM glycine, 20% methanol.
24. Anti-HA primary antibody.
25. 3 kDa MWCO spin concentrators.
26. Desalting columns (PD-10 with Sephadex G-25 resin, or suitable alternative).
27. BCA or equivalent protein estimation kit.
28. 250 mM Propargylamine stock.

2.2 Cell/Tissue Lysis for HA-Ub-PA Labeling

1. Buffer C: 50 mM Trizma, 5 mM $MgCl_2$, 0.5 mM EDTA, 250 mM sucrose, pH 7.5 (adjusted with HCl).
2. 1 M dithiothreitol (DTT) – (can be stored at −20 °C).
3. Acid-washed glass beads (0.1 mm).
4. BCA or equivalent protein estimation kit.

2.3 HA-Ub-PA Immunoprecipitation for LC-MS/MS

1. 5% SDS (sodium dodecyl sulfate) stock.
2. 10% NP-40 (also known as IGEPAL CA-630) stock.
3. Buffer D: 50 mM Trizma base, 0.5% (v/v) NP-40, 150 mM NaCl, 20 mM $MgCl_2$, adjust to pH 7.4 with HCl.
4. Anti-HA-agarose slurry.
5. 2× Laemmli buffer: 20% glycerol, 4% SDS, 10 mM DTT, 100 mM Tris-HCl (pH 6.8).
6. S-trap micro kit.
7. 200 mM DTT stock.
8. 200 mM Iodoacetamide stock.
9. 12% phosphoric acid.
10. 0.2% formic acid.
11. 50% acetonitrile.
12. 90% methanol, 100 mM triethylammonium bicarbonate (TEAB) (pH 7.55).
13. 50 mM triethylammonium bicarbonate (TEAB), pH 8.5.
14. Trypsin.
15. Vacuum concentrator (SpeedVac).
16. Buffer E: 98% H_2O, 2% ACN, 0.1% TFA.
17. Glass mass spectrometry vials.

2.4 HPLC High-pH Fractionation

1. Waters Acquity UPLC Peptide BEH C18 column.
2. Buffer F: H_2O, pH 10 adjusted with ammonium hydroxide.
3. Buffer G: 90% ACN, 10% H_2O, pH 10 adjusted with ammonium hydroxide.

2.5 HA-Ub-PA Immunoprecipitation for LC-MS/MS Using High-Throughput Bravo Agilent Liquid Handling Platform

1. 5% SDS (sodium dodecyl sulfate).
2. 10% NP-40 (also known as IGEPAL CA-630).
3. Buffer D: 50 mM Trizma base, 0.5% (v/v) NP-40, 150 mM NaCl, 20 mM $MgCl_2$, adjust to pH 7.4 with HCl.
4. Protein A (PA-W) cartridges (Agilent, G5496-60000).
5. Anti-HA antibody.
6. Phosphate buffered saline (PBS).
7. 0.15% trifluoroacetic acid (TFA).

8. Trypsin.
9. 100 mM triethylammonium bicarbonate (TEAB) (pH 8.5).
10. 20% formic acid.

2.6 LC-MS/MS Standard/Fractionated Samples

1. 50 cm EASY-Spray column (2 μm beads, 75 μm ID).
2. Buffer H: acetonitrile, 0.1% formic acid, 5% DMSO.

2.7 LC-MS/MS TimsTOF

1. Evotips.
2. Isopropanol.
3. Buffer I: acetonitrile, 0.1% formic acid.
4. Buffer J: H_2O, 0.1% formic acid.
5. PepSep 8 cm analytical C18 column (3 μm beads, 100 μm ID).

3 Methods

3.1 HA-Ub-PA Probe Synthesis

3.1.1. Transform the pTYB2 HA-ubiquitin (76delG) plasmid vector (or a suitable alternative) into competent BL21 (DE3) *E. coli* using the aseptic technique. pTYB2 contains intein and a chitin-binding domain (CBD), which is fused to the C-terminal of ubiquitin for purification and subsequent reaction with the warhead.

1. Add 1 pg-100 ng of the vector to 50 μL of the competent *E. coli*, and place on ice for 30 minutes.
2. Heat shock for 10 seconds at 42 °C.
3. Return to the ice for 5 minutes.
4. Add 950 μL of sterile SOC media, and shake at 37 °C, 250 RPM, for 60 minutes.
5. Mix the cells and carry out 5 × 1 in 10 serial dilutions.
6. Spread 100 μL of each serial dilution onto separate sterile LB agar plates containing 100 μg/mL ampicillin (or different antibiotic + concentration in the case of alternative selection gene) (*see* **Note 1**).

3.1.2 Liquid Culture of Transformed *E. coli* Using the Aseptic Technique (*see* **Note 2**)

1. From the transformation, inoculate 2 × 40 mL of LB media containing 100 μg/mL ampicillin in a 250 mL conical flask, each with a single colony, and incubate at 37 °C 200 RPM overnight.
2. Optional: After an overnight growth period, glycerol stocks can be made to 20% glycerol and kept at −80 °C for future use.
3. Add 40 mL to 2 × 1 L of sterile LB in a 2 L conical flask containing 100 μg/mL ampicillin, and incubate at 37 °C 200 RPM until an OD_{600} of 0.8–0.9 is reached.

4. Induce protein expression by adding 0.4 mM IPTG, and incubate cultures overnight at 17 °C, 200 RPM.
5. Collect bacterial pellets by centrifugation at 6800 × g for 12 minutes, remove supernatant, and store at −80 °C.

3.1.3 HA-Ub-Mesna Purification from HA-Ub-Intein-CBD

1. For steps with *, remove 200 μL of the sample to analyze purification efficiency by SDS-PAGE gel and Western blot after purification.
2. Resuspend bacterial pellet in 20 mL of buffer A, and sonicate at an amplitude of 35% on ice for 10 × 30 s, with 30 s breaks in between.
3. Centrifuge the lysate at 30,000 × g for 30 minutes at 4 °C.
4. Transfer the supernatant to a fresh tube, and disinfect the tube with pellet*.
5. Fill a 1.5 × 10 mL chromatography column with 10 mL of chitin bead slurry using a clamp stand.
6. Wash the column with 10 mL of buffer A four times.
7. Add the lysate supernatant to the column and collect the flow-through.
8. Reapply the flow-through to the column and collect the flow-through again*.
9. Wash the column with 10 mL of buffer A, and collect the flow-through*.
10. Wash the column with a further 30 mL of buffer A and collect the flow-through*.
11. Store all flow-throughs and 200 μL aliquots at −20 °C.
12. Close the column and fill it with buffer B.
13. Seal the column with parafilm to ensure no liquid can leak out, and incubate at 37 °C with 180 RPM shaking overnight.
14. Let the resin settle and collect the flow-through*. At this point, the flow-through will be HA-Ub-Mesna.
15. Wash the column with 15 mL buffer A*.
16. Wash the column with a further 45 mL buffer A*.
17. Aliquot HA-Ub-Mesna-containing eluates and store the eluates at −80 °C.

3.1.4 SDS-PAGE Gel and Western Blot to Assess HA-Ub-Mesna Formation

1. Take 40 μL of 200 μL aliquots, add 10 μL of 5× Laemmli buffer, and boil at 90 °C for 5 minutes.
2. Run an acrylamide gel with 25 μL of each sample, in the order shown in Table 1, twice. Cut the gel in half, and use one side for Coomassie staining and one side for Western blotting.

Table 1
Example SDS-PAGE and a Western blot sample list

Lane	Sample
1	Molecular weight marker
2	Total lysate
3	Flow-through 1
4	Wash 1 flow-through
5	Wash 2 flow-through
6	HA-Ub-Mesna
7	HA-Ub-Mesna wash 1
8	HA-Ub-Mesna wash 2

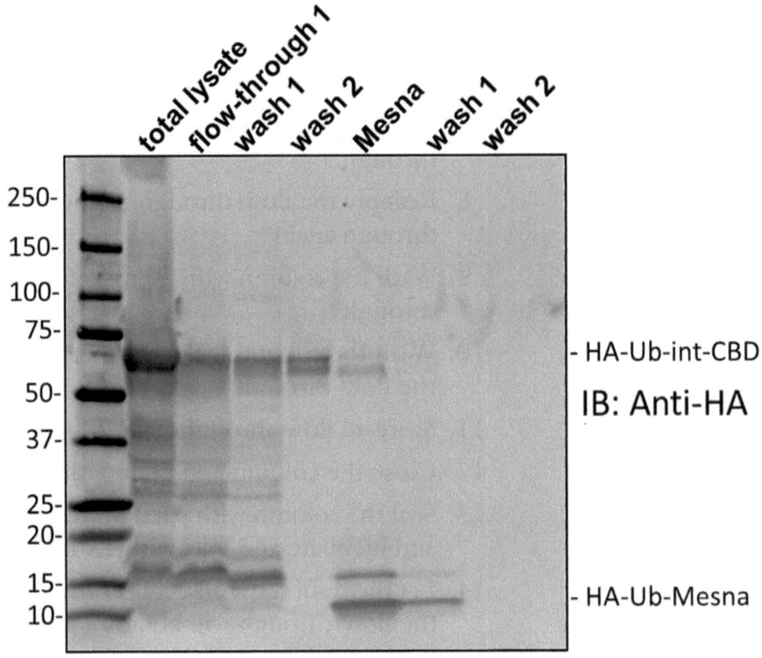

Fig. 3 Example anti-HA Western blot with samples from different stages of the HA-Ub-PA purification. From left to right: total lysate, flow-through 1, wash 1 flow-through, wash 2 flow-through, HA-Ub-Mesna, HA-Ub-Mesna wash 1, HA-Ub-Mesna wash 2

3. Assess purity of HA-Ub-Mesna by Coomassie staining/destaining the SDS-PAGE gel. HA-Ub-PA size is expected to be ~10 kDa.
4. Check for the presence of HA-Ub-Mesna by anti-HA immunoblotting the Western blot (Fig. 3).

5. Spin concentrate samples of pure HA-Ub-Mesna (typically HA-Ub-Mesna and HA-Ub-Mesna wash 1). A tenfold concentration with a 3 kDa MWCO filter should be sufficient.
6. Desalt the HA-Ub-Mesna using PD-10 desalting columns into buffer A.
7. Check the concentration of the HA-Ub-Mesna by BCA or equivalent protein estimation kit. Typically, the concentration is 4–6 mg/mL.

3.1.5 Formation of HA-Ub-PA from HA-Ub-Mesna

1. Dilute the HA-Ub-Mesna stock to 1.5 mg/mL using buffer A.
2. Add 250 mM propargylamine to the HA-Ub-Mesna, and leave for 1 hour on the rocking shaker at room temperature.
3. Desalt using PD-10 columns, and ascertain concentration using BCA or equivalent protein estimation kit.
4. Aliquot and store at –80 °C long-term or –20 °C short-term.
5. HA-Ub-PA probe formation can be assessed by intact protein LC-MS/MS.

3.2 Cell or Tissue Lysis for HA-Ub-PA Labeling

1. Keep lysate on ice throughout.
2. Add fresh DTT to buffer C to a final concentration of 1 mM (*see* **Note 3**).
3. Resuspend cell pellet in a volume of buffer C to yield a protein concentration of between 2 and 10 mg/mL (*see* **Notes 4** and **5**).
4. Add acid-washed glass beads to the lysate at a 1:2 v/v ratio of beads to buffer.
5. Vortex the lysate with the beads for 30 seconds, and return lysate to the ice for 1 minute. Repeat this a further nine times (*see* **Note 6**).
6. To pellet cell debris and glass beads, clarify the lysate at 14,000 ×g, 4 °C for 25 minutes (*see* **Note** 7).
7. Remove the supernatant without disturbing the pellet.
8. Quantify the protein concentration of the supernatant using a BCA kit or alternative.

3.3 HA-Ub-PA Labeling Optimization

1. HA-Ub-PA:lysate protein concentration ratios must be optimized for each batch.
2. Concentration dependence of 0.1–5 μL of HA-Ub-PA usually demonstrates that ~0.5 μL of HA-Ub-PA is sufficient to fully label 50 μg of lysate protein, with a 37 °C, 45 minutes incubation. Include an additional control sample without HA-Ub-PA labeling. An anti-HA Western blot will show a band at ~10 kDa, indicating saturating concentrations of HA-Ub-PA (Fig. 4).

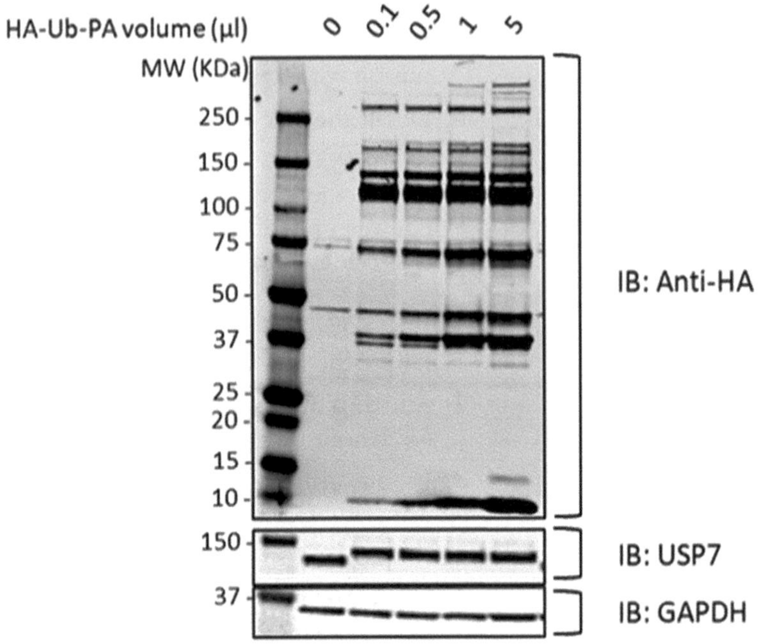

Fig. 4 Example Western blot for an HA-Ub-PA concentration dependence with 50 μg of MCF-7 lysate protein per lane. USP7 and anti-HA immunoblotting demonstrate binding saturation with 0.1 μL of HA-Ub-PA per 50 μg of lysate protein

3. Many DUBs are fully labeled with much shorter incubation times. In cases where a specific DUB is to be studied by Western blot, with non-covalent inhibitors, a time dependence of probe labeling should also be carried out to minimize probe incubation times/concentrations and, therefore, potential inhibitor displacement while ensuring complete DUB labeling.

3.4 HA-Ub-PA Immunoprecipitation for LC-MS/MS: Standard

1. Follow cell lysis outlined in 3.2, allowing for 500 μg lysate protein per condition. Include an additional control sample without HA-Ub-PA labeling.
2. Dilute with buffer C to 300 μL (minus the volume of inhibitor (optional) and HA-Ub-PA (required volume ascertained at 3.3)).
3. Incubate with HA-Ub-PA at 37 °C for 45 minutes.
4. Quench the labelling with 0.4% of SDS (from a 5% stock) and 0.5% NP-40 (from a 10% stock).
5. Dilute to 1 mL, with the addition of 661 μL of buffer D.
6. Remove 50 μL for a Western blot to check for immunoprecipitation efficiency.
7. Pause point: The HA-Ub-PA labeled lysates can be stored short term at −80 °C.
8. Aliquot 150 μL of anti-HA-agarose slurry per sample, using a cut tip to ensure reproducible pipetting of slurry.

9. Wash the slurry by adding 750 μL of buffer D, vortex briefly, and centrifuge at 2000 × g for 1 minute at room temperature to pellet the beads. Remove and discard the supernatant without disturbing the beads.
10. Repeat **step 9** a further three times.
11. After removing buffer D from the agarose beads for a final time, add the HA-Ub-PA labeled lysates to the beads, and incubate overnight at 4 °C on an end-over-end rotator.
12. Centrifuge the beads at 2000 × g for 1 minute, and collect and retain the supernatant. Remove a 50 μL aliquot for a Western blot to check the immunoprecipitation efficiency, and store the rest in case the immunoprecipitation has not worked.
13. Wash the slurry by adding 750 μL of buffer D, vortex briefly, and centrifuge at 2000 × g for 1 minute to pellet the beads at room temperature. Remove and discard the supernatant without disturbing the beads.
14. Repeat **step 13** a further three times.
15. Add 110 μL of 2× Laemmli buffer, vortex briefly, and boil at 90 °C for 5 minutes to elute proteins.
16. Remove a 10 μL aliquot of the eluate for the Western blot to check that the immunoprecipitation has worked efficiently.
17. Pause point: All samples can be stored at –20 °C short term or 80 °C long term.
18. Run samples on a Western blot, and blot for HA to ensure that HA-Ub-PA has labeled the lysate compared to the control and ensure that the signal is missing in the initial immunoprecipitation supernatant and enriched in the immunoprecipitation eluate.
19. Prepare the samples for LC-MS/MS analysis using the S-trap micro-column protocol, which includes reduction, alkylation, and digestion steps.
20. SpeedVac samples and resuspend in 20 μL of buffer E. Use vortexing and a sonicating water bath to ensure peptides are fully resuspended. Centrifuge at 17,000 × g for 10 minutes, and transfer the supernatant to glass mass spectrometry vials.

3.5 HPLC High pH Fractionation for Increased DUBome Depth

1. Degas buffer F + G with 10 minutes in a sonicating water bath or alternative apparatus.
2. With the C18 column attached and relevant pressure alarm set, flush system with 98% buffer F and 2% buffer G for 10 minutes at 0.2 mL/min.
3. Prime syringe and inject sample from the end of **step 3.3** onto the column.
4. For the following 65 minutes gradient, collect fractions every 1 minute for the first 60 minutes at a flow rate of 100 μL/min.

5. Start with a 15 minutes wash of 98% buffer F and 2% buffer G.
6. Set a gradient to increase to 65% buffer F and 35% buffer G over 30 minutes.
7. Finish the gradient by switching to 5% buffer F and 95% buffer G in 0.1 minutes.
8. Wash with 5% buffer F and 95% buffer G for 9.9 minutes.
9. Return to 98% buffer F and 2% buffer G in 0.1 minutes.
10. Wash with 98% buffer F and 2% buffer G for 9.9 minutes.
11. Concatenate 16.6 μL of each sample from 60 fractions into 10 using the following example interval:

    Sample 1: F1 + F11 + F21 + F31 + F41 + F51.

    Sample 2: F2 + F12 + F22 + F32 + F42 + F52
12. SpeedVac samples and resuspend in 20 μL of buffer E. Use vortexing and a sonicating water bath to ensure peptides are fully resuspended. Centrifuge at 17,000 × g for 10 minutes, and transfer to glass mass spectrometry vials.

3.6 HA-Ub-PA Immunoprecipitation for LC-MS/MS: High-Throughput Using Agilent Bravo Liquid Handling Platform (ABPP-HT)

3.6.1 Label Cellular DUBs with HA-Ub-PA and Quench the Reaction with SDS and NP-40

Perform HA-Ub-PA labelling as in Subheading 3.3 with the volume alterations as follows.

1. Dilute the required amount of lysate protein (*see Note 8*) with buffer C to 75 μL (minus the volume of inhibitor (optional) and HA-Ub-PA (required volume ascertained at 3.2)).
2. Incubate with HA-Ub-PA at 37 °C for 45 minutes.
3. Quench the labelling with 0.4% of SDS (from a 5% stock) and 0.5% NP-40 (from a 10% stock).
4. Dilute to a final volume of 250 μL with buffer D.
5. Remove an appropriate volume to check for immunoprecipitation efficiency by Western blot.
6. Pause point: At this point, the HA-Ub-PA-labeled lysates can be stored short term at −80 °C.

*3.6.2 Immobilize Anti-HA Antibody on Protein a Agilent Cartridges (PA-W) (see **Note 9**) Using the App for Immobilization (Agilent Sample Prep WorkBench v3.0.0)*

1. Prime cartridges with 100 μL of PBS at 300 μL/min.
2. Equilibrate cartridges with 50 μL of PBS at 10 μL/min.
3. Dilute antibody to a concentration of 2 mg/mL using PBS.
4. Load 100 μg (50 μL) of antibody onto cartridges at 3 μL/min.
5. Carry out a cup wash with 50 μL with PBS at 10 μL/min.
6. Carry out an internal cartridge wash of 100 μL with PBS at 10 μL/min.
7. Re-equilibrate the cartridges with 50 μL of PBS at 10 μL/min.

3.6.3 Immunoprecipitate Samples Using the Affinity Purification App (Agilent Sample Prep WorkBench v3.0.0)

1. After immobilization of the anti-HA antibody on the protein A cartridges, prime the cartridges with 100 μL of buffer D at 300 μL/min.
2. Equilibrate the cartridges with 50 μL of buffer D at 10 μL/min.
3. Load the sample at 1 μL/min.
4. Carry out a cup wash with 50 μL of buffer D at 10 μL/min.
5. Carry out an internal cartridge wash with 50 μL of buffer D at 10 μL/min.
6. Elute bound proteins using 50 μL of 0.15% TFA in water.

3.6.4 Trypsin Digestion of Samples

1. Neutralize eluates with 180 μL of 100 mM TEAB.
2. Digest samples with 1 μg of trypsin overnight at 37 °C.
3. Acidify samples with formic acid (final concentration 1%).

3.7 LC-MS/MS: Standard and Fractionated Samples

1. Samples are run using a liquid chromatography-tandem mass spectrometer setup. The type of mass spectrometer depends on what is available. We have previously used a Dionex UltiMate 3000 LC and a quadrupole-orbitrap mass spectrometer.
2. The LC method – Separate samples with a 50 cm EASY-Spray column (2 μm beads, 75 μm ID), or alternative on a 60 minutes gradient of 2–35% of acetonitrile in 0.1% formic acid and 5% DMSO (buffer H), with a flow rate of 250 nL/min.
3. Data-dependent acquisition (DDA) method – Acquire the MS1 spectra with a resolution of 70,000 and AGC target of 3e6 ions for a maximum injection time of 100 ms. Set the top 15 abundant peaks to be fragmented after isolation with a mass window of 1.6 Th at a resolution of 17,500 with a maximum injection time of 128 ms. Set normalized collision energy to 28% (HCD) (*see* **Note 10**).

3.8 LC-MS/MS: TimsTOF

To match the high-throughput methodology outlined in Subheading 3.6, an Evosep LC and timsTOF MS/MS can be run at a throughput of 100 samples per day. Short gradients on other mass spectrometers may also be tractable. However, the methodology outlined in Subheading 3.6 was optimized using an Evosep-timsTOF Pro setup, with the methodology of this setup detailed below.

1. Activate Evotips by soaking them with 100 μL of isopropanol.
2. Add 20 μL of buffer I and centrifuge at 700 × g for 1 minute.
3. Soak tips in isopropanol again.
4. Add 20 μL of buffer J and centrifuge at 700 × g for 1 minute.
5. Load 200 ng of peptide in 20 μL and centrifuge at 700 × g for 1 minute.

6. Add 20 μL of buffer J and centrifuge at 700 × g for 1 minute.
7. Add 100 μL of buffer J and centrifuge at 700 × g for 10 seconds.
8. Separate peptides on a PepSep 8 cm analytical C18 column (3 μm beads, 100 μm ID), with the Evosep One pre-set 100 samples per day gradient [19].
9. Data-dependent acquisition (DDA) method – Acquire data in PASEF mode (oTOF control 6.2.105/HyStar 5.1.8.1) in a mass range of 100–1700 m/z with four PASEF frames (three cycles overlap). Set the ion mobility window from 1/k0 0.85 to 1.3 V/cm^2, ramp time 100 ms with a locked duty cycle (*see* **Note 10**).

3.9 Data Processing

A number of software packages are available free of charge for the analysis of proteomic data. For data-dependent acquisition, as outlined here, we have previously used MaxQuant [20] and FragPipe [21, 22] for analysis. Both MaxQuant and FragPipe have extensive online tutorials. Briefly outlined below are the search settings for ABPP LC-MS/MS experimentation. Unless specified below, leave all settings as default for both MaxQuant or FragPipe. Download a recent UniProt FASTA database for the experimental species.

3.9.1 FragPipe Analysis

1. Load the "LFQ-MBR" workflow (assuming the sample size is >1).
2. Allocate the RAM and set the parallelism as appropriate.
3. Add the appropriate raw files to the workflow.
4. Set the data type as DDA.
5. Input the experiment and replicate the name to generate a multi-experiment report.
6. Under the database tab, load the UniProt FASTA file, ensure 50% decoys are present, and add decoys if they are not already included.
7. Ensure the default setting lists methionine oxidation and protein N-terminal acetylation as variable modifications and cysteine carbamidomethylation as a fixed modification.
8. Select "IM-MS" for timsTOF PASEF data and "Regular MS" for other data.
9. Select RUN.

3.9.2 MaxQuant Analysis

1. Load the appropriate raw files, and input proper experimental names and fractions if applicable.
2. Ensure the default setting lists methionine oxidation and protein N-terminal acetylation as variable modifications and cysteine carbamidomethylation as a fixed modification.
3. Set the label-free quantitation to LFQ.
4. Set the appropriate instrumentation.

5. Under sequences, select the location of the FASTA sequence file.
6. Under Identification, select match between runs.
7. Under folder locations, set a preferred location for the "combined output folder."
8. Under Performance, allocate an appropriate number of processors and select "start."

3.10 Data Analysis

1. Western blotting – A semi-quantitative analysis can be performed by measuring the densitometry of bands with imaging software such as ImageJ or Image studio. To identify the IC_{50} of an inhibitor, the intensity of the probe-labeled band can be plotted as a percentage of both probe and no probe bands. When normalized to a probe-labeled sample without inhibitor, the activity of a DUB with a specific concentration of inhibitor can be identified as a percentage (Fig. 5). Quantified DUB activity can then be fitted using software such as GraphPad to extract IC_{50} values.
2. LC-MS/MS – DUBs can be identified from the list given in Table 2. The enrichment of the DUBs should be compared to a no probe control to ensure that they have bound to the probe and not bound to the antibody/beads during the immunoprecipitation. If they are enriched, then intensities from a probe sample can be assumed to represent 100% DUB activity where labeling was complete, and immunoprecipitation was efficient. Any reduction in intensity with the addition of an inhibitor can then be plotted as % activity when normalized against the probe control sample.

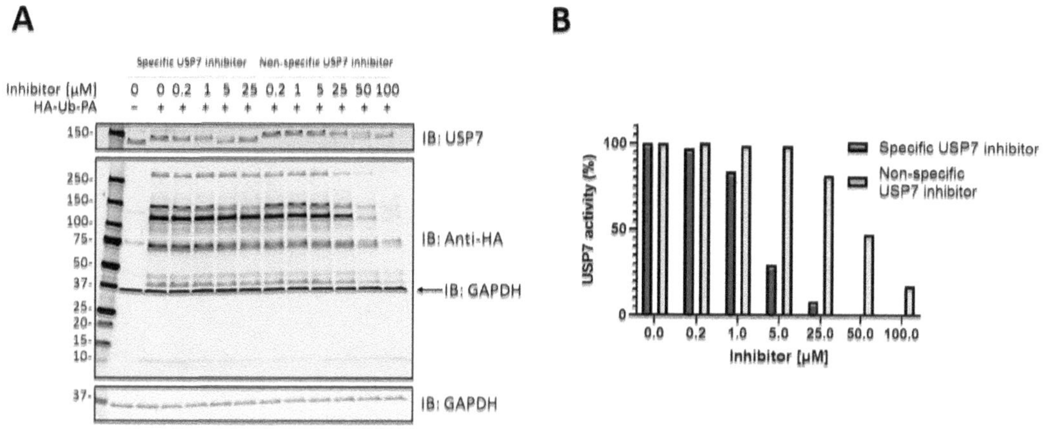

Fig. 5 Example Western blot for a specific and non-specific USP7 inhibitor. (**a**) USP7 immunoblotting shows the concentration dependence of the inhibitor preventing HA-Ub-PA binding; anti-HA immunoblotting shows that the inhibitor on the left is specific to USP7 up to a concentration where USP7 is fully inhibited. In contrast, the inhibitor on the right is widely cross-reacting and preventing HA-Ub-PA binding with multiple DUBs. (**b**) Example of densitometric quantification of the USP7 Western blot, with the top USP7 band as a percentage of both USP7 bands, with samples normalized as a percentage of the control

Table 2
List of cysteine active DUBs, with those previously shown to react with HA-Ub-PA from MCF-7 lysates using the fractionated deep DUBome methodology highlighted in grey shading

UniProt identifier	Gene name	DUB family
Q9BXU7	**USP26**	USP
Q9HBJ7	**USP29**	USP
P35125	**USP6** HRP1 TRE2	USP
Q8NFA0	**USP32** USP10	USP
K7EKZ1	**USP32**	USP
Q9P2H5	**USP35** KIAA1372 USP34	USP
Q8NB14	**USP38** KIAA1891	USP
O94782	**USP1**	USP
Q9Y5T5	**USP16** MSTP039	USP
Q70EL2	**USP45**	USP
Q86UV5	**USP48** USP31	USP
Q70CQ4	**USP31** KIAA1203	USP
Q70EL4	**USP43**	USP
O94966	**USP19** KIAA0891 ZMYND9	USP
P51784	**USP11** UHX1	USP
Q13107	**USP4** UNP UNPH	USP
Q9Y4E8	**USP15** KIAA0529	USP
Q70EL1	**USP54** C10orf29	USP
Q70EK8	**USP53** KIAA1350	USP
Q504Q3	**PAN2** KIAA0710 USP52	USP
Q9NQC7	**CYLD** CYLD1 KIAA0849 HSPC057	USP
Q9NVE5	**USP40**	USP
Q93009	**USP7** HAUSP	USP
Q96K76	**USP47**	USP
Q70CQ2	**USP34** KIAA0570 KIAA0729	USP
Q9UPU5	**USP24** KIAA1057	USP
Q93008	**USP9X** DFFRX FAM USP9	USP
O00507	**USP9Y** DFFRY	USP
Q9UHP3	**USP25** USP21	USP
Q96RU2	**USP28** KIAA1515	USP
Q9Y2K6	**USP20** KIAA1003 LSFR3A VDU2	USP
Q8TEY7	**USP33** KIAA1097 VDU1	USP
Q86T82	**USP37** KIAA1594	USP
P45974	**USP5** ISOT	USP
Q92995	**USP13** ISOT3	USP
Q53GS9	**USP39** CGI-21 HSPC332 PRO2855	USP
Q9H0E7	**USP44**	USP
Q70CQ1	**USP49**	USP
Q9P275	**USP36** KIAA1453	USP
Q9H9J4	**USP42**	USP

(continued)

Table 2
(continued)

UniProt identifier	Gene name	DUB family
Q0WX57	USP17L24 **USP17** USP17H USP17I USP17J USP17K USP17L USP17M; USP17L25; USP17L26; USP17L27; USP17L28; USP17L29; USP17L30	USP
Q6R6M4	USP17L2 DUB3 USP17 USP17H USP17I USP17J USP17K **USP17L** USP17M	USP
Q70CQ3	**USP30**	USP
Q14694	**USP10** KIAA0190	USP
Q9UMW8	**USP18** ISG43	USP
Q3LFD5	**USP41**	USP
P40818	**USP8** KIAA0055 UBPY	USP
Q9UK80	**USP21** USP23 PP1490	USP
P54578	**USP14** TGT	USP
Q70EL3	**USP50**	USP
Q70EK9	**USP51**	USP
Q9UPT9	**USP22** KIAA1063 USP3L	USP
A6NNY8	**USP27X** USP22L USP27	USP
Q9Y6I4	**USP3**	USP
O75604	**USP2** UBP41	USP
O75317	**USP12** UBH1 USP12L1	USP
P62068	**USP46**	USP
Q7RTZ2	USP17L1 **USP17L** USP17L1P	USP
H0YMI4	**USP3**	USP
P09936	**UCHL1**	UCHL
P15374	**UCHL3**	UCHL
Q9Y5K5	**UCHL5** UCH37 AD-019 CGI-70	UCHL
Q92560	**BAP1** KIAA0272 hucep-6	UCHL
Q7L8S5	**OTUD6A** DUBA2	OTU
Q8N6M0	**OTUD6B** DUBA5 CGI-77	OTU
Q5VVQ6	**YOD1** DUBA8 HIN7 OTUD2 PRO0907	OTU
Q5VV17	**OTUD1** DUBA7 OTDC1	OTU
Q96G74	**OTUD5**	OTU
Q5T2D3	**OTUD3** KIAA0459	OTU
Q96JH7	**VCPIP1** KIAA1850 VCIP135	OTU
Q01804	**OTUD4** HIN-1 KIAA1046	OTU
Q6GQQ9	**OTUD7B** ZA20D1	OTU
Q8TE49	**OTUD7A** C15orf16 CEZANNE2 OTUD7	OTU
Q9UGI0	**ZRANB1** TRABID	OTU
P21580	**TNFAIP3** OTUD7C	OTU
Q96BN8	**OTULIN** FAM105B	OTU

(continued)

Table 2
(continued)

UniProt identifier	Gene name	DUB family
Q96FW1	**OTUB1** OTB1 OTU1 HSPC263	OTU
Q96DC9	**OTUB2** C14orf137 OTB2 OTU2	OTU
Q9NP73	**ALG13** CXorf45 GLT28D1 MDS031	OTU
Q9NUU6	OTULINL **FAM105A**	OTU
Q8N5J2	**MINDY1** FAM63A KIAA1390	MINDY
Q8NBR6	**MINDY2** FAM63B KIAA1164	MINDY
Q9H8M7	**MINDY3** C10orf97 CARP DERP5 FAM188A MSTP126 My042	MINDY
Q4G0A6	**MINDY4** C7orf67 FAM188B	MINDY
Q5D1E8	**ZC3H12A** MCPIP MCPIP1	MCPIP
Q15040	**JOSD1** JSPH1 KIAA0063	Josephin
Q8TAC2	**JOSD2** SBBI54	Josephin
P54252	**ATXN3** ATX3 MJD MJD1 SCA3	Josephin
Q9H3M9	**ATXN3L** ATX3L MJDL	Josephin

4 Notes

1. To make agar plates, combine LB with 1.5% agar (V/W) and autoclave to sterilize. Add antibiotic from a pre-dissolved stock when the media bottle is cool enough to touch but before it solidifies. At this point, the plates can be poured.
2. Volumes for bacterial growth can be scaled depending on the required yield of HA-Ub-PA.
3. DTT oxidizes rapidly and should be thawed and kept on ice, with short-term storage of concentrated stocks at −20 °C.
4. Protein concentrations will vary depending on cell type; therefore, lysis buffer volumes will need to be adjusted accordingly.
5. If necessary, a buffer with detergent can be used. Anecdotally, freeze-thawing of lysates with detergent leads to less efficient HA-Ub-PA probe labeling.
6. The vortex time can be shortened by using a bead beater if available. If analyzing tissues, either homogenize the material first with a glass douncer and continue by vortexing with the glass beads, or homogenize and lyse using a bead beater.
7. This centrifugation step is appropriate if the DUB of interest is cytosolic. Consider lower centrifugation speeds to retain particular organelles depending on the DUB of interest.

8. Because of variation in protein estimation kits and probe batches, it is best first to do a lysate concentration dependence to ascertain an optimal lysate protein concentration for use with the Agilent Bravo liquid handling platform. Optimization of lysate protein concentration will maximize the number of DUBs identified by LC-MS/MS while ensuring that the column is not saturated, which is necessary for accurate IC_{50} extractions and cross-reactivity data. We have previously found that 250 μg of lysate protein was optimal, with a concentration dependence ranging from 50 μg to 500 μg, usually encompassing the optimal concentration. Western blot can assess immunoprecipitation efficiency as outlined in 3.3 by checking for anti-HA immunoblotting signal depletion in immunoprecipitation flow-through and signal enrichment in the eluate.
9. These cartridges are specific to the Bravo liquid handling platform. Protein G cartridges also work comparably.
10. Data can also be collected via a data-independent acquisition (DIA) mode. If this methodology is applied, DIA-NN software can be used for processing.

Acknowledgments

We thank members of the Kessler group for their helpful discussions. We are grateful to Roman Fischer and Iolanda Vendrell, leads of the Discovery Proteomics Facility, for expert help with mass spectrometry analysis. H.B.L.J. was supported by a Bristol Myers Squibb award to B.M.K. A.P.F. and B.M.K were supported by the Chinese Academy of Medical Sciences (CAMS) Innovation Fund for Medical Science (CIFMS), China (grant number: 2018-I2M-2-002), and by Pfizer.

References

1. Harrigan JA, Jacq X, Martin NM et al (2018) Deubiquitylating enzymes and drug discovery: emerging opportunities. Nat Rev Drug Discov 17:57–78
2. Pinto-Fernández A, Davis S, Schofield AB et al (2019) Comprehensive landscape of active deubiquitinating enzymes profiled by advanced chemoproteomics. Front Chem 7:592
3. Hanpude P, Bhattacharya S, Dey A et al (2015) Deubiquitinating enzymes in cellular signaling and disease regulation. IUBMB Life 67:544–555
4. Borodovsky A, Ovaa H, Kolli N et al (2002) Chemistry-based functional proteomics reveals novel members of the deubiquitinating enzyme family. Chem Biol 9:1149–1159
5. Altun M, Kramer HB, Willems LI et al (2011) Activity-based chemical proteomics accelerates inhibitor development for deubiquitylating enzymes. Chem Biol 18:1401–1412
6. Hewings DS, Flygare JA, Bogyo M et al (2017) Activity-based probes for the ubiquitin conjugation-deconjugation machinery: new chemistries, new tools, and new insights. FEBS J 284:1555–1576
7. Shanmugham A, Ovaa H (2008) DUBs and disease: activity assays for inhibitor

development. Curr Opin Drug Discov Devel 11:688–696

8. Ekkebus R, Flierman D, Geurink PP et al (2014) Catching a DUB in the act: novel ubiquitin-based active site directed probes. Curr Opin Chem Biol 23:63–70
9. Weber A, Elliott PR, Pinto-Fernandez A et al (2017) A linear diubiquitin-based probe for efficient and selective detection of the deubiquitinating enzyme OTULIN. Cell Chem Biol 24:1299–1313.e7
10. Ekkebus R, van Kasteren SI, Kulathu Y et al (2013) On terminal alkynes that can react with active-site cysteine nucleophiles in proteases. J Am Chem Soc 135:2867–2870
11. Hewings DS, Heideker J, Ma TP et al (2018) Reactive-site-centric chemoproteomics identifies a distinct class of deubiquitinase enzymes. Nat Commun 9:1162
12. Conole D, Mondal M, Majmudar JD et al (2019) Recent developments in cell permeable deubiquitinating enzyme activity-based probes. Front Chem 0:876
13. Ward JA, McLellan L, Stockley M et al (2016) Quantitative chemical proteomic profiling of ubiquitin specific proteases in intact cancer cells. ACS Chem Biol 11:3268–3272
14. Cho J, Park J, Kim EE et al (2020) Assay systems for profiling deubiquitinating activity. Int J Mol Sci 21:5638
15. Leestemaker Y, de Jong A, Ovaa H (2017) Profiling the activity of deubiquitinating enzymes using chemically synthesized ubiquitin-based probes. Methods Mol Biol 1491:113–130
16. Kramer HB, Nicholson B, Kessler BM et al (2012) Detection of ubiquitin–proteasome enzymatic activities in cells: application of activity-based probes to inhibitor development. Biochim Biophys Acta 1823:2029–2037
17. Skelly M, Spoel S (2021) Characterising plant deubiquitinases with in vitro activity-based labelling and ubiquitin chain disassembly assays. Bio Protoc 11:e4015
18. Jones HBL, Heilig R, Fischer R et al (2021) ABPP-HT – high-throughput activity-based profiling of deubiquitylating enzyme inhibitors in a cellular context. Bio Protoc 9:640105
19. Bache N, Geyer PE, Bekker-Jensen DB et al (2018) A novel LC system embeds analytes in pre-formed gradients for rapid, ultra-robust proteomics. Mol Cell Proteomics 17:2284–2296
20. Tyanova S, Temu T, Cox J (2016) The MaxQuant computational platform for mass spectrometry-based shotgun proteomics. Nat Protoc 11:2301–2319
21. Kong AT, Leprevost FV, Avtonomov DM et al (2017) msFragger: ultrafast and comprehensive peptide identification in mass spectrometry-based proteomics. Nat Methods 14:513
22. Leprevost FV, Haynes SE, Avtonomov DM et al (2020) Philosopher: a versatile toolkit for shotgun proteomics data analysis. Nat Methods 17:869

Chapter 8

High-Throughput MALDI-TOF Mass Spectrometry-Based Deubiquitylating Enzyme Assay for Drug Discovery

Virginia De Cesare and Paul Davies

Abstract

Ubiquitylation (or ubiquitination) is the reversible conjugation of a 76-amino-acid polypeptide (ubiquitin) to a target protein to modulate various biological processes. Deubiquitylating enzymes (DUBs) are a class of enzymes that specifically remove ubiquitin from a substrate. In recent years DUBs have garnered significant attention as a new class of targets in multiple therapeutic areas. The recent development of high-throughput Matrix-Assisted Laser Desorption/Ionization-Time of Flight mass spectrometry (MALDI-TOF MS) has provided new tools to perform drug discovery screening. Here we present a facile and high-throughput step-by-step protocol of the MALDI-TOF MS-based DUB assay for screening the activity of DUBs in vitro. In a MALDI-TOF DUB assay, we quantitate the amount of mono-ubiquitin generated by the in vitro cleavage of ubiquitin chains. The presented protocol takes advantage of nanoliter dispensing robotics and automated MALDI-TOF MS analysis to screen large and diverse compound libraries.

Key words Ubiquitin, Deubiquitylating enzyme, DUBs, MALDI-TOF mass spectrometry, Drug discovery, High-throughput screening

1 Introduction

Ubiquitylation is a posttranslational modification that impacts multiple cellular processes. Removal of ubiquitin from its substrate is the activity of specific ubiquitin proteases, called deubiquitylating enzymes (DUBs). The human genome encodes approximately 100 DUBs that are divided into seven families [1–3]: the USPs family, with 56 members, 17 ovarian tumor proteases (OTUs), 12 JAB1/MPN/MOV34 Zn-dependent metalloprotease (JAMMs), 4 Machado-Josephin domain proteases (MJD), 4 ubiquitin c-terminal hydrolases (UCHs), 5 motif interacting with ubiquitin-containing novel DUB family (MINDYs), and 1 zinc finger-containing ubiquitin peptidase 1 (ZUP1).

In 2014, Ritorto et al. [4] employed MALDI-TOF MS to characterize the in vitro activity and substrate specificity of a panel

Julie A. Maupin-Furlow and Mariola J. Edelmann (eds.), *Deubiquitinases: Methods and Protocols*,
Methods in Molecular Biology, vol. 2591, https://doi.org/10.1007/978-1-0716-2803-4_8,

of 32 DUBs. This method, named the MALDI-TOF DUB assay, relies on detecting the free ubiquitin signal generated by the activity of a DUB toward ubiquitin dimers. Quantification of DUB activity is achieved by using heavy labelled ^{15}N ubiquitin as an internal standard. The method was initially performed in standard 0.5 mL tubes and was therefore low-throughput in nature. Here, we report a step-by-step protocol of a high-throughput version of the original MALDI-TOF DUB assay, adapted to employ a nanoliter pipetting system, fast automatic detection, and semi-automatic data analysis [5–9].

Initially, compounds are deposited on 384-well plates using acoustic delivery systems and incubated with the DUB of interest. The reaction is started by adding a ubiquitin dimer of suitable linkage and stopped by adding trifluoroacetic acid. Positive hits are then confirmed by dose-response analysis, allowing calculation of IC50 values. Cherry-picked positive "hits" are then tested against a panel of different DUBs to assess their specificity within the same phylogenetic family and across six out of the seven DUB families. The 384-plate format is typically employed for extensive compound screening, and up to 3072 data points can be acquired daily.

2 Materials

Prepare all solutions using Milli-Q water and mass spectrometry grade solvents (ethanol, acetonitrile, and isopropanol) for all stages. Low-protein-binding tubes are recommended for the handling and storage of ubiquitin substrates.

2.1 Solutions and Reagents

1. Prepare MALDI-TOF matrix by weighing 7.6 mg of 2, 5-dihydroxyacetophenone (DHAP) in a 1.5 mL tube. Add 375 μL of 100% ethanol. In a separate tube, weight 25 mg diammonium hydrogen citrate, and resuspend in 1 mL Milli-Q water. Add 125 μL of the aqueous diammonium hydrogen citrate to the ethanol/DHAP matrix solution. Shake the mixture for at least an hour. Store the matrix solution for up to 1 week at room temperature (20° – 25 ° C). Do not store at +4° or −20 °C.
2. Enzyme buffer: 40 mM Tris-HCl pH 7.5, BSA 0.01%, 1 mM dithiothreitol (DTT), or tris(2-carboxyethyl)phosphine (TCEP). To prepare 10 mL of enzyme buffer, mix 0.4 mL of 1 M Tris-HCl stock solution and 10 μL of 10% bovine serum albumin (BSA). Bring to volume with Milli-Q water. Keep on ice. Add 20 μL of 0.5 M TCEP or DTT before using this solution to dilute the enzyme of choice.
3. Substrate solution: 40 mM Tris-HCl pH 7.5, 0.01% BSA. Use enzyme buffer (without the DTT or TCEP) to dilute the substrate of choice to a final concentration of 0.1 mg/mL.

To be kept on ice until use. Substrate solution can be stored at 4 °C for several days in low binding tubes.

4. Stopping solution: 6% trifluoroacetic acid (TFA). To prepare 100 mL of stopping solution, add 6 mL of 100% TFA to 94 mL of Milli-Q water. Agitate gently and keep on ice until use. The stopping solution can be stored at 4 °C for up to 2 weeks. *See* **Note 1** regarding TFA and the care taken when handling this material.
5. ^{15}N ubiquitin solution: produce and purify the ^{15}N ubiquitin as previously described [8], and then dilute in Milli-Q water to a final concentration of 4 μM.
6. MALDI target cleaning solution: to be prepared under chemical fume hood. For 1 L: mix 300 mL of LC-MS grade acetonitrile and 699 mL of Milli-Q water. Add 1 mL of 100% trifluoroacetic acid (30% acetonitrile, 0.1% TFA (v/v) solution). Refer to **Note 1** for TFA risks and handling.
7. High-purity nitrogen gas.
8. Recombinant deubiquitylase enzyme (in vitro active; see Table 1 and **Note 2**).
9. Ubiquitin dimer (or trimer/tetramer), produced and purified as previously described [10, 11]. *See* **Note 2**.

2.2 Equipment, Supplies, and Software

1. MTP AnchorChip 1536 BC (Part Number # 8280787).
2. 384-well microplate, PS, small volume, hibase, white (Greiner Bio-One, Cat. number 784904).
3. 384-well LSVD plate, non-sterile polypropylene (TTP Labtech, Cat. number 4150-05828)
4. Microplate 384/V-PP, Protein LoBind (Eppendorf, Order no 0030 624.300).
5. SILVERseal sealer aluminum foil.
6. Rapiflex MALDI-TOF mass spectrometer (Bruker – equipped with Compass software for FlexSeries 2.0).
7. FlexControl software (Bruker).
8. FlexAnalysis software (Bruker).
9. ZERO_Script file (*see* **Note 3** for code availability), to be imported into the Flex Analysis Methods folder.
10. UW Ultrawave QS18 ultrasonic cleaning bath.
11. 20 cm × 13 cm × 4 cm stainless steel box for sonicating MALDI target.
12. FluidX, XRD-384 reagent dispenser. To be tested for accuracy ahead of experiment, *see* **Note 4**.
13. TTP Labtech Mosquito HTS equipped with five positions plate deck.
14. 16 Channel Electronic VIAFLO Pipette, 2–50 μL volume range and appropriate tips.

Table 1
In vitro active deubiquitylases currently available for selectivity profiling in the MRC-PPU DUB profiling service

Enzyme	Length	Accession	Tag	DU number	Expression system	Substrate
USP1 G670A G671A/WDR48	1-785/1-732	O94782.1/ Q8TAF3-1	His6/ None	DU23056	Insect cells	K63
USP2	1-396	O75604-4	GST	DU13025	*E. coli*	K63
USP4	1-963	Q13107-1	His6	DU14350	*E. coli*	K48
USP5	1-858	P45974-1	His6	DU15641	*E. coli*	K63
USP6	529-1406	P35125-1	GST	DU37745	Insect cells	K63
USP7	1-1102	Q93009-1	His6	DU15644	*E. coli*	K63
USP8	1-1118	P40818-1	His6	DU15645	Insect cells	K48
USP9x	1553-1995	Q93008-3	GST	DU20628	*E. coli*	K11
USP9y	1550-2000	O00507-1	GST	DU21846	*E. coli*	K11
USP10	1-798	Q14694-1	His6	DU15647	Insect cells	K63
USP11	1-963	P51784-1	GST	DU20016	*E. coli*	K48
USP12/WDR48	1-370/1-732	O75317-1/ Q8TAF3-1	His6/ None	DU24373	Insect cells	K48
USP15	1-952	Q9Y4E8-2	GST	DU37753	Insect cells	K48
USP16	1-823	Q9Y5T5-1	His6	DU25349	*E. coli*	K63
USP19	1-1318	O94966-1	GST	DU37789	Insect cells	K63
USP20	1-914	Q9Y2K6-1	GST	DU15664	*E. coli*	K63
USP21	196-565	Q9UK80-1	GST	DU22385	*E. coli*	K48
USP25	1-1055	Q9UHP3-2	GST	DU21610	*E. coli*	K48
USP27x	1-438	A6NNY8-1	GST	DU21193	*E. coli*	K11
USP28	1-1077	Q96RU2-1	GST	DU20233	*E. coli*	K48
USP30	57-517	Q70CQ3-1	GST Clvd	DU36294	*E. coli*	K48
USP35	390-978	Q9P2H5-1	GST	DU67487	*E. coli*	K63
USP36	81-461	Q9P275-1	GST	DU4944	*E. coli*	K11
USP38	1-1042	Q8NB14-1	GST	DU35473	*E. coli*	K11
USP45	1-814	Q70EL2-1	GST	DU15681	*E. coli*	K48
USP46/WDR48	1-366/1-732	P62068-1/ Q8TAF3-1	His6/ None	DU24347	Insect cells	K48

(continued)

Table 1
(continued)

Enzyme	Length	Accession	Tag	DU number	Expression system	Substrate
USP47	1-1355	Q96K76-4	His6	DU15682	Insect cells	K48
USP48	1-1053	Q86UV5-1	GST	DU27626	Insect cells	K11
CYLD	2-956	Q9NQC7-1	His6	DU1834	*E. coli*	K63
OTULIN	1-352	Q96BN8-1	GST Clvd	DU43487	*E. coli*	M1
OTU1	1-348	Q5VVQ6-1	GST	DU36559	*E. coli*	K11
OTUD1	270-481	Q5VV17-1	His6	DU25080	*E. coli*	K63
OTUD3	1-398	Q5T2D3-1	GST	DU21323	*E. coli*	K11
OTUD5 phospho Ser 177	1-571	Q96G74-1	GST Clvd	DU21450	*E. coli*	K63
OTUB1	1-271	Q96FW1-1	GST	DU19741	*E. coli*	K48
OTUB2	1-234	Q96DC9-1	GST	DU32795	*E. coli*	K63
VCPIP1	25-561	Q96JH7-1	GST	DU44386	*E. coli*	K48
vOTU	1 - 183	3ZNH_A[a]	GST	DU45351	*E. coli*	K63
TRABID	245-697	Q9UGI0-1	His	DU22468	*E. coli*	K63
A20	1-366	P21580-1	GST	DU32912	*E. coli*	K48
Cezanne	1-843	Q6GQQ9-1	GST	DU20899	*E. coli*	K11
UCHL1	1-223	P09936-1	His	DU15693	*E. coli*	Ub-W
UCHL3	1-230	P15374-1	GST	DU21015	*E. coli*	Ub-W
UCHL5	1-329	Q9Y5K5-1	GST	DU12810	*E. coli*	Ub-W
BAP1	1-240	Q92560-1	GST	DU63658	*E. coli*	Ub-W
JOS2	1-188	Q8TAC2-1	His	DU20941	*E. coli*	K11
AMSH	256-424	O95630-1	GST	DU15719	*E. coli*	K63
AMSH-LP	1-436	Q96FJ0-1	GST	DU15780	*E. coli*	K63
MINDY1	1-469	Q8N5J2-1	MBP	DU59325	*E. coli*	K48 (4mer)
MINDY2	1-621	Q8NBR6-1	GST	DU63404	*E. coli*	K48
MINDY3	1-445	Q9H8M7-1	GST	DU47870	*E. coli*	K48

[a]*vOTU* viral homologue of the ovarian tumor protease superfamily from the Crimean-Congo hemorrhagic fever virus. Substrate: ubiquitin dimer of the indicated linkage or alternative ubiquitin substrate. 4mer: tetramer

3 Methods

Carry out all procedures at room temperature unless otherwise specified.

3.1 Preliminary Preparations

1. Clean the MALDI target. To do so, place the MALDI target (384 or 1536 AnchorChip) in a stainless steel box of adequate size. Pour 100% isopropanol onto the target to completely cover it, and sonicate for 3 minutes (using the UW Ultrawave QS18 Ultrasonic Cleaning Bath). Remove isopropanol and add 30% acetonitrile and 0.1% TFA (v/v) solution, sonicate for further 3 minutes, and then dry the MALDI target plate using a stream of high-purity nitrogen.

3.2 MALDI-TOF DUB Assay for HTS Screening

Compound libraries are usually delivered by non-contact acoustic transfer on 384 well plates of choice. The final concentration of the compounds varies depending on the type of library in use. 10–50 μM final is a commonly used range. Depending on the material starting concentration, 5–500 nL can be deposited on the assay plate. Higher amounts can also be spotted. However, the final DMSO concentration should not exceed 3% (compounds are usually resuspended in 100% DMSO).

1. Aliquot compounds into assay plate (384-well microplates) using non-contact acoustic delivery. Reserve two columns for both positive (DMSO only) and negative (no enzyme) controls (see Fig. 1 and **Notes 5** and **6**).
2. Using FluidX, XRD-384 reagent dispenser, aliquot 3 μL of enzyme solution into each assay plate column except for the negative control column (see Fig. 1).
3. Using FluidX, XRD-384 reagent dispenser, aliquot 3 μL of enzyme buffer into the negative control column (see Fig. 1).
4. Cover the plate with aluminum sealing foil to prevent evaporation, and incubate at room temperature for 5–30 minutes.
5. Using FluidX, XRD-384 reagent dispenser, add 3 μL of the substrate to each column of the entire assay plate.
6. Cover the assay plate with aluminum sealing foil to prevent evaporation and incubate at room temperature for the time suggested by initial optimization experiments (*see* **Note** 7).
7. Stop the reaction by adding 3 μL TFA to a final 2% (v/v) concentration to each column (*see* Fig. 1).

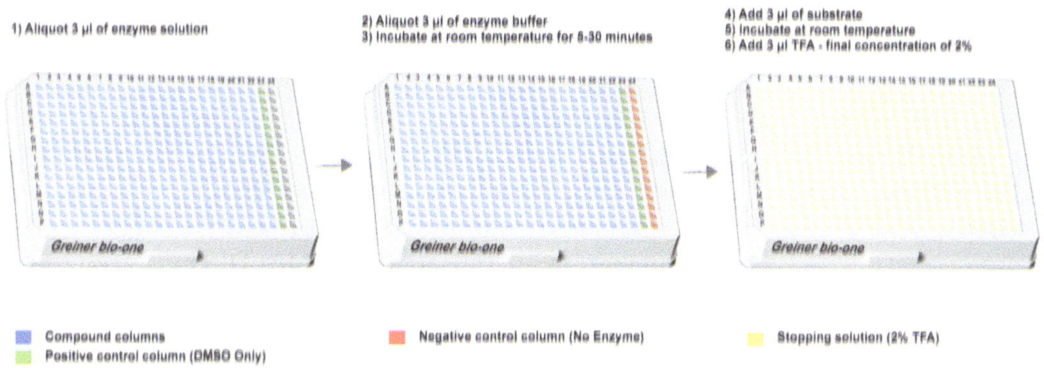

Fig. 1 MALDI-TOF MS HTS workflow

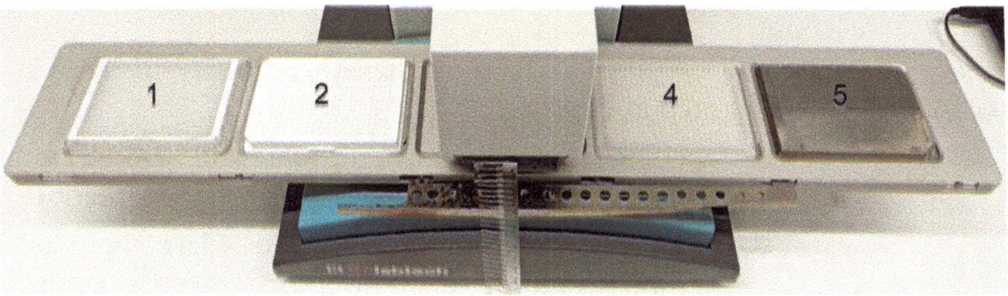

Fig. 2 TTP Labtech Mosquito HTS plate disposition. Reservoir plate with ^{15}N Ubiquitin and DHAP matrix (position 1). Assay plate (position 2). Mixing plate (position 4). MALDI target (position 5)

3.3 Spot the Assay Plates on the MTP AnchorChip 1536 MALDI Target

This step is performed using a TTP Labtech Mosquito HTS (equipped with five positions plate deck). The instrument aliquots 1050 nL of sample from the assay plate into the mixing plate (384-well LSVD plate, non-sterile polypropylene). The sample will be extensively mixed with 1200 nL of DHAP matrix (taken from reservoir plate) and spotted on the MTP AnchorChip 1536 MALDI target (*see* Fig. 2, **Notes 8** and **9**).

1. Aliquot a sufficient amount of DHAP and ^{15}N ubiquitin into a reservoir plate (microplate 384/V-PP), and cover the plate with aluminum sealing foil to prevent evaporation.
2. On the TTP Labtech Mosquito HTS, position the reservoir plate on deck position 1, assay plate on deck position 2, mixing plate on deck position 4 and the MTP AnchorChip 1536 BC MALDI target on deck position 5.
3. Aliquot 1050 nL of assay plate column 1 into mixing plate column 1.
4. Add 500 nL of ^{15}N ubiquitin to mixing plate column 1.
5. Add 1.2 µL of DHAP matrix to mixing plate column 1.
6. Mix thoroughly using 2× 10 mixing cycles of 1000 nL.
7. Spot 260 nL on the MTP AnchorChip 1536 MALDI target.

8. Repeat **steps 3** to 7 for all subsequent assay plate columns.
9. Allow the MALDI target to dry completely at room temperature.

3.4 Automatic MALDI-TOF MS Analysis

1. Open Flex Control and select a fingerprint (.par file) Reflectron Positive (RP) acquisition (acquisition window 8–9 kDa).
2. Insert and dock the freshly spotted MALDI target into the instrument, and select appropriate MALDI target geometry.
3. Wait for the instrument to reach an adequate vacuum level.
4. Calibrate the instrument using the ^{15}N ubiquitin signal (8669.47 *m/z*).
5. In the Automation run, select "New" and start a new run using the automatic run wizard.

3.5 Statistical Analysis for HT MALDI-TOF DUB Assay

1. Download and allocate scripts necessary for analysis as indicated in **Note 3**.
2. Open the Flex Analysis software, and select one spectrum within the folder containing the raw data of interest.
3. Open the "Method" menu, and select the Flex Analysis method (extension. FAMSMethod) "ZERO" script.
4. Open the "Method" menu and select "Edit Script."
5. Modify the destination folder (strOutputDir) and result file name (StrOutputFileName) as desired, and save the modified script.
6. Open the flexAnalysis Batch Process tool and select spectra to process using the previously modified "ZERO" script. The zero script will export mass areas of each spot from the raw data into a .csv column format. For the negative controls, the zero script considers the area of the background noise detectable in the ubiquitin *m/z* window.
7. Copy and paste the zero script output columns (Spectrum, Compound, m/z, S/N, Resolution and Area) into the Input sheet of the Baseline data check file.
8. Convert Output zero deleted sheet (the last sheet of Baseline data check file) to .txt plain format.
9. Drag text file onto the Grid .exe file. The script will produce a new file with GRID included in the file name. The GRID file will select ubiquitin and ^{15}N ubiquitin area values and report them in the same position as spotted on the MALDI plate.
10. Calculate ubiquitin over ^{15}N ubiquitin area ratio and % of substrate consumed (*see* **Note 10**).
11. Evaluate data quality by calculating Z' scores, *see* **Notes 11** and **12**.

4 Notes

1. TFA should be handled in a chemical fume hood using protective eyewear and gloves. TFA is used as a component of the stopping solution and for cleaning the MALDI target.
2. High-throughput screening campaigns should be performed using the same batch of both enzyme and substrate, to be partitioned prior to use into single-use aliquots. If this is not possible, new protein batches should be tested side by side before the screening.
3. The Zero Script in The in-house scripts are publicly available in GitHub at https://github.com/Vdecesare/GRID-script.git and https://github.com/Vdecesare/Zero-Script.git). Locate the FAMSMethod in the Data\Methods\flexAnalysisMethods.
4. All liquid handling systems should be tested for accuracy before starting the main procedure. Quality control of the FluidX, XRD-384 reagent dispenser should report a coefficient of variation (CV%) ≤5. If the CV% is >5, a new liquid handling head might be required. Accurate calibration of the TTP Labtech Mosquito HTS is also fundamental to ensure optimal mixing and spotting on the MALDI target.
5. DTT and TCEP are strong nucleophiles and may therefore compete for the compound with the active cysteine of the DUBs. Carefully evaluate DTT or TCEP presence in the reaction buffer depending on the DUB and compound library in

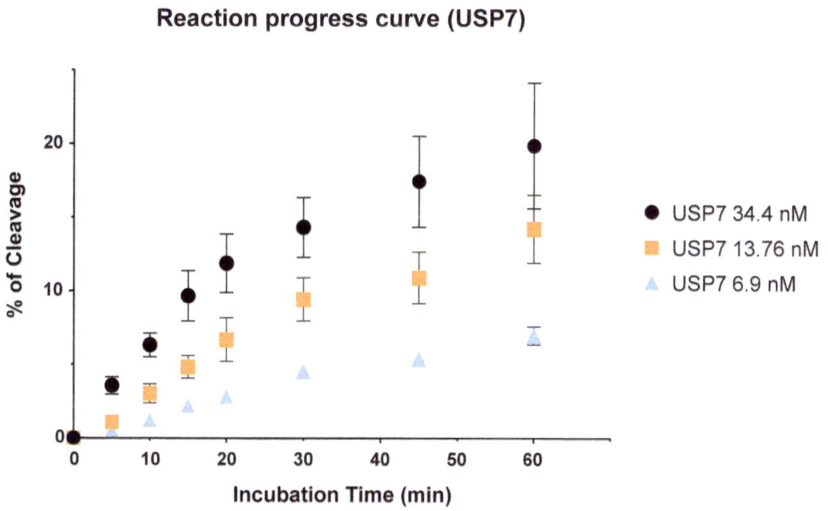

Fig. 3 Reaction progress curve of USP7. Different enzymatic concentrations are tested over time. The linearity of the reaction is evaluated to identify the ideal enzyme concentration and incubation time

use. Enzymatic activity should be evaluated in the presence or absence of a reducing agent in a side-by-side comparison.

6. Both positive (DMSO only) and negative controls (no enzyme or presence of strong DUB inhibitor) are necessary to calculate the reliability and quality of data from each plate. A positive control indicates the maximum signal achieved without any interference, while the negative control determines the background signal. In absence of ubiquitin signal, as for in the negative control and in case of strong inhibition, the "Zero script" will integrate the signal background area to allow Z' Prime calculations.
7. Before starting an HTS campaign, it is fundamental to identify ideal enzyme and substrate concentrations, incubation times, and temperatures. This optimization can be done by testing several enzyme concentrations over a time course experiment (reaction progress curve; *see* Fig. 3). Data points should be collected at the beginning of the reaction to identify active compounds when an excess of the substrate results in linear product formation. In Fig. 3, representative results of a reaction progress curve of USP7 tested at three different concentrations (34.4 nM, 13.76 nM, and 6.9 nM). The percentage of a cleaved substrate (% of cleavage) is obtained by applying the equation in **Note 9**. At the final concentration of 34.4 nM, the reaction is linear for up to 15 minutes. The reaction incubation time can be extended to 20 minutes by reducing the final concentration of the enzyme down to 13.76 nM. Further dilution to 6.9 nM results in a nonlinear product formation curve.
8. Extensive mixing cycles of sample and DHAP matrix are required to activate the formation of sample/DHAP matrix crystals and ensure optimal MALDI-TOF MS signal. The number of mixing cycles and the volume can be adjusted depending on sample concentration and complexity.
9. The DHAP matrix is prepared with 75% ethanol; the evaporation of mixed samples is time-sensitive. The mixture of a sample and matrix must be immediately spotted onto the MALDI target before the next steps can take place. Once spotted, the reaction mixture should be analyzed as soon as possible to avoid degradation of the sample.
10. Ubiquitin/^{15}N ubiquitin ratio can be translated into % of substrate consumed (% of cleavage) using the previously reported equation [4]:

$$x = \left(\frac{\text{AreaUbi}}{\text{Area }^{15}N\text{ Ub}} \times \left[^{15}N\text{ Ubi}\right]\right)/2 \Big) / [\text{DiUbi}] \times 100 \quad (1)$$

11. The aim of HTS statistical analysis is the robust identification of positive hits. The Z-Prime (or Z', *See* Eq. 2), extensively used in the HTS field, takes into account both the signal window between positive and negative controls and their statistical variability [12]. Z-Prime is defined as three times the sum of positive and negative control standard deviations (σ_p, σ_n) divided by the difference of the averages of positive and negative controls (μ_p, μ_n):

$$\text{Z} - \text{Prime} = 1 - \frac{3\left(\sigma_{p+\sigma_n}\right)}{\left|\mu_{p-\mu_n}\right|} \tag{2}$$

12. Z-Prime scores higher than 0.5 are considered acceptable.

Contributions

V.DC. and P.D. wrote the manuscript.
The data that support the anticipated results are available from the corresponding author upon reasonable request. Other enquiries can also be addressed to https://dub-screen.mrc.ac.uk/.

References

1. Clague MJ, Barsukov I, Coulson JM et al (2013) Deubiquitylases from genes to organism. Physiol Rev 93(3):1289–1315. https://doi.org/10.1152/physrev.00002.2013
2. Nijman SM, Luna-Vargas MP, Velds A et al (2005) A genomic and functional inventory of deubiquitinating enzymes. Cell 123(5): 773–786. https://doi.org/10.1016/j.cell.2005.11.007
3. Clague MJ, Urbe S, Komander D (2019) Breaking the chains: deubiquitylating enzyme specificity begets function. Nat Rev Mol Cell Biol 20(6):338–352. https://doi.org/10.1038/s41580-019-0099-1
4. Ritorto MS, Ewan R, Perez-Oliva AB et al (2014) Screening of DUB activity and specificity by MALDI-TOF mass spectrometry. Nat Commun 5:4763. https://doi.org/10.1038/ncomms5763
5. Huguenin-Dezot N, De Cesare V, Peltier J et al (2016) Synthesis of isomeric phosphoubiquitin chains reveals that phosphorylation controls deubiquitinase activity and specificity. Cell Rep 16(4):1180–1193. https://doi.org/10.1016/j.celrep.2016.06.064
6. Magiera K, Tomala M, Kubica K et al (2017) Lithocholic acid hydroxyamide destabilizes cyclin D1 and induces G0/G1 arrest by inhibiting deubiquitinase USP2a. cell. Chem Biol 24(4):458–70 e18. https://doi.org/10.1016/j.chembiol.2017.03.002
7. Kategaya L, Di Lello P, Rouge L et al (2017) USP7 small-molecule inhibitors interfere with ubiquitin binding. Nature 550(7677): 534–538. https://doi.org/10.1038/nature24006
8. De Cesare V, Moran J, Traynor R et al (2020) High-throughput matrix-assisted laser desorption/ionization time-of-flight (MALDI-TOF) mass spectrometry-based deubiquitylating enzyme assay for drug discovery. Nat Protoc 15(12):4034–4057. https://doi.org/10.1038/s41596-020-00405-0
9. De Cesare V, Carbajo Lopez D, Mabbitt PD et al (2021) Deubiquitinating enzyme amino acid profiling reveals a class of ubiquitin esterases. Proc Natl Acad Sci U S A 118(4). https://doi.org/10.1073/pnas.2006947118
10. Pickart CM, Raasi S (2005) Controlled synthesis of polyubiquitin chains. Methods Enzymol

399:21–36. https://doi.org/10.1016/S0076-6879(05)99002-2

11. Bremm A, Komander D (2012) Synthesis and analysis of K11-linked ubiquitin chains. Methods Mol Biol 832:219–228. https://doi.org/10.1007/978-1-61779-474-2_15

12. Zhang JH, Chung TD, Oldenburg KR (1999) A simple statistical parameter for use in evaluation and validation of high throughput screening assays. J Biomol Screen 4(2):67–73. https://doi.org/10.1177/108705719900400206

Chapter 9

Bioinformatical Approaches to the Discovery and Classification of Novel Deubiquitinases

Thomas Hermanns and Kay Hofmann

Abstract

Deubiquitinating enzymes (DUBs) are active at multiple levels of the eukaryotic ubiquitin system. DUBs are important for ubiquitin activation and maintaining cellular ubiquitin levels but can also edit or dissolve ubiquitin chains or deconjugate ubiquitin from substrates. Eukaryotic DUBs can be grouped into seven molecular classes, most of which enzymes are cysteine proteases assuming the papain fold. In recent years, an ever-increasing number of pathogen-encoded DUBs have been characterized, which are active inside the host cell and help the pathogens to evade the defense response. At first sight, bacterial and viral DUBs appear to be very different from their eukaryotic counterparts, making them hard to identify by bioinformatic methods. However, apart from very few exceptions, bacterial and viral DUBs are distantly related to eukaryotic DUB classes and possess several hallmarks that can be used to identify high-confidence DUB candidates from pathogen genomes – even in the complete absence of biochemical or functional annotation. This chapter addresses bioinformatical DUB discovery approaches based on a previously published analysis of DUB evolution. The core set of bioinformatical tools required for this endeavor are freely accessible and do not require a particular bioinformatics infrastructure.

Key words Ubiquitin, Deubiquitinating enzymes, Evolution, Papain-fold cysteine proteases

1 Introduction

Ubiquitination regulates a wide variety of cellular processes, including protein stability, vesicular trafficking, autophagy, inflammation, and the cellular defense against invading pathogens [1]. Deubiquitinating enzymes (DUBs) are proteases, which specifically cleave peptide or isopeptide bonds formed by the C-terminus of ubiquitin and have several important roles in the eukaryotic ubiquitin system. Among other tasks, DUBs are crucial for ubiquitin activation and for maintaining cellular ubiquitin levels, but they also break ubiquitin chains and can edit or even remove ubiquitin conjugates from their substrates [2]. Deubiquitinases – like other components of the ubiquitin system – are found in all eukaryotes. In addition, many DUBs are also encoded by intracellular pathogens such as

Julie A. Maupin-Furlow and Mariola J. Edelmann (eds.), *Deubiquitinases: Methods and Protocols*,
Methods in Molecular Biology, vol. 2591, https://doi.org/10.1007/978-1-0716-2803-4_9,

Table 1
DUB classes and their coverage by domain databases

Type	DUB	PDB	InterPro	Pfam	Detection of non-eukaryotic members
Metalloprotease	JAMM	2ZNV	IPR000555	PF01398	Bacterial members not active as DUBs
Papain-fold cysteine proteases	UCH	3KW5	IPR001578	PF01088	None known
	USP	2Y5B	IPR001394	PF00443	Bacterial and viral members typically missed by HMMs
	OTU	5LRV 4I6L	IPR003323 IPR019400	PF02338 PF10275	Few bacterial members covered by HMM, many missed
	Josephin	2JRI	IPR006155	PF02099	None known
	MINDY	5JQS	IPR033979 IPR025257	PF04424 PF13898	None known
	ZUFSP	6EI1	IPR012462	PF07910	None known
	SENP/ULP	2IO0	IPR003653	PF15499	Few bacterial members covered by HMM, many missed
	RavD	6NJD	Not covered	not covered	Only described in *Legionella pneumophila*
	Tegument	2J7Q	IPR006928	PF04843	Only described in herpesviruses

The columns labeled “PDB,” “InterPro,” and “Pfam” indicate the accession number of the DUB class in the respective databases. The OTU class has two subtypes that are covered by different database entries. The rightmost column indicates if non-eukaryotic members of these DUB classes are known and, if yes, whether those members are covered by the domain databases (Procedure 4.1)

bacteria and viruses, which do not have a ubiquitin system on their own. Bacterial and viral DUBs serve as effector molecules after secretion into the host cell, where they help the pathogens evade the ubiquitin-mediated defense response [3, 4].

Eukaryotic DUBs belong to seven classes, which are structurally and evolutionarily distinct and show important differences in their specificity. One class (JAMM) contains metalloproteases. The other six DUB classes, including UCH, USP, OTU, Josephin, MINDY, and ZUFSP, are papain-fold cysteine proteases [2]. Members of all seven eukaryotic DUB classes have been characterized structurally, typically in complex with their ubiquitin substrate; representative examples are listed in Table 1. Initially, many DUB classes have been discovered by classical genetics or by reactivity toward DUB-directed activity-based probes [5], while others have been discovered by bioinformatical methods similar to those described in this chapter [6–10]. Today, it appears unlikely that fundamentally new eukaryotic DUB classes still await their discovery. Since all known eukaryotic DUBs are clearly related to one of the seven classes mentioned above, which are covered by protein domain and family databases such as Pfam and InterPro [11, 12], the identification of DUBs from newly sequenced eukaryotic genomes is straightforward.

Unlike eukaryotic deubiquitinases, bacterial DUBs do not always belong to the seven established classes [3]. While many bacterial DUBs are clearly related to the eukaryotic OTU family [3, 13], a larger group belongs to an enzyme family usually referred to as the "CE-clan," which also contains eukaryotic proteases specific for the ubiquitin-related modifiers SUMO and NEDD8, but not for ubiquitin itself [14]. While most known bacterial DUBs belong to either the OTU or CE-clan families, two exceptions are known. TssM from *Burkholderia* is related to the eukaryotic USP class, and *Legionella* RavD does not appear to be related to any known enzyme class. Viruses also code for deubiquitinating enzymes, which are often related to the eukaryotic USP and OTU classes and typically recognize – besides ubiquitin itself – the ubiquitin-related modifier ISG15, which is important for antiviral signaling [4, 15, 16]. In addition, one unusual group of deubiquitinases is found exclusively in herpesviruses: the tegument-DUBs, which appear unrelated to any other DUB class and often recognize the ubiquitin-like modifier NEDD8 besides ubiquitin [17, 18].

All known bacterial and viral deubiquitinases have been shown to be papain-fold cysteine proteases and resemble the eukaryotic DUBs in their catalytic mechanism and mode of ubiquitin recognition. Nevertheless, identifying unknown bacterial or viral DUBs from genome sequences is more problematic than their eukaryotic counterparts. Even those bacterial and viral DUBs with relationship to established DUB classes or CE-clan enzymes tend to be very divergent members, which are often not found by the family-specific hidden Markov models (HMMs) used by the Pfam and InterPro databases [11, 12] (Table 1). Moreover, the fact that Legionella RavD and the herpesvirus-specific tegument proteases are not visibly related to any known DUB family suggests that other pathogen-derived DUBs might also elude bioinformatical detection by their high divergence or by belonging to entirely new enzyme families.

In a recent study addressing the evolution of deubiquitinases, we noted a distant but statistically significant similarity between several DUB classes, suggesting a shared ancestry [8]. We also noted several sequence and structural features shared by multiple DUB classes, which can be used to discern bona fide deubiquitinases from other types of cysteine proteases. One hallmark of DUBs and other GlyGly cleaving proteases – but absent from other papain-fold enzymes – is the "aromatic gatekeeper motif" adjacent to the catalytic histidine residue [8]. Other features shared by multiple DUB classes concern the recognition mode of the outgoing (S1) ubiquitin substrate, mainly the stabilization of the ubiquitin C-terminal R-x-R motif by one or two salt bridges, and the recognition of the Ile-44 surface patch of ubiquitin. Since recognizing these features usually requires structural information on the enzyme::ubiquitin complex, they were deemed less useful for DUB discovery from sequence material alone [8]. However, with the increasing availability of high-quality structure prediction software such as the recently published AlphaFold [19], incorporating structural criteria into a bioinformatical DUB discovery pipeline has become feasible.

2 General Approach

The following section describes the steps necessary to test a candidate sequence for its likelihood to encode a deubiquitinating activity. An outline of the procedure is shown in Fig. 1. In general, a protein sequence is considered a good DUB candidate if it fulfills one of the following three criteria:

(i) A portion of the sequence yields a significant match to an established DUB domain descriptor (sequence-to-family search, using Pfam, InterPro, or other related classification tools).

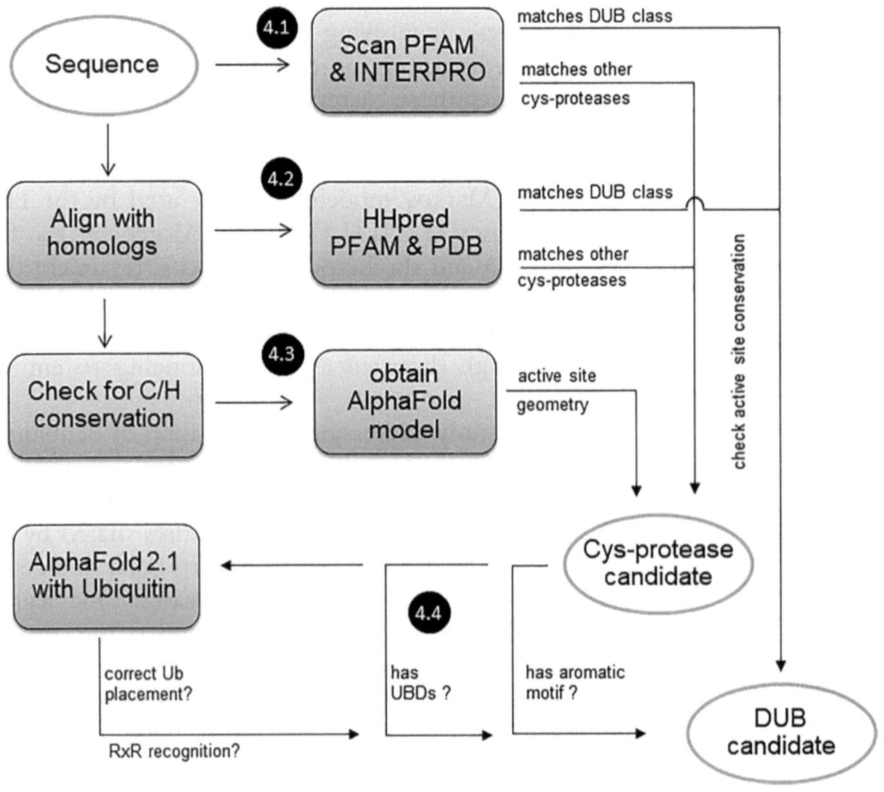

Fig. 1 Flow diagram of the DUB discovery pipeline. The black circles labeled "4.1" to "4.4" refer to the section of the text where the analyses are described

(ii) A portion of the sequence is distantly related to an established DUB domain family, as detected by family-to-family searches using HHPRED. At the same time, this similarity should score higher than any similarity to non-DUB families.

(iii) A portion of the sequences is predicted to encode a papain-fold cysteine protease, and the sequence fulfills additional criteria supporting a DUB function over a generic protease function. The papain fold can be established by distant similarity to other papain-fold proteases (family-to-family searches) or by prediction of the 3D structure. The additional criteria include the presence of an aromatic gatekeeper motif, the presence of predicted ubiquitin-binding domains, or a promising outcome of a 3D model of the enzyme/substrate complex.

In all cases, "reality checks" should be performed before a – possibly tedious – experimental validation. Similarity-based criteria will also find inactive members of protein families, so the conservation of active-site residues must be verified. Several of the described criteria also apply to proteases for ubiquitin-like modifiers. The chances of a candidate sequence to be active against ubiquitin rather than SUMO, ISG15, or NEDD8 depend on the context. For example, pathogenic bacteria are most likely to encode DUBs, while RNA viruses can be expected to possess enzymes deconjugating ISG15.

3 Software

Most bioinformatical procedures described in this chapter can be performed using freely accessible worldwide web (WWW) servers and do not require any bioinformatical infrastructure at the user's site.

3.1 Sequence-Against-Sequence Searches

Sequence-against-sequence searches are required to find homologs of a given sequence, such as orthologs and closely related paralogs. The most widely used tool is BLAST [20], for which a web server is available at http://blast.ncbi.nlm.nih.gov/Blast.cgi. Another recommended tool, somewhat more sensitive than BLAST, is JackHMMER [21]; a web server is available at https://www.ebi.ac.uk/Tools/hmmer/search/jackhmmer.

3.2 Multiple Sequence Alignment and Visualization

The recommended tool for aligning several related protein sequences (i.e., a sequence family) is MAFFT [22]. Web servers running MAFFT are available at http://www.ebi.ac.uk/Tools/msa/mafft and http://mafft.cbrc.jp/alignment/server. This software is also available for download from http://mafft.cbrc.jp/alignment/software.

Most multiple alignment servers support viewing/editing the resulting alignment with Jalview [23], which is also separately available from https://www.jalview.org/.

3.3 Domain Detection (Sequence-Against-Family Searches)

Multiple software tools are available to scan a user-specified query sequence against a collection of predefined protein or domain families. Recommended web servers for this purpose include Pfam [11] and InterPro [12]. Web servers are available at http://pfam.xfam.org/search and http://www.ebi.ac.uk/interpro, respectively. If the query sequence is contained in the UniProt database (https://www.uniprot.org), it is advisable to query Pfam and InterPro with the UniProt accession number, rather than the sequence, to save processing time. Both databases also offer download of the family/domain descriptors (encoded as hidden Markov models) and the necessary software, which is useful for running high-throughput searches.

3.4 Distant Similarity (Family-Against-Family) Searches

The sensitivity and specificity of sequence database searches can be greatly enhanced by using information from a protein family, rather than a single protein, as the query and running the search against a database of pre-compiled protein or protein domain families. Usually, the alignment information of the query and target families are stored as hidden Markov models, converting the task to an HMM-against-HMM search. The recommended tool is HHPRED [24, 25]; a web server is available at https://toolkit.tuebingen.mpg.de/tools/hhpred. Of crucial importance is selecting the target database, which has to comprise all major DUB classes in a recognizable form. Suitable options are the databases called "Pfam-A" (build from all Pfam domains) and "PDB_mmCIF70" (build from all PDB structure entries and their homologs). Both databases contain all relevant enzyme classes and use accession numbers, as shown in Table 1.

3.5 Prediction of Enzyme Structures and Complexes

The last years have seen a dramatic improvement in tertiary structure modeling of "difficult proteins" beyond the application of homology-based modeling. Several tools for de novo modeling of protein structures are available, including RoseTTAfold [26], accessible as a web server at https://robetta.bakerlab.org/submit.php, and AlphaFold [19], available for download from https://github.com/deepmind/alphafold. At the time of writing, no publicly accessible web server running AlphaFold is available. However, many pre-computed AlphaFold predictions for the entire proteomes of important model organisms have been made available at https://alphafold.ebi.ac.uk; these predictions are also linked from the UniProt entries of the corresponding proteins.

Since predictions by both RoseTTAfold and AlphaFold are computationally expensive, it is usually advisable to run the folding on the sequence region suspected to contain the DUB domain (maybe with some flanking sequences on either side), rather than on the entire sequence – in particular, if the sequences exceed a length of 800 amino acids. The downloadable AlphaFold software (version 2.1) supports the simultaneous modeling of multiple sequences known or suspected to form a complex. In our experience, this mode – although certainly not perfect – is often helpful for "guessing" DUB::ubiquitin contact sites.

3.6 Structure Visualization and Comparison

Structural models downloaded from UniProt or obtained by modeling programs can be visualized by tools such as PyMOL (available from https://pymol.org) or Discovery Studio Visualizer (available from https://discover.3ds.com/discovery-studio-visualizer-download). These tools are also helpful in identifying and extracting folding units (domains) from larger protein structures.

For the comparison of structures to structural databases and the superposition of related structures, the most widely used (and probably most sensitive) tool is Dali [27], which is available as a web server at http://ekhidna2.biocenter.helsinki.fi/dali. A much faster but less sensitive tool is Foldseek, available at https://search.foldseek.com/search.

4 Procedure

The main procedure described in this section starts from a single protein sequence of unknown functionality and asks whether the encoded protein might have DUB activity. In most situations, there will be no definitive answer coming from bioinformatical analysis alone; an experimental validation is always necessary. However, the procedure described here is much faster and easier than an experimental screen on many candidate proteins, like the one leading to the discovery of RavD [28]. Subheading 4.5 addresses the possibility and challenges of performing proteome-wide screens.

4.1 Scanning Domain Databases for DUB Domains

1. Prepare a text file of the query sequence (one-letter amino acid abbreviations, no numbers, or other symbols; spaces are acceptable), and paste the sequence into the query window of the Pfam or InterPro web servers (Subheading 3.3). These servers also allow to specify the UniProt accession number rather than the sequence. In many cases, this will speed up the response considerably since Pfam and InterPro can use pre-computed results.
2. On the result page, check if any of the identified domains or families are associated with DUB or other cysteine protease activities. Scanning the result for possible ubiquitin-binding domains is also advisable, as this information will be helpful at a later stage (Subheading 4.4).
3. If the result indicates a significant similarity to a DUB domain, check for conservation of the active site residues (*see* **Note 1**). If the active site residues are conserved, you can consider your query a high-confidence DUB candidate.

4. If the results indicate a significant relationship to a (non-DUB) cysteine protease family, in particular a family with poorly defined function, it might still be a DUB – but most likely will cleave something else. It is advisable to check for conservation of the active site residues (*see* **Note 1**) and continue with the additional tests described in Subheading 4.4.
5. If most of the query sequence is matched by domains functionally unrelated to DUB activity (*see* **Note 2**), the sequence is most likely not a DUB, and further analyses are unlikely to succeed.
6. Otherwise, continue with the analyses of Subheading 4.2. If some parts of the query sequence were matched by irrelevant domains (*see* **Note 2**), it is advisable to continue further analyses selectively with the sequence portion without those matches.

4.2 Searching for Distant DUB Similarities

1. If the analyses of Subheading 4.1 did not uncover any association with DUBs or other cysteine proteases, but did not suggest a non-DUB function for the entire protein, continue. Use the entire sequence, or (preferably) individual sequence regions not covered by irrelevant domain matches, to establish a sequence family and generate multiple alignments and subsequent HHPRED search.
2. Since the HHPRED server (Subheading 3.4) can take care of all required steps, the simplest option is to paste the query sequence into the search box of the HHPRED toolkit, select "Pfam-A" and "PDB_mmCIF70" as the search databases, and leave all other parameters at their defaults. Without any intervention by the user, the server will perform a sequence-to-sequence database search, generate multiple alignments of the significant hits, create a hidden Markov model from this alignment, and run this HMM against the specified HMM databases. For more control over these processes, an alternative method is described in **Note 3**. The HHPRED output page sorts the hits according to their significance (*see* **Note 4**). Unlike Pfam, multiple overlapping hits are possible. Some hits will be to Pfam families (names corresponding to Pfam accession number, linking to their Pfam page), while others will link to structure entries from the PDB database. Hits representing DUB domains can be recognized by their name tag or correspondence to the accession numbers listed in Table 1.
3. If the top hit of the HHPRED search indicates a significant similarity to a DUB family (either by Pfam or PDB hit), go to the alignment section at the bottom of the result page, and check for conservation of the active site residues (*see* **Note 5**). If the active site residues are conserved, you can consider your query a high-confidence DUB candidate. Other higher-scoring

hits to irrelevant families, which cover a different part of the query sequence, do not invalidate high-scoring DUB hits (*see* **Note 4**).

4. If the relevant top hits reported by the HHPRED server belong to other cysteine proteases rather than established DUBs, the query sequence might still have DUB activity. It is advisable to check for conservation of the active site residues (*see* **Note 5**) and continue with the additional tests described in Subheading 4.4.
5. If none of the significant hits reported by HHPRED hint toward a DUB or other cysteine protease activities, it is likely that the query protein does not have DUB activity. However, there is still the possibility of the query protein belonging to a new cysteine protease family, unrelated to anything stored in Pfam and PDB. To investigate this possibility, continue with Subheading 4.3.

4.3 De Novo Discovery of DUBs and Cysteine Proteases

1. For the discovery of fundamentally new cysteine protease families, we can exploit the fact that active site residues tend to be highly conserved. To that end, search for homologs of your query sequence by BLAST (Subheading 3.1), and create multiple alignments of the query and all significant hits by MAFFT (Subheading 3.2). Alternatively, the JackHMMER program (Subheading 3.1) can be used for the combined search and alignment step. Inspect the resulting alignment for highly conserved cysteine and histidine residues within an appropriate distance from each other (*see* **Note 6**). If this is the case, check if a highly conserved aromatic residue follows the conserved histidine.
2. If the previous step resulted in a promising alignment with conserved residues that might constitute an active site, use the query sequence for generating an AlphaFold model (Subheading 3.5). If the query sequence is long (>800 aa), it is advisable to model only the region harboring the predicted active site, with 50 flanking residues added at both ends. After modeling, use a structure visualizer (Subheading 3.6) to check the positions of the conserved cysteine and histidine residues. If the conserved cysteine forms the N-terminal end of an alpha-helix, the conserved histidine lies within an adjacent beta-strand, and their side chain distance is less than 4 Å (measured from the cysteine sulfur atom to a side chain nitrogen atom of histidine), the predicted fold could be a papain-like protease.
3. Further information on the nature of the predicted fold can be obtained by structure comparison searches using Dali or Foldseek (Subheading 3.6).

4. If the active site is conserved and the fold is confirmed to be papain-like, the query sequence might belong to a novel DUB class – or another type of papain-fold (*see* **Note 7**). Continue with the additional tests described in Subheading 4.4

4.4 Tests to Discriminate DUBs from Other Proteases

The methods described here should be applied whenever previous analyses suggest the query protein to assume a papain fold with conserved active site residues, but no specific similarity to a known DUB class could be established.

1. Check the multiple alignments of the query sequence with its homologs for the residue following the catalytic histidine. If this residue is aromatic (Phe, Tyr, Trp) and is highly conserved in the alignment, this is a good indication that the papain-fold cysteine protease will cleave after GlyGly motifs – a property shared by DUBs and deconjugating enzymes for ubiquitin-like modifiers (*see* **Note 8**).
2. Check the query sequence for the presence of known ubiquitin-binding domains. This information is usually provided by the domain database search described in Subheading 4.1. The association of ubiquitin-binding domains with cysteine proteases is observed in many DUBs and strongly supports DUB prediction.
3. Use AlphaFold (Subheading 3.5) to predict the structure of a two-protein complex including the predicted protease (or its catalytic region) and mono-ubiquitin. Visualize the resulting complex (Subheading 3.6), and check if the C-terminus of ubiquitin is placed adjacent to the active site residues.
4. If the modeled protease behaves like a typical DUB, the following structural features should be present: (i) the aromatic residue behind the catalytic histidine should be in contact with the penultimate glycine residue of ubiquitin, (ii) Arg-72 and/or Arg-74 of ubiquitin should form a salt bridge with an acidic residue of the protease, and (iii) Ile-44 of ubiquitin should have extensive contacts with parts of the protease. Not all DUBs will fulfill all of these criteria, but most do. Each criterion will support the prediction of the query sequence as DUB.
5. If the structural model does show specific contacts for the C-terminal arginine residues or the Ile-44 surface of ubiquitin, check the contact residues for their conservation. If these residues are conserved (beyond the average conservation level of the family), this fact will provide further support for the importance of these contacts and thus for the DUB prediction.

4.5 High-Throughput Searches

The procedures described in Subheading 4.1, 4.2, 4.3, and 4.4 assume that a single query sequence is tested for potential DUB activity. Except for the AlphaFold models, all necessary steps can be performed through web-based services without requiring the local installation of software and databases. Another relevant question is the automated search of a large number of query sequences – possibly the entire proteome of a pathogen – for putative DUB candidates. The upscaling of the procedures presented above is not easy since several of the steps require human interventions. A feasible approach to analyzing sequences in high-throughput mode involves automatizing filtering steps, which are restrictive enough to leave only a small number of candidates for subsequent manual inspection. All software tools described in Subheading 3.1, 3.2, 3.3, 3.4, and 3.5 can be downloaded for local installation on a Linux-based computer; many of the tools also require the download of associated databases. The following steps outline a possible filtering pipeline:

1. As an optional first step, filter out all sequences that have obvious start-to-end similarity to proteins from organisms not expected to encode DUBs. This step will remove about half of all sequences for typical bacterial genomes.
2. For each sequence in the remaining query set, perform an automated search against a local version of the Pfam database (Subheading 3.3). Select all hits to known DUB domains, and add these hits to the candidate list. For all other sequences, mask the regions covered by non-DUB domains.
3. For the remaining sequence regions, gather homologs from other species using BLAST or HMMER (Subheading 3.1), and create multiple alignments for each sequence family (Subheading 3.2).
4. Convert the multiple alignments to hidden Markov models, and search HMM collections of Pfam- and PDB-derived sequence families. These steps can be performed by scripting a locally installed HHPRED package (Subheading 3.4). Scan the output for significant hits to DUB-related HMMs (e.g., based on the accession numbers listed in Table 1). Sequences with significant hits to DUB-related HMMs can be added to the candidate list.
5. It may be advantageous (but more challenging to implement) to collect sequences with significant or nearly significant hits to other papain-fold protease domains, test them for conservation of the active site residues, and subject them to AlphaFold modeling (Subheading 3.5).
6. Proteins on the candidate list can then be subjected to the manual evaluation described in Subheading 4.1, 4.2, 4.3, and 4.4.

5 Notes

1. The present implementation of Pfam and InterPro web servers does not display an alignment of the domain consensus with the query sequence – a feature that was available in earlier versions and is supported by local installations of these databases. In Pfam, the conservation of active site residues is indicated by color markers displayed above the graphical domain display: red markers indicate sites annotated in the protein entry and supported by literature, and purple markers indicate sites predicted by the sequence-to-domain alignment.
2. Most domain matches reported by Pfam and InterPro have a functional annotation. Any domain match incompatible with DUB activity indicates that the region covered by this domain can be excluded from further analysis. However, some matches are labeled as "domain of unknown function." Regions covered by those domains cannot formally be excluded but are nevertheless unlikely to include active DUBs.
3. While HHPRED can be used with a single query sequence, the web server also supports the use of multiple alignment as a query. To obtain such an alignment, homologs can be searched by BLAST and aligned by MAFFT. By doing so, it becomes possible to edit the alignment or to remove questionable sequences before submitting them to HHPRED.
4. The main significance indicator of HHPRED is the "E-value," which indicates how many hits of a certain quality can be expected from a random database. Thus, low E-values are better, and values below 0.01 are considered highly significant. Note that the hits are sorted by their "overall" significance, irrespective of the sequence region covered by the hit. Interesting DUB-related hits may be found in the lower portions of the result list, e.g., if several highly significant hits to other regions of the query sequence are dominant. The graphical display at the top of the result pages shows both the (color-coded) quality of the hits and their position relative to the input sequence.
5. Unlike Pfam, HHPRED displays at the bottom of the result page the alignment of the query sequence to the consensus sequence of the matched domains. However, the information on the position of the active site residues is not contained in this display and has to be obtained elsewhere, e.g., by following the link to the domain description.
6. There are no hard limits on DUB domain size and the spacing of the active site residues. Both Cys-before-His and His-before-Cys arrangements are common. Among known DUBs, the eukaryotic USP class has particularly great catalytic domains due to the presence of large insertions; in this class,

the Cys-to-His distance can be up to 800 residues. In most typical DUBs, including bacterial and viral types, the domain size ranges from 150 to 350 residues, and the active site residues tend to be close to the domain borders, with a distance of 100–300 residues between them.

7. Except for the JAMM-metalloproteases, all DUBs follow the papain fold. Conversely, most papain-fold enzymes are proteases, but not necessarily DUBs. However, several families of papain-fold enzymes have non-proteolytic activities. Among them are transpeptidases, glutaminases, and specific acetyltransferases belonging to the CE-clan.
8. Note that the aromatic residue following the catalytic histidine is conserved in DUB families and other papain-fold proteases cleaving after GlyGly. Moreover, other proteases might "accidentally" carry an aromatic residue at this position. Thus, it is essential to evaluate the conservation of the aromatic residue rather than its presence in the query sequence. Also, note that a few DUBs (e.g., ZUFSP and tegument-type) lack this aromatic motif altogether.

Acknowledgments

This work was supported by the DFG grant HO 3783/3-1.

References

1. Rape M (2018) Ubiquitylation at the crossroads of development and disease. Nat Rev Mol Cell Biol 19(1):59–70. https://doi.org/10.1038/nrm.2017.83
2. Clague MJ, Urbe S, Komander D (2019) Breaking the chains: deubiquitylating enzyme specificity begets function. Nat Rev Mol Cell Biol 20(6):338–352. https://doi.org/10.1038/s41580-019-0099-1
3. Hermanns T, Hofmann K (2019) Bacterial DUBs: deubiquitination beyond the seven classes. Biochem Soc Trans 47(6):1857–1866. https://doi.org/10.1042/BST20190526
4. Proulx J, Borgmann K, Park IW (2021) Role of virally-encoded deubiquitinating enzymes in regulation of the virus life cycle. Int J Mol Sci 22(9). https://doi.org/10.3390/ijms22094438
5. Borodovsky A, Ovaa H, Kolli N, Gan-Erdene T, Wilkinson KD, Ploegh HL, Kessler BM (2002) Chemistry-based functional proteomics reveals novel members of the deubiquitinating enzyme family. Chem Biol 9(10):1149–1159
6. Scheel H, Tomiuk S, Hofmann K (2003) Elucidation of ataxin-3 and ataxin-7 function by integrative bioinformatics. Hum Mol Genet 12(21):2845–2852. https://doi.org/10.1093/hmg/ddg297
7. Maytal-Kivity V, Reis N, Hofmann K, Glickman MH (2002) MPN+, a putative catalytic motif found in a subset of MPN domain proteins from eukaryotes and prokaryotes, is critical for Rpn11 function. BMC Biochem 3:28. https://doi.org/10.1186/1471-2091-3-28
8. Hermanns T, Woiwode I, Guerreiro RF, Vogt R, Lammers M, Hofmann K (2020) An evolutionary approach to systematic discovery of novel deubiquitinases, applied to Legionella. Life Sci Alliance 3(9). https://doi.org/10.26508/lsa.202000838
9. Hermanns T, Pichlo C, Woiwode I, Klopffleisch K, Witting KF, Ovaa H, Baumann U, Hofmann K (2018) A family of unconventional deubiquitinases with modular chain specificity determinants. Nat Commun 9(1):799. https://doi.org/10.1038/s41467-018-03148-5

10. Abdul Rehman SA, Kristariyanto YA, Choi SY, Nkosi PJ, Weidlich S, Labib K, Hofmann K, Kulathu Y (2016) MINDY-1 is a member of an evolutionarily conserved and structurally distinct new family of deubiquitinating enzymes. Mol Cell 63(1):146–155. https://doi.org/10.1016/j.molcel.2016.05.009
11. Finn RD, Mistry J, Tate J, Coggill P, Heger A, Pollington JE, Gavin OL, Gunasekaran P, Ceric G, Forslund K, Holm L, Sonnhammer EL, Eddy SR, Bateman A (2011) The Pfam protein families database. Nucleic Acids Res 38(Database issue):D211–D222
12. Hunter S, Apweiler R, Attwood TK, Bairoch A, Bateman A, Binns D, Bork P, Das U, Daugherty L, Duquenne L, Finn RD, Gough J, Haft D, Hulo N, Kahn D, Kelly E, Laugraud A, Letunic I, Lonsdale D, Lopez R, Madera M, Maslen J, McAnulla C, McDowall J, Mistry J, Mitchell A, Mulder N, Natale D, Orengo C, Quinn AF, Selengut JD, Sigrist CJ, Thimma M, Thomas PD, Valentin F, Wilson D, Wu CH, Yeats C (2009) InterPro: the integrative protein signature database. Nucleic Acids Res 37(Database issue):D211–D215
13. Schubert AF, Nguyen JV, Franklin TG, Geurink PP, Roberts CG, Sanderson DJ, Miller LN, Ovaa H, Hofmann K, Pruneda JN, Komander D (2020) Identification and characterization of diverse OTU deubiquitinases in bacteria. EMBO J 39(15):e105127. https://doi.org/10.15252/embj.2020105127
14. Pruneda JN, Durkin CH, Geurink PP, Ovaa H, Santhanam B, Holden DW, Komander D (2016) The molecular basis for ubiquitin and ubiquitin-like specificities in bacterial effector proteases. Mol Cell 63(2):261–276. https://doi.org/10.1016/j.molcel.2016.06.015
15. Bailey-Elkin BA, Knaap RCM, Kikkert M, Mark BL (2017) Structure and function of viral deubiquitinating enzymes. J Mol Biol 429(22):3441–3470. https://doi.org/10.1016/j.jmb.2017.06.010
16. Zong Z, Zhang Z, Wu L, Zhang L, Zhou F (2021) The functional deubiquitinating enzymes in control of innate antiviral immunity. Adv Sci (Weinh) 8(2):2002484. https://doi.org/10.1002/advs.202002484
17. Gastaldello S, Hildebrand S, Faridani O, Callegari S, Palmkvist M, Di Guglielmo C, Masucci MG (2010) A deneddylase encoded by Epstein-Barr virus promotes viral DNA replication by regulating the activity of cullin-RING ligases. Nat Cell Biol 12(4):351–361. https://doi.org/10.1038/ncb2035
18. Schlieker C, Weihofen WA, Frijns E, Kattenhorn LM, Gaudet R, Ploegh HL (2007) Structure of a herpesvirus-encoded cysteine protease reveals a unique class of deubiquitinating enzymes. Mol Cell 25(5):677–687. https://doi.org/10.1016/j.molcel.2007.01.033
19. Jumper J, Evans R, Pritzel A, Green T, Figurnov M, Ronneberger O, Tunyasuvunakool K, Bates R, Zidek A, Potapenko A, Bridgland A, Meyer C, Kohl SAA, Ballard AJ, Cowie A, Romera-Paredes B, Nikolov S, Jain R, Adler J, Back T, Petersen S, Reiman D, Clancy E, Zielinski M, Steinegger M, Pacholska M, Berghammer T, Bodenstein S, Silver D, Vinyals O, Senior AW, Kavukcuoglu K, Kohli P, Hassabis D (2021) Highly accurate protein structure prediction with AlphaFold. Nature 596(7873):583–589. https://doi.org/10.1038/s41586-021-03819-2
20. Altschul SF, Gish W, Miller W, Myers EW, Lipman DJ (1990) Basic local alignment search tool. J Mol Biol 215(3):403–410. https://doi.org/10.1016/S0022-2836(05)80360-2
21. Potter SC, Luciani A, Eddy SR, Park Y, Lopez R, Finn RD (2018) HMMER web server: 2018 update. Nucleic Acids Res 46(W1):W200–W204. https://doi.org/10.1093/nar/gky448
22. Katoh K, Standley DM (2013) MAFFT multiple sequence alignment software version 7: improvements in performance and usability. Mol Biol Evol 30(4):772–780. https://doi.org/10.1093/molbev/mst010
23. Waterhouse AM, Procter JB, Martin DM, Clamp M, Barton GJ (2009) Jalview Version 2–a multiple sequence alignment editor and analysis workbench. Bioinformatics 25(9):1189–1191. https://doi.org/10.1093/bioinformatics/btp033
24. Soding J (2005) Protein homology detection by HMM-HMM comparison. Bioinformatics 21(7):951–960. https://doi.org/10.1093/bioinformatics/bti125
25. Zimmermann L, Stephens A, Nam SZ, Rau D, Kubler J, Lozajic M, Gabler F, Soding J, Lupas AN, Alva V (2018) A completely reimplemented MPI bioinformatics toolkit with a new HHpred server at its core. J Mol Biol 430(15):2237–2243. https://doi.org/10.1016/j.jmb.2017.12.007
26. Baek M, DiMaio F, Anishchenko I, Dauparas J, Ovchinnikov S, Lee GR, Wang J, Cong Q, Kinch LN, Schaeffer RD, Millan C, Park H, Adams C, Glassman CR, DeGiovanni A, Pereira JH, Rodrigues AV, van Dijk AA, Ebrecht AC, Opperman DJ, Sagmeister T, Buhlheller C, Pavkov-Keller T, Rathinaswamy MK, Dalwadi U, Yip CK, Burke JE, Garcia KC,

Grishin NV, Adams PD, Read RJ, Baker D (2021) Accurate prediction of protein structures and interactions using a three-track neural network. Science 373(6557):871–876. https://doi.org/10.1126/science.abj8754

27. Holm L (2019) Benchmarking fold detection by DaliLite v.5. Bioinformatics 35(24): 5326–5327. https://doi.org/10.1093/bioinformatics/btz536

28. Wan M, Wang X, Huang C, Xu D, Wang Z, Zhou Y, Zhu Y (2019) A bacterial effector deubiquitinase specifically hydrolyses linear ubiquitin chains to inhibit host inflammatory signalling. Nat Microbiol 4(8):1282–1293. https://doi.org/10.1038/s41564-019-0454-1

Chapter 10

Application of Archaea in Deubiquitinase-Like Enzyme Discovery and Activity Assay

Nathaniel L. Hepowit and Julie A. Maupin-Furlow

Abstract

Archaea can be used as microbial platforms to discover new types of deubiquitinase-like (DUB-like) enzymes and to produce ubiquitin/ubiquitin-like (Ub/Ubl) protein conjugates as substrates for DUB/DUB-like activity assays. Here we outline how to use archaea to synthesize, purify, and assay the activity of DUB-like enzymes with unusual properties, including catalytic activity in hypersaline conditions, organic solvents, and high temperatures. We also outline the application of archaea in forming Ub/Ubl isopeptide linkages that include the covalent attachments of diverse archaeal and eukaryotic Ub/Ubls to target proteins. Archaea form these Ub/Ubl-linked protein conjugates in vivo, and the resulting products are found to serve as useful DUB substrates for in vitro assays.

Key words Archaea, SILAC, Mass spectrometry, Proteomics, *Haloferax volcanii*, Oxidative stress

1 Introduction

Archaea are microorganisms, distinct from bacteria and eukaryotes, that can be used as platforms for discovering deubiquitinase-like (DUB-like) enzymes and activity assays. While archaea are single-celled prokaryotes that appear visually like bacteria, advances in molecular biology now suggest that archaea are descendants of organisms that gave rise to the eukaryote cell [1]. Because of this evolutionary relationship, many biological systems central to eukaryotic cell function are conserved in archaea yet more streamlined. This relationship provides an opportunity to use archaea to investigate the basic structure and function of eukaryotic molecular systems, such as the ubiquitin–proteasome and associated deubiquitinases (DUBs) [2, 3]. Proteins of the JAMM metalloprotease family are identified in archaea that are structurally related to the Rpn11 subunit of the eukaryotic 26S proteasome [4, 5] and that catalyze DUB-like activity [3]. These "archaeal DUBs" cleave a broad spectrum of ubiquitin-like proteins (Ubls) from protein

Julie A. Maupin-Furlow and Mariola J. Edelmann (eds.), *Deubiquitinases: Methods and Protocols*,
Methods in Molecular Biology, vol. 2591, https://doi.org/10.1007/978-1-0716-2803-4_10,

conjugates, including the removal of Ubls that are isopeptide-linked to lysine residues or linear-linked to the N-terminus of the modified protein [3, 6, 7]. The Ub/Ubl proteins removed from the target protein by the archaeal DUBs include archaeal Ubls and eukaryotic Ubls or ubiquitin (Ub), thus revealing the broad range spectrum of substrates cleaved by these enzymes [8]. Here we outline how to identify, synthesize, purify, and assay the activity of archaeal DUBs. We also clarify how to use archaea to form Ub/Ubl isopeptide linkages, thus offering a method to form Ub/Ubl-linked protein conjugates that can serve as DUB substrates through simple microbial culture.

2 Materials

1. *E. coli* strains: TOP10 (for cloning) and Rosetta (DE3) (for expression of recombinant HvJAMM1).
2. *Haloferax volcanii* DS2 genomic DNA ordered from ATCC (ATCC29605) or prepared from whole cells by osmotic lysis (*see* **Note 1**).
3. pET15b or other related pET plasmid vectors.
4. PCR primer pair: HvJAMM1 NdeI up 5'-TTCCGGGcatatgACTTCGAGTAGGCTCTCTCTC-3′ and HvJAMM1 BlpI stop 5'-AATgctcagcTCACTCGACGGCGACGGAGAG-3′.
5. PCR reaction reagents: nuclease-free water, 5× HF buffer, 10-mM dNTP mix, 10-μM primer (forward and reverse), Phusion DNA Polymerase (for high-fidelity PCR-based cloning), Taq DNA polymerase (for colony screening).
6. PCR cleanup kit (e.g., Monarch PCR & DNA Cleanup Kit, 5 μg; New England Biolabs).
7. Microvolume spectrophotometer (e.g., Take3 microvolume plate with Gen5 software, BioTek Instruments).
8. 25× TAE stock solution: 121-g Tris base, 28.5-mL glacial acetic acid, 50-mL EDTA (pH 8.0). Add H_2O to 1 L.
9. Agarose gel electrophoresis: 1× TAE buffer (running buffer, diluted from 25× stock), agarose gel (0.8% agarose in 1× TAE buffer).
10. DNA agarose gel electrophoresis equipment.
11. Restriction digestion reagents (NEB): 10× CutSmart Buffer, NdeI, BlpI.
12. DNA gel extraction kit (e.g., QIAquick from Qiagen).
13. DNA ligation reagents: 10× T4 DNA ligase buffer, T4 DNA ligase.
14. DNA plasmid miniprep kit (e.g., PureLink HiPure Plasmid Miniprep Kit, Invitrogen).

15. Luria–Bertani (LB) medium (per liter): 10-g tryptone, 5-g yeast extract, and 10-g NaCl. Include 15-g agar for plates. Autoclave for 20 min and cool in 60 °C water bath for 30 min.

    Aseptically supplement the medium with ampicillin (100 mg/L) and chloramphenicol (30 mg/L) as needed. For plates, mix the medium, avoiding bubbles, and distribute into plates (20 mL per plate).
16. *E. coli* pET15b-based expression medium: LB supplemented with ampicillin (100 mg/L) and chloramphenicol (30 mg/L).
17. Incubator for orbital shaking of cultures.
18. 1-M isopropyl β-D-1-thiogalactopyranoside (IPTG): make fresh in sterile deionized water (dH_2O).
19. *E. coli* lysis buffer A: 20-mM Tris (pH 7.5), 150-mM NaCl, 40-mM imidazole, 1-mM PMSF, 1-mM DTT.
20. *E. coli* lysis buffer B: 20-mM Tris (pH 7.5), 4-M NaCl, 40-mM imidazole, 1-mM PMSF, 1-mM DTT.
21. Chromatography columns: HisTrap Ni^{2+}-NTA column (MilliporeSigma).
22. TBS buffer: 50-mM Tris–HCl (pH 7.5), 150-mM NaCl.
23. French press and French press cell.
24. 0.2-μM syringe filters, PVDF, or other low protein binding membrane.
25. HisTrap wash buffer: 20-mM Tris (pH 7.5), 2-M NaCl, 40-mM imidazole, 1-mM DTT.
26. HisTrap elution buffer: 20-mM Tris (pH 7.5), 2-M NaCl, 400-mM imidazole, 1-mM DTT.
27. Tris–salt buffer: 20-mM Tris (pH 7.5), 2-M NaCl.
28. HEPES–salt buffer: 20-mM HEPES (pH 7.5), 2-M NaCl.
29. *Haloferax volcanii* H26 or DS70: strain curve of pHV2 [9, 10].
30. Plasmids that express N-terminal Flag-tagged Ubl proteins in *H. volcanii*, including pJAM947 (Flag–SAMP1), pJAM949 (Flag–SAMP2), and pJAM977 (Flag–SAMP3) [11, 12] and pJAM1314 (separately expresses Flag–SAMP1 and MoaE–StrepII) (where the SAMPs are archaeal Ubl proteins) (*see* **Note 2**).
31. ATCC974 medium (100 mL): solution A (75 mL)—12.5-g NaCl, 5.0-g $MgCl_2{\cdot}6H_2O$, 0.5-g K_2SO_4, 0.02-g $CaCl_2{\cdot}6H_2O$, deionized water up to 75 mL; adjust pH to 6.8 with 1-M KOH. Solution B (25 mL): 0.5-g tryptone, yeast extract 0.5 g, deionized water up to 25 mL; adjust pH to 6.8 with 1-M KOH. Sterilize separately and mix aseptically. The medium was supplemented with 15-g agar for plates and novobiocin at 0.2 μg/mL and 50-mM DMSO as indicated in the method.

32. *H. volcanii* lysis buffer: 20-mM Tris–HCl (pH 7.4), 1-M NaCl, 1× EDTA-free protease inhibitor (Mini-Protean, Roche), 1-mM PMSF, 5-mM EDTA, 0.1% Tween-20.
33. Anti-Flag M2 agarose (Sigma; product number A2220).
34. Plastic minicolumns for purification with small quantities of resin (100–250 μL).
35. Flag IP wash buffer: 20-mM Tris–HCl (pH 7.4), 1-M NaCl.
36. Flag elution buffer: 20-mM Tris–HCl (pH 7.4), 1-M NaCl, 3× Flag peptide (0.1 mg/mL).
37. Strep-Tactin column (Qiagen).
38. Protein concentration kits: Bradford-based protein assay kit and micro-BCA kit.
39. Centrifugal filter unit: 10-kDa MWCO (15-mL sample), 10-kDa MWCO (0.5-mL sample).
40. Dialysis set: dialysis membrane (10-kDa MWCO), magnetic dialysis clamp, magnetic stirrer, midi-dialysis tube (10-kDa MWCO, 50–800-μL sample).
41. Resolving gel for 12% SDS-PAGE (15 mL): 5-mL dH_2O, 3.75-mL Tris–HCl (1 M, pH 8.8), 150-μL 10% SDS, 50-μL 10% ammonium persulfate, 6-mL acrylamide/bis (30% 37.5:1; Bio-Rad), 14-μL tetramethylethylenediamine (TEMED).
42. 4% stacking gel for SDS-PAGE (5 mL): 3-mL dH_2O, 1.25-mL Tris–HCl (0.5 M, pH 6.8), 50-μL 10% SDS, 30-μL 10% ammonium persulfate, 0.675-mL acrylamide/bis (30% 37.5: 1; Bio-Rad), 8-μL TEMED.
43. Reducing sample loading buffer (for SDS-PAGE): 2.5-mL 0.5-M Tris–HCl (pH 6.8), 1-mL β-mercaptoethanol, 1.5-mL 1% bromophenol blue, 5-mL 10% SDS, 1.5-mL 75% glycerol.
44. 10× SDS-PAGE buffer (1 L): 30.3-g Tris base, 144-g glycine, 10-g SDS. Dissolve Tris base and glycine in 700-mL H_2O. Add H_2O to 1 L and add SDS.
45. Protein electrophoresis and transblot equipment.
46. PVDF membranes.
47. Biosafe Coomassie blue stain (e.g., SimplyBlue Biosafe Stain, Novex, Life Technologies).
48. Sypro Ruby stain (e.g., Invitrogen).
49. GelDoc analyzer (e.g., Bio-Rad system with blue and white trays).
50. Tris–glycine transfer buffer: 25-mM Tris base, 192-mM glycine, 20% methanol. Store at 4 °C.
51. TBST buffer: 150-mM NaCl, 20-mM Tris (pH 7.5), 0.05% Tween 20.
52. Blocking solution: 3% nonfat milk powder in TBST buffer.

53. Mouse monoclonal anti-Flag M2 antibody.
54. Rabbit anti-StrepII polyclonal antibody (GenScript).
55. Anti-mouse and anti-rabbit, horse radish peroxidase (HRP)-linked antibodies.
56. ECL Western blotting substrate (e.g., Pierce).
57. iBright imager or X-ray film (Hyperfilm; Amersham Biosciences) with X-ray film developer.

3 Methods

3.1 Cloning, Expression, and Purification of a Recombinant Archaeal DUB

1. Download the DNA and protein sequences associated with the DUB/DUB-like protein of interest. For this example, you can use the UCSC Archaeal Genome Browser [13] to retrieve the archaeal DUB sequences by selecting the *Haloferax volcanii* DS2 genome browser and entering the HVO_2505 gene locus tag number corresponding to HvJAMM1.
2. Perform 3D modeling of the deduced DUB/DUB-like protein to predict the proximity of the active site to the N- and C-terminal tails where an affinity tag may be added. The affinity tag will be used to purify the protein. For this example, model the 3D structure of HvJAMM1 using Phyre2 [14]. Visualize the 3D model using ChimeraX [15], and use this visualization to examine the proximity of the N- and C-termini to the putative active site residues and other structural folds that may be relevant to the DUB/DUB-like activity. In this case, the N- and C-terminal tails are at a significant distance from the active site suggesting the affinity tag may be positioned at either end of HvJAMM1 (Fig. 1a). A 3D structural model of HvJAMM1 generated by AlphaFold [16] that is available at UniProtKB D4GTS4 [16] may also be used for this purpose.
3. Design PCR primers to clone the DNA fragment carrying the DUB/DUB-like gene of interest into a pET vector (*see* **Note 3**). The design should be such that the PCR product can be inserted into the multiple cloning site of the pET vector in frame with the coding sequence of the affinity tag of interest. Include the restriction enzyme sites on the 5′ ends of the PCR primers compatible with the vector. Add a few extra random nucleotides on the 5′ ends of the primers to enable the restriction enzymes to cleave the PCR product. For this example, design the PCR primers with NdeI and BlpI sites to insert the PCR product into plasmid pET15b to express the archaeal DUB (HvJAMM1) with an N-terminal His-tag (Fig. 1b, c).

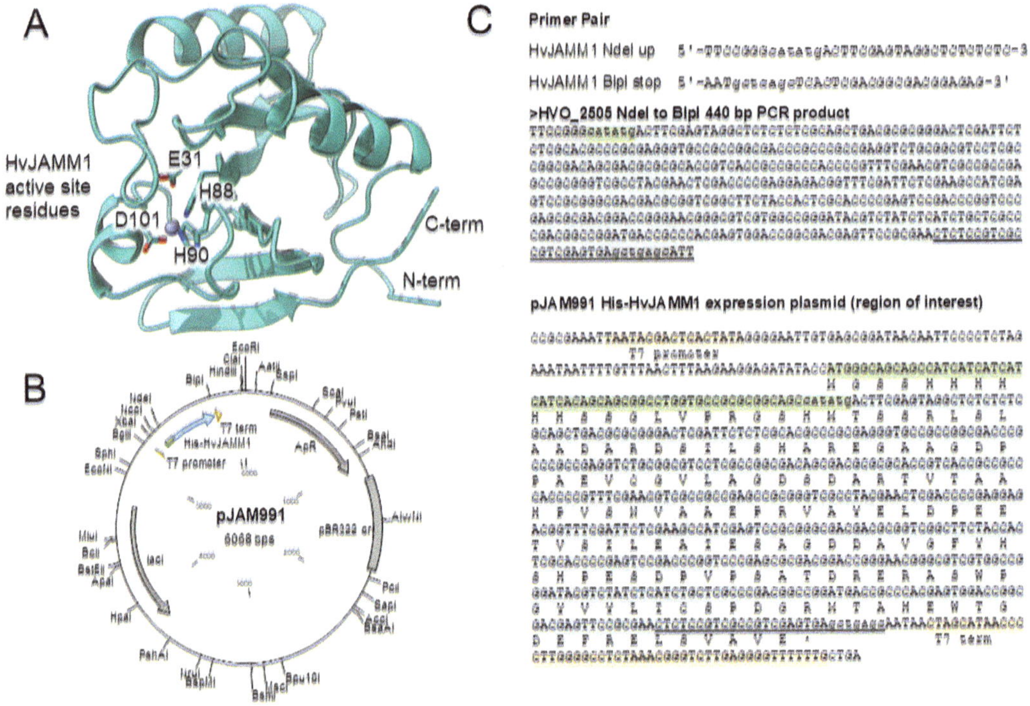

Fig. 1 Strategy for generating a plasmid to synthesize an N-terminal His-tag fusion of a DUB/DUB-like enzyme (HvJAMM1) in recombinant *E. coli*. (**a**) Example of a 3D model of a DUB/DUB-like enzyme (HvJAMM1) to examine the active site residues in relation to the N- and C-termini and guide decisions regarding the positioning of the affinity tag. (**b**) Example of a pET-based plasmid used to express an archaeal DUB with an affinity tag (His-HvJAMM1). (**c**) The primer pair used to isolate the archaeal DUB gene (*hvo_2505* encoding HvJAMM1) by PCR, the PCR product, and a zoom-in of the region of interest on the His-HvJAMM1 expression plasmid pJAM991. Lowercase, NdeI and BlpI sites. Yellow, T7 promoter and terminator. Green, His-tag and thrombin cleavage site sequence encoded by the pET vector. Blue, HvJAMM1 protein sequence. Gray, BlpI site, and random sequences are added to the 5′ ends of the primers to facilitate restriction enzyme digest

4. Use PCR to amplify the DUB/DUB-like gene using the genomic DNA of the organism of interest as the template and the primer pair (designed in **step 3**). For this example, use HvJAMM1 NdeI up and HvJAMM1 BlpI stop primers and *Haloferax volcanii* genomic DNA to generate a 440-bp DNA fragment using the reaction mixture and thermal cycling setting as indicated (Table 1).
5. Using a PCR cleanup kit, remove the excess primers and other impurities from the PCR product.
6. Determine the purity and DNA concentration of the PCR product sample by measuring the A260:A280 ratio using a microvolume spectrophotometer.
7. Digest the PCR product and plasmid vector (pET15b) at 37 °C for 2 h using the NdeI and BlpI restriction enzymes, and buffer in a reaction mixture as recommended by the supplier (Table 2).

Table 1
Example of PCR conditions used for isolation of an archaeal DUB gene (*hvo_2505*, HvJAMM1)

PCR reaction mixture	*(100 μL)*
dH_2O	67.0 μL
5× HF buffer	20.0 μL
10-mM dNTPs	2.0 μL
100-μM HvJAMM1 FWD primer	2.5 μL 5'-TTCCGGGcatatgACTTCGAGTAGGCTCTCTCTC-3' (with *NdeI* cut site)
100-μM HvJAMM1 REV primer	2.5 μL 5'-AATgctcagcTCACTCGACGGCGACGGAGAG-3' (with *BlpI* cut site)
Haloferax volcanii genomic DNA (200 ng/mL)	5.0 μL
Phusion DNA polymerase (2,000 U/mL)	1.0 μL (2 units)
Thermal cycle:	
98 °C	1.0 min
98 °C 65 °C 72 °C	0.5 min 0.5 min } 40 cycles 0.5 min
72 °C	10 min
25 °C	Hold

Table 2
Example of restriction enzyme digest to cleave the insert and vector DNA and generate compatible ends for T4 DNA ligation

Restriction digestion mixture	*(100 μL)*
10× CutSmart buffer	10 μL
DNA (PCR product or pET15b)	100 ng to 1 μg
NdeI	1 μL (10 units)
BlpI	1 μL (10 units)
dH_2O	*To 100 μL*
37 °C for 2 h	

Table 3
Example of a DNA ligation reaction to generate a plasmid encoding an affinity-tagged DUB/DUB-like enzyme for expression in recombinant *E. coli*

T4 DNA ligation mixture	*(10 μL)*
10× T4 DNA ligase buffer	1 μL
pET15b (vector, cut NdeI and BlpI)	50 ng
hvo_2505 DNA (insert, cut with NdeI and BlpI)	50 ng
T4 DNA ligase	0.5 μL
dH_2O (nuclease-free)	*To 10 μL*
16 °C overnight	
T4 DNA ligation mixture control	*(10 μL)*
10× T4 DNA ligase buffer	1 μL
pET15b (vector, cut NdeI and BlpI)	50 ng
T4 DNA ligase	0.5 μL
dH_2O (nuclease-free)	*To 10 μL*
16 °C overnight	

8. Run the digested DNA samples on a 0.8% agarose gel with 1× TAE buffer at 100 V for 30 min.
9. Excise the DNA bands (0.4 kb for the PCR product and 5.7 kb for the pET15b vector). Purify the DNA from the gel slices using a DNA gel extraction kit as recommended by the supplier eluting into 20-μL nuclease-free dH_2O.
10. Ligate the purified DNA fragments overnight (16–20 h) at 16 °C, including a negative vector control alone as indicated (Table 3).
11. Transform the 10-μL ligation product into 100-μL suspension of *E. coli* XL10-Gold ultracompetent cells (Stratagene) as recommended by the product protocol.
12. Allow the transformants to grow in pre-heated (37 °C) LB broth for 45 min.
13. Centrifuge the culture at 3500 × *g* for 3 min and spread harvested cells onto LB agar plate containing ampicillin (100 μg/mL).
14. After overnight incubation at 37 °C, select and streak colonies onto fresh LB plate (with ampicillin), and incubate overnight at 37 °C. If the number of colony-forming units (CFUs) per plate is higher on the control plate when compared to the test ligation plate (*see* **Note 4**).

15. Screen cells that harbor the pET15b-His-HvJAMM1 plasmid (pJAM991) by diagnostic PCR using the cloning primers (HvJAMM1 NdeI up and HvJAMM1 BlpI stop primers).
16. Confirm the fidelity of the plasmid construct by Sanger DNA sequencing using T7 promoter FWD and T7 terminator REV primers.
17. Purify the HvJAMM1-pET15b plasmid from XL10-Gold cells using a DNA plasmid miniprep kit as recommended by the product purification protocol.
18. Transform the HvJAMM1-pET15b plasmid into *E. coli* Rosetta (DE3), and plate cells onto LB agar plate with ampicillin (100 μg/mL) and chloramphenicol (30 μg/mL). Incubate the plate overnight (16–20 h) at 37 °C.
19. Pool five to ten colonies and grow in 10-mL LB–ampicillin–chloramphenicol broth at 37 °C for 2 h with agitation (200 rpm, orbital shaking).
20. Upscale the culture into 500-mL LB–ampicillin–chloramphenicol broth with a starting OD_{600} of 0.02 unit, and allow cells to grow at 37 °C with agitation at 200 rpm.
21. When cells reach 0.1–0.2 OD_{600} unit, reduce the temperature of shaking incubator to 20 °C.
22. At OD_{600} of 0.5–0.6 unit, set aside a 1-mL culture for SDS-PAGE (pre-induction control sample). Collect cells from the 1-mL aliquot by centrifugation (13,800 × *g*) for 1 min. Lyse cells by adding the sample loading buffer and boil for 10 min. To compute the volume (in μL) of the sample loading buffer to be added into the cell pellet harvested from a 1-mL culture aliquot, multiply the OD_{600} of cells by 154. A 10-μL sample as loading volume for SDS-PAGE is equivalent to 0.065 OD_{600} of cells. Freeze this pre-induction control sample at −20 °C until use.
23. Induce the remaining cells to express the recombinant His-HvJAMM1 by adding 0.5 mL of 1-M IPTG into the 500-mL culture.
24. After 5 h of IPTG induction, set aside 1-mL culture for SDS-PAGE (post-induction sample). Process cells as described in **step 22** (Fig. 2a) (*see* **Note 5**).
25. Harvest the remaining cells by centrifugation (3500 × *g*) for 10 min at 4 °C. (Pre-weigh the conical centrifuge tube.)
26. Wash cell pellet with ice-chilled TBS buffer and collect cells by centrifugation (3500 × *g*) for 10 min at 4 °C. Weigh the cell pellet. (Optional: After harvest, store the cell pellet at −80 °C until use.)

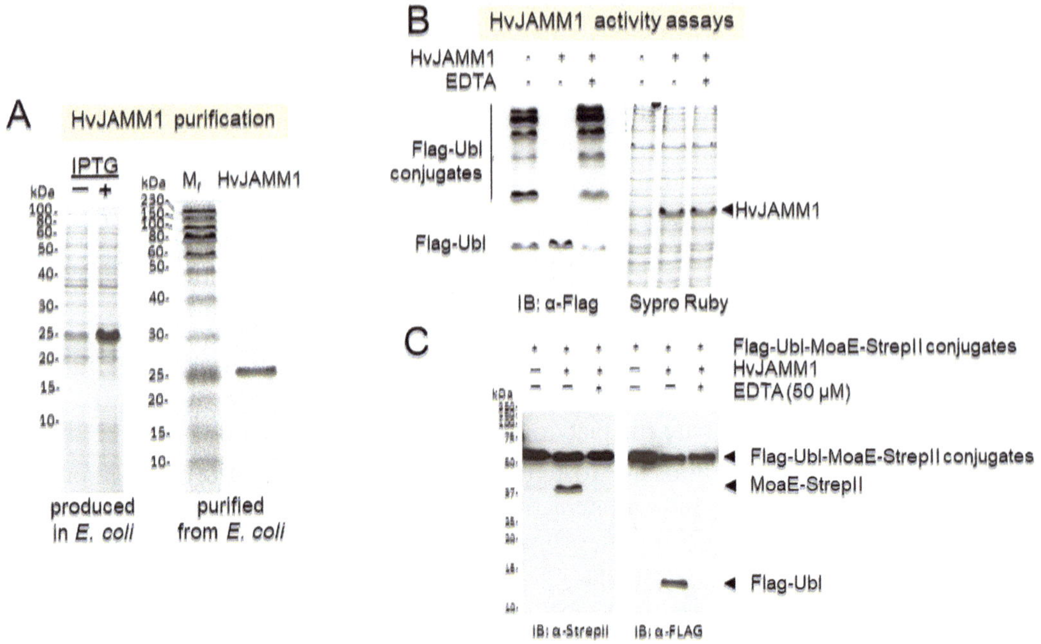

Fig. 2 Purification and activity assay of DUB/DUB-like enzymes using Ubl conjugates synthesized by *H. volcanii* as substrates. (**a**) Archaeal DUB (His-HvJAMM1) expressed and purified by Ni^{2+}-affinity chromatography from recombinant *E. coli* Rosetta (DE3)-pJAM991 as indicated. (**b**) Assay of His-HvJAMM1 with endogenous protein isopeptide linked to Flag–Ubl (Flag–Ubl conjugates, where the Ubl is SAMP2) and (**c**) with a single protein target (MoaE–StrepII) isopeptide linked to Flag–Ubl (Flag–SAMP1) and then tandem affinity purified prior to activity assay. All Ubl conjugates were synthesized in *H. volcanii*. (Hepowit et al. [3])

27. Resuspend the cell pellet in ice-chilled *E. coli* lysis buffer A (5-mL lysis buffer A per 1-g wet weight cells) by gentle pipetting in and out.
28. Lyse cells by passing the cells through a French press (pre-chilled) thrice at 2000 psi.
29. Add an equal volume of ice-chilled *E. coli* lysis buffer B. Clarify the lysate by centrifugation (9200 × *g* at 4 °C for 10 min) and filtration (0.2-μM syringe filter). Keep the lysate on ice.
30. Using an FPLC, wash the HisTrap column with ice-chilled HisTrap wash buffer in a 1-mL/min flow rate until the A280 remains constant at a baseline level.
31. Load the clarified lysate (conical tube on ice) into the HisTrap column at a 0.5-mL/min flow rate.
32. Wash the HisTrap column with ice-chilled HisTrap wash buffer at a flow rate of 1 mL/min until the A280 remains at a constant level.
33. Elute the resin-captured His-HvJAMM1 with HisTrap elution buffer in a 0.5 mL/min flow rate until the A280 returns to baseline level. Collect the elution flow-through at 1 mL per fraction.

34. Using the Bradford-based protein assay as a guide, pool all fractions with a protein concentration of 0.5 mg/mL and above.
35. Buffer exchange by dialyzing (10-kDa MWCO, clamped dialysis membrane) the eluted His-HvJAMM1 with 1-L HEPES–salt buffer for 3 h at 4 °C with gentle rotation using a magnetic stirrer.
36. Transfer the dialysis bag into fresh 4-L HEPES–salt buffer, and continue the dialysis overnight (16–20 h) at 4 °C with gentle rotation. (Dialysis removes the imidazole.)
37. Measure the concentration of His-HvJAMM1 using a Micro BCA Protein Assay Kit and a microvolume spectrophotometer.
38. Determine the purity of the HisTrap-isolated His-HvJAMM1 by running a 1-μg aliquot on SDS-PAGE gel followed with Coomassie blue and/or Sypro Ruby staining (Fig. 2a). Include the pre- and post-induction samples on separate wells of the SDS-PAGE gel.
39. Filter-sterilize (0.2 μM) the purified recombinant His-HvJAMM1 and store at 4 °C. (In a high-salt buffer, the His-HvJAMM1 is still active even for a year at 4 °C storage. For more extended storage, add 10% glycerol, aliquot the enzyme [50 μL/microfuge tube], flash-freeze with liquid N_2, and store at −80 °C. Avoid refreezing the thawed enzyme.)

3.2 Formation and Enrichment of Isopeptide-Linked Ubl-Conjugated Proteins Using the Archaeon H. volcanii *as a Microbial Platform*

1. Transform *H. volcanii* H26 with a plasmid that constitutively expresses an N-terminal Flag-tagged Ubl protein. The transformation is by a polyethylene glycol 600 method as previously described [17] (*see* **Notes 6–8**).
2. Spread the transformants onto ATCC974 agar plates supplemented with novobiocin (0.2 μg/mL), and incubate at 42 °C for 3–5 days.
3. As a startup culture, pool three to five colonies into a 5-mL ATCC974 medium with novobiocin, and incubate at 42 °C for 24 h with agitation (200 rpm).
4. To upscale the culture, transfer the 5-mL culture into a 250-mL flask that contains 500 mL of fresh ATCC974 medium with novobiocin (0.2 μg/mL) and 50-mM DMSO. Incubate the culture flask at 42 °C with agitation (200 rpm).
5. Harvest cells at stationary phase (after 48–72 h of incubation) by centrifugation at 3500 × *g* for 10 min at room temperature.
6. Wash the cell pellet with 20-mL ice-cold Tris–salt buffer by pipetting in and out. Remove the wash buffer by centrifugation.
7. Resuspend cells in 10-mL ice-chilled Tris lysis buffer (4 mL per 1-g wet weight cells), and lyse cells by passing through a French press thrice at 2000 psi.

8. Clarify the lysate by centrifugation (9200 × *g* at 4 °C for 10 min) and filtration (0.2-μM syringe filter).
9. Pack a minicolumn with 200-μL anti-Flag resin.
10. Condition the anti-Flag column with 1 mL of Tris–salt buffer thrice. Apply air pressure using a rubber bulb for proper flow-through.
11. Gently load the clarified lysate into the anti-Flag column without forming any bubble.
12. Wash the column with 1-mL Tris–salt buffer thrice.
13. Elute the resin-bound protein isopeptide linked to the Flag–Ubl with 0.6-mL 3× Flag peptide (0.1-mg/mL Tris–salt buffer). Allow the eluent to pass through the column by gravity flow.
14. Buffer exchange by dialyzing (10-kDa MWCO midi-dialysis tube) the eluted proteins with 1-L HEPES–salt buffer for 3 h at 4 °C with gentle rotation using a magnetic stirrer.
15. Transfer the dialysis bag into fresh 1-L HEPES–salt buffer, and continue the dialysis overnight (16–20 h) at 4 °C with gentle rotation. (Dialysis also removes the 3× Flag peptide.)
16. Concentrate the Flag–Ubl protein conjugates to 100–200 μL using a centrifugal filter unit (10-kDa MWCO) by spinning at 10,000 × *g* at 4 °C.
17. Measure the concentration of Flag–Ubl protein conjugates using the Micro BCA Protein Assay Kit and a microvolume spectrophotometer.
18. Confirm the enriched Flag–Ubl protein conjugates by Western blot using an anti-Flag antibody.
19. Store the eluted Flag–Ubl protein conjugates at −80 °C until use (*see* **Note 9**).

3.3 Tandem Affinity Purification of Ubl-Modified Proteins from the Archaeon H. volcanii

1. Transform *H. volcanii* H26 with pJAM1314, which expresses the Ubl protein with an N-terminal Flag tag and the target protein separately. This example uses Flag–SAMP1 as the Ubl protein and MoaE–StrepII as the target protein.
2. Spread a dilution series of the transformants onto ATCC974 agar plates that contain novobiocin (0.2 μg/mL) to ensure isolated colonies. Incubate the plates at 42 °C for 3–5 days.
3. As a startup culture, pool three to five colonies into a 5-mL ATCC974 medium with novobiocin, and incubate at 42 °C for 24 h with agitation (200 rpm).
4. To upscale the culture, transfer the 5-mL culture into a 250-mL flask that contains 500 mL of fresh ATCC974 medium with novobiocin. Incubate the culture flask at 42 °C with agitation (200 rpm).

5. Harvest cells at stationary phase (after 48–72 h of incubation) by centrifugation at 3500 × *g* for 10 min.
6. Wash the cell pellet with 20-mL ice-chilled Tris–salt buffer by pipetting in and out. Remove the wash buffer by centrifugation.
7. Resuspend cells in 10-mL ice-chilled Tris lysis buffer (4 mL per 1-g wet weight cells), and lyse cells by passing through a French press thrice at 2000 psi.
8. Clarify the lysate by centrifugation (9200 × *g* at 4 °C for 10 min) and filtration (0.2-μM syringe filter). Keep the lysate on ice.
9. Equilibrate a 1-mL Strep-Tactin column with ice-chilled Tris–salt buffer using an FPLC with 0.3-mL/min flow rate.
10. Auto-load the lysate into the column using an FPLC with a 0.3-mL/min flow rate. Keep the lysate on ice while being pumped into the column.
11. Wash the column with ice-chilled Tris–salt buffer until the UV–vis signal goes to baseline level.
12. Elute the MoaE–StrepII and the isopeptide-linked Flag–SAMP1–MoaE–StrepII from the Strep-Tactin column with Tris–salt buffer supplemented with 5-mM *d*-desthiobiotin.
13. Pack a minicolumn with 200-μL anti-Flag resin.
14. Condition the anti-Flag column with 1 mL of Tris–salt buffer thrice. Apply air pressure using a rubber bulb for proper flow-through.
15. Gently load the clarified lysate into the anti-Flag column without forming any bubble.
16. Wash the column with 1-mL Tris–salt buffer thrice to remove any unmodified MoaE–StrepII.
17. Elute the resin-bound Flag–SAMP proteins with 0.6-mL 3× Flag peptide (0.1-mg/mL Tris–salt buffer). Allow the eluent to pass through the column by gravity flow.
20. Buffer exchange by dialyzing (10-kDa cutoff mini-dialysis tube) the eluted proteins with 1-L HEPES–salt buffer for 3 h at 4 °C with gentle rotation using a magnetic stirrer.
21. Transfer the dialysis bag into fresh 1-L HEPES–salt buffer, and continue the dialysis overnight (16–20 h) at 4 °C with gentle rotation. (Dialysis also removes the 3× Flag peptide.)
22. Concentrate the Flag–SAMP–MoaE conjugates to 100–200 μL using a centrifugal filter unit (10 MWCO) by spinning at 10,000 × *g* at 4 °C.
23. Measure the concentration of isopeptide-linked Flag–SAMP1–MoaE–StrepII using the Micro BCA Protein Assay Kit and a microvolume spectrophotometer.

24. Confirm the purity of enriched Flag–SAMP1–MoaE–StrepII by Western blot using anti-Flag and anti-StrepII antibodies.
25. Store the isopeptide-linked Flag–SAMP1–MoaE–StrepII at −80 °C until use.

3.4 Assay the Isopeptidase Activity of the Purified DUB/DUB-Like Enzyme Using the Ubl Conjugates Formed in the Archaeon

To monitor the activity of purified DUB/DUB-like enzymes, an in vitro assay can be performed using the Ubl conjugates formed in *H. volcanii* as substrates. In this example, the "salt-loving" Zn^{2+} metalloprotease HvJAMM1 is assayed for isopeptidase activity. Thus, the buffers are supplemented with 2-M NaCl to maintain enzyme stability and activity. EDTA is included to strip the Zn^{2+} from the activity site of the archaeal DUB as a negative control.

1. To monitor enzyme activity, mix the following reagents in a 50-μL PCR tube using Table 4 as a guide. The 20-μL reaction mixture (tube #2) contains 2-M NaCl, 100-μM $ZnCl_2$, 10-μM His-HvJAMM1, and 1-μg Flag–Ubl modified substrate (Flag–SAMP2 conjugates enriched from *H. volcanii*). Negative controls include no enzyme (tube #1), heat-denatured enzyme (tube #3), and EDTA treatment (tube #4).
2. Using a thermal cycler, incubate the PCR tubes containing the reaction mixture at 50 °C for 2 h. A time-course should be performed to determine optimal detection of enzyme activity.
3. Stop the reaction by adding 20 ul of 2× sample buffer and boiling for 10 min.
4. Run the reaction mixtures on a 12% SDS-PAGE gel and visualize by immunoblotting analysis (Fig. 2b) (detailed procedure in Subheading 3.5.3).

Table 4
Example of the setup for analyzing DUB/DUB-like activity using Ubl-modified proteins produced in *H. volcanii* as a substrate

PCR tube	HEPES–salt buffer	DUB/DUB-like enzyme[a](100 μM)	Ubl-modified substrate[b](100 ng/μL)	$ZnCl_2$[c] (1 mM)	EDTA (1 M)
1	8 μL	–	10 μL	2 μL	–
2	6 μL	2 μL	10 μL	2 μL	–
3	6 μL	2 μL (heat-killed)	10 μL	2 μL	–
4	5 μL	2 μL	10 μL	2 μL	1 μL

[a]His-HvJAMM1 purified from recombinant *E. coli*
[b]Flag–SAMP1 conjugates or isopeptide-linked Flag–SAMP1–MoaE–StrepII enriched from *H. volcanii*
[c]In HEPES–salt buffer

3.5 Protein Visualization by Staining and Immunoblotting Analysis

1. Load 10 μL of protein sample (in loading buffer) into each well of a 12% SDS-PAGE gel submerged in ice-chilled 1× SDS-PAGE running buffer.
2. Using vertical PAGE electrophoresis equipment, run the SDS-PAGE gel at 150 V for 70 min at room temperature.

3.5.1 Biosafe Coomassie Blue Staining

1. Transfer the gel into a tray containing 15 mL (or enough to soak the gel) of Biosafe Coomassie blue stain.
2. Rock the tray at room temperature for 1 h.
3. Discard the stain and quickly wash the gel thrice with dH_2O.
4. Soak the gel with dH_2O and allow the gel to destain by rocking at room temperature overnight. Submerge a Kimwipes napkin into the water. Blue stain in water attaches to the napkin.
5. Visualize the gel using a GelDoc analyzer (white tray).

3.5.2 Sypro Ruby Staining

1. Transfer the gel into a black staining container (with lid) with 15 mL (or enough to soak the gel) of Sypro Ruby stain.
2. Stain the gel for 3 h at room temperature by gentle rocking.
3. Transfer the gel into a clean black staining container, and wash with 30% methanol/7% acetic acid solution by gentle rocking for 30 min.
4. Visualize the gel using a GelDoc analyzer (purple tray).

3.5.3 Immunoblotting (Western Blotting) Analysis

1. On the black side of the gel holder cassette, place the following in order (from bottom to top): pre-wet foam pad, pre-wet Whatman filter paper, SDS-PAGE gel (containing the proteins), PVDF membrane (pre-soaked in methanol), pre-wet Whatman filter paper, and pre-wet foam pad. Close the holder cassette to form a sandwich.
2. Place the whole sandwich into the transblot tank on a proper orientation (black side of the gel holder cassette facing the black side of the electrode assembly).
3. Transfer at 4 °C for 90 min at 100 V with stirring.
4. Take the PVDF membrane out and soak with 20-mL blocking buffer in a tray. Rock for 1 h at room temperature.
5. Add the primary antibody (anti-Flag or anti-StrepII) to a final concentration of 1:10,000 dilution, and incubate with gentle rocking at room temperature for 1 h.
6. Quickly wash the PVDF membrane thrice with TBST. Incubate the membrane in 20-mL TBST (fourth wash) for 5 min with rocking.
7. Discard the TBST, and add 20 mL of blocking buffer containing 1:10,000 dilution of the secondary antibody (anti-mouse

or anti-rabbit, HRP-linked). Incubate with gentle rocking at room temperature for 1 h.

8. Quickly wash the PVDF membrane thrice with TBST. Incubate the membrane in 20-mL TBST (fourth wash) for 30 min with rocking.
9. Fix the PVDF membrane on a cassette. Treat the membrane with ECL Western blotting substrate as specified by the product protocol.
10. Cover the PVDF membrane with a plastic film and fix it with adhesive tape.
11. In the darkroom, overlay the X-ray film on the fixed PVDF membrane and quickly seal the cassette.
12. Take X-ray images of films exposed for 5 s, 1 min, 15 min, and 30 min to obtain the best exposure.

4 Notes

1. *H. volcanii* genomic DNA can easily be prepared from whole cells for use as a template in PCR as follows: Harvest 1 mL of *H. volcanii* culture freshly grown to log to early stationary phase (OD600 nm of 0.5–0.8) by centrifuging for 1 min using the maximum setting in a microcentrifuge, at room temperature. Remove the supernatant, add 400-μL nuclease-free water, and disrupt the cell pellet using a pipettor. The "salt-loving" cells will immediately lyse by osmotic lysis. Inactivate DNases by heating the sample to 70–75 °C for 10 min. If preparing the genomic DNA in this manner, we recommend using 1–4 μL of the DNA sample per 50–100-μL PCR.
2. The *H. volcanii* strains and plasmids can be obtained from the Maupin-Furlow Lab of the University of Florida. For details on plasmid, *see* [3, 11, 12].
3. A similar cloning strategy can be used to synthesize DUB/DUB-like enzymes in recombinant *H. volcanii* with the pJAM shuttle plasmids, thus allowing similar primers and cloning sites (NdeI to BlpI for generation of an N-terminal His-tag fusion protein) [3]. C-terminal StrepII and His-tag expression plasmids and strategies are also available for expression in *H. volcanii* and are compatible with the high-salt buffers used to maintain halophilic protein activity [18].
4. Prior to the T4 DNA ligation reaction, the vector can be treated with Antarctic phosphatase to remove its 5′ phosphoryl groups and, thus, minimize vector self-ligation as needed. We recommend incubating the vector with the restriction enzymes as outlined in Table 2. After the 2 h incubation, add the

following reagents directly to the 100-μL reaction mixture: 11-ul 10× Antarctic phosphatase reaction buffer and 1-ul Antarctic phosphatase (5 units/ul, final 5 units per pmol ends of vector DNA). Incubate the sample at 37 °C for 30 min. Directly apply the mixture to the DNA gel (no need to heat inactivate).

5. Expression of the affinity-tagged protein in recombinant *E. coli* may occur in the absence of IPTG (as observed in Fig. 2). If this non-induced expression is high, an evident detection of the recombinant protein in the induced culture may be complicated. In this case, the empty vector can be used as a negative control to confirm protein synthesis.
6. An N-terminal Flag–His6-tag can also be added to the coding sequence of the Ubl and used to generate Ubl conjugates in *H. volcanii* [8]. This tandem tag allows for purifying the Ubl conjugates by anti-Flag or Ni^{2+}-affinity chromatography.
7. If expressing a foreign gene, we recommend synthesizing the mature form of the Ubl with the coding sequence codon-optimized for translation in *H. volcanii* using the Codon Optimization On-Line (COOL) web platform. NdeI/BlpI restriction sites can be added to the sequence 5′ and 3′ ends to facilitate downstream cloning. The entire sequence can be synthesized by GenScript and ligated into a pUC57 plasmid. This design provides a plasmid that carries the Flag–Ubl or Flag–His6–Ubl gene on a region that can be released by NdeI/BlpI digestion and ligated into an *E. coli–H. volcanii* shuttle plasmid (pJAM202) to create a Flag–His6–Ubl or Flag–Ubl expression plasmid. This shuttle plasmid is selected for transformation in *E. coli* by ampicillin resistance and in *H. volcanii* by novobiocin resistance. Once the construct is confirmed in a general cloning strain such as *E. coli* TOP10, the plasmid is passaged through a methylation-deficient strain of *E. coli* (GM2163) prior to transformation into *H. volcanii.*
8. A *Halohandbook* is available as a valuable reference for general molecular biology experiments with the halophilic archaea [17].
9. Ubl conjugates are also found to serve as substrates for HvJAMM1 when prepared from the *H. volcanii* Flag–Ubl expressing strains as follows:
   (a) Grow the cells to stationary phase in 100 mL of GMM-Ala with novobiocin (0.2 μg/mL).
   (b) Harvest cells and wash twice in 20-mL ice-chilled HEPES–salt buffer.
   (c) Resuspend the cells in the same buffer (1 mL per 1-g wet weight).

(d) Sonicate the cell suspension (3 × 30 to 60 s at 140 W on ice).

(e) Clarify the lysate by centrifugation (9200 × *g* for 15 min at 4 °C).

(f) Boil the supernatant for 20 min and further clarify by centrifugation (9200 g for 15 min at 4 °C).

(g) Adjust total protein concentration to 5 mg/mL, aliquot, and store at −80 °C until further use.

Acknowledgments

Funds were awarded to JMF to advance archaeal bioenergy through the US Department of Energy, Office of Basic Energy Sciences, Division of Chemical Sciences, Geosciences and Biosciences, Physical Biosciences Program (DOE DE-FG02-05ER15650) and to provide evolutionary insight into biological systems through the National Institutes of Health, General Medical Sciences (2R01GM057498-18).

References

1. Liu Y, Makarova KS, Huang WC, Wolf YI, Nikolskaya AN, Zhang X, Cai M, Zhang CJ, Xu W, Luo Z et al (2021) Expanded diversity of Asgard archaea and their relationships with eukaryotes. Nature 593(7860):553–557
2. Fu X, Liu R, Sanchez I, Silva-Sanchez C, Hepowit NL, Cao S, Chen S, Maupin-Furlow J (2016) Ubiquitin-like proteasome system represents a eukaryotic-like pathway for targeted proteolysis in archaea. *mBio* 7(3): e00379–e00316
3. Hepowit NL, Uthandi S, Miranda HV, Toniutti M, Prunetti L, Olivarez O, De Vera IM, Fanucci GE, Chen S, Maupin-Furlow JA (2012) Archaeal JAB1/MPN/MOV34 metalloenzyme (HvJAMM1) cleaves ubiquitin-like small archaeal modifier proteins (SAMPs) from protein-conjugates. Mol Microbiol 86(4):971–987. https://doi.org/10.1111/mmi.12038. Epub 2012 Oct 9. PMID: 22970855; PMCID: PMC3558616
4. Tran HJ, Allen MD, Löwe J, Bycroft M (2003) Structure of the Jab1/MPN domain and its implications for proteasome function. Biochemistry 42(39):11460–11465
5. Ambroggio XI, Rees DC, Deshaies RJ (2004) JAMM: a metalloprotease-like zinc site in the proteasome and signalosome. PLoS Biol 2(1):E2
6. Cao S, Hepowit N, Maupin-Furlow JA (2015) Ubiquitin-like protein SAMP1 and JAMM/MPN+ metalloprotease HvJAMM1 constitute a system for reversible regulation of metabolic enzyme activity in Archaea. PLoS One 10(5): e0128399
7. Cao S, Engilberge S, Girard E, Gabel F, Franzetti B, Maupin-Furlow JA (2017) Structural insight into ubiquitin-like protein recognition and oligomeric states of JAMM/MPN+ proteases. Structure 25(6):823–833. e826
8. Jiang B, Adams Z, Moonah S, Shi H, Maupin-Furlow J, Moskovitz J (2020) The antioxidant enzyme methionine sulfoxide reductase A (MsrA) interacts with Jab1/CSN5 and regulates its function. Antioxidants (Basel) 9(5): E452
9. Allers T, Ngo HP, Mevarech M, Lloyd RG (2004) Development of additional selectable markers for the halophilic archaeon *Haloferax volcanii* based on the *leuB* and *trpA* genes. Appl Environ Microbiol 70(2):943–953
10. Wendoloski D, Ferrer C, Dyall-Smith ML (2001) A new simvastatin (mevinolin)-resistance marker from *Haloarcula hispanica* and a new *Haloferax volcanii* strain cured of plasmid pHV2. Microbiology 147(Pt 4): 959–964
11. Humbard MA, Miranda HV, Lim JM, Krause DJ, Pritz JR, Zhou G, Chen S, Wells L,

Maupin-Furlow JA (2010) Ubiquitin-like small archaeal modifier proteins (SAMPs) in *Haloferax volcanii*. Nature 463(7277):54–60

12. Miranda HV, Antelmann H, Hepowit N, Chavarria NE, Krause DJ, Pritz JR, Bäsell K, Becher D, Humbard MA, Brocchieri L et al (2014) Archaeal ubiquitin-like SAMP3 is isopeptide-linked to proteins via a UbaA-dependent mechanism. Mol Cell Proteomics 13(1):220–239
13. Schneider KL, Pollard KS, Baertsch R, Pohl A, Lowe TM (2006) The UCSC archaeal genome browser. Nucleic Acids Res 34(Database issue): D407–D410
14. Kelley LA, Mezulis S, Yates CM, Wass MN, Sternberg MJ (2015) The Phyre2 web portal for protein modeling, prediction and analysis. Nat Protoc 10(6):845–858
15. Pettersen EF, Goddard TD, Huang CC, Meng EC, Couch GS, Croll TI, Morris JH, Ferrin TE (2021) UCSF ChimeraX: structure visualization for researchers, educators, and developers. Protein Sci 30(1):70–82
16. Jumper J, Evans R, Pritzel A, Green T, Figurnov M, Ronneberger O, Tunyasuvunakool K, Bates R, Žídek A, Potapenko A et al (2021) Highly accurate protein structure prediction with AlphaFold. Nature 596(7873):583–589
17. Dyall-Smith M (2009) The halohandbook: protocols for halobacterial genetics v.7.2. http://www.haloarchaea.com/resources/halohandbook/Halohandbook_2009_v7.2mds.pdf
18. Humbard MA, Zhou G, Maupin-Furlow JA (2009) The N-terminal penultimate residue of 20S proteasome α1 influences its N^{α} acetylation and protein levels as well as growth rate and stress responses of *Haloferax volcanii*. J Bacteriol 191(12):3794–3803

Chapter 11

Characterization of the Ubiquitin and ISG15 Deconjugase Activity of SARS-CoV-1 and SARS-CoV-2 Papain-Like Protease

Digant Nayak, Priscila dos Santos Bury, Marcin Drag, Alfred N. Nijkerk, Farid El Oualid, and Shaun K. Olsen

Abstract

Both severe acute respiratory syndrome coronavirus 1 and 2 (SARS-CoV-1 and SARS-CoV-2) encode a papain-like protease (PLpro), which plays a vital role in viral propagation. PLpro accomplishes this function by processing the viral polyproteins essential for viral replication and removing the small proteins, ubiquitin and ISG15 from the host's key immune signaling proteins, thereby preventing the host's innate immune response. Although PLpro from both SARS-CoV-1 and SARS-CoV-2 are structurally highly similar (83% sequence identity), they exhibit functional variability. Hence, to further elucidate the mechanism and aid in drug discovery efforts, the biochemical and kinetic characterization of PLpro is needed. This chapter describes step-by-step experimental procedures for evaluating PLpro activity in vitro using activity-based probes (ABPs) along with fluorescence-based substrates. Herein we describe a step-by-step experimental procedure to assess the activity of PLpro in vitro using a suite of activity-based probes (ABPs) and fluorescent substrates and how they can be applied as fast and yet sensitive methods to calculate kinetic parameters.

Key words Deubiquitinase, deISGylase, DUB, PLpro, COVID-19, ISG15, Ubiquitin, Activity-based probes (ABPs), Activity assay, Papain-like protease, SARS-CoV-1, SARS-CoV-2

1 Introduction

After the discovery of severe acute respiratory syndrome coronavirus 2 (SARS-CoV-2) in Wuhan, China [1, 2] around 2019 December, the virus has already spread across all the continents of the world with more than 100 million confirmed cases and more than five million fatalities till date [3]. SARS-CoV-2 is the causative agent for novel coronavirus disease 2019 (COVID-19) with influenza-like symptoms including fever or chills, cough, headache, diarrhea, etc., with the most prominent indicators being shortness of breath and loss of taste or smell. The emergence

Julie A. Maupin-Furlow and Mariola J. Edelmann (eds.), *Deubiquitinases: Methods and Protocols*, Methods in Molecular Biology, vol. 2591, https://doi.org/10.1007/978-1-0716-2803-4_11,

of COVID-19 as a global pandemic serves as the third example of causing acute disease in humans, with SARS-CoV-1 outbreak being the first, beginning in 2002, followed by the appearance of the Middle East respiratory syndrome (MERS-CoV) around 2012 [4]. Although SARS-CoV-2 shares high sequence similarity and disease physiology with SARS-CoV-1, SARS-CoV-2 has fewer case fatality and increased infectivity, which leads to quick and worldwide transmission [5, 6]. While there have been recent developments with emergency approvals of vaccines, which have reduced the spread, there have been cases of new variants which may bypass the immune responses developed by vaccines. Hence there is a dire need to develop antiviral drugs with enhanced efficacies to combat COVID-19.

Both SARS-CoV-1 and SARS-CoV-2 encode for a papain-like cysteine protease (PLpro) [7], which has been a significant focus for drug development. In the host, this enzyme promotes viral replication by processing the viral encoded single-chain nonstructural proteins (NSPs) into individual subunits via proteolytic hydrolysis of the peptide bonds [8]. These individual proteins (termed NSP1, NSP2, etc.) help in viral genome replication, among other functions [9]. It was also later identified that PLpro has deubiquitinating and deISGylating activity. Here, PLpro cleaves the isopeptide bond between host cellular proteins and the small proteins ubiquitin (Ub) and ISG15 (Interferon Stimulating gene 15), respectively [10–12]. Since posttranslational modifications with Ub and ISG15 play a crucial role in the innate immune response to viral infection, PLpro-induced de-conjugation of these modifiers promotes viral pathogenesis. Moreover, PLpro is also responsible for inhibiting key innate immune molecules like cytokines and chemokines [13–15]. Hence, it is no surprise that PLpro is one of the key targets for therapeutic intervention.

Various structures for PLpro from both SARS-CoV-1 and recently from SARS-CoV-2 have been reported, in both apo form and in complex with suitable substrates. Also, many structural studies have identified small molecules that impede PLpro function [7, 16, 17]. However, prerequisite to structural studies pertaining to small molecules, it is imperative to set up a platform to assess the activity of PLpro in vitro for high-throughput studies. Also, it was vital for PLpro/substrate structures to stabilize the complex because of the low affinity between them. Although the PLpro structures provided a mechanistic understanding of the activity of this enzyme, they did not shed light on their kinetics with respect to various substrates (like Ub versus ISG15). However, with technical advances, these substrates were modified, which could be used to capture or measure the activities of PLpro and other deubiquitinating/deISGylating enzymes [18–20]. One of the modifications included the generation of activity-based probes (ABPs) [21].

ABPs are substrate mimetics that present an electrophilic "warhead" to the active site nucleophile to produce a stable, covalent adduct. Application of these ABPs was indispensable in capturing the PLpro/substrate complex for structural studies [22–25]. In this chapter we will focus on ABPs equipped with the vinyl pentynyl sulfone (VPS) warhead, an electrophile that is also suitable for further functionalization by click chemistry [26]. The described ABP protocol is also applicable to Ub and ISG15 ABPs equipped with other electrophiles [23]. Despite all the merits of these ABPs, they are inadequate when it comes to kinetic information, which is important in understanding substrate specificity. Various (quenched) fluorescent substrates are commercially available. The most known substrates are equipped with 7-amido-4-**m**ethylcoumarin (**AMC**) or Rhodamine110Gly (Fig. 1). Although AMC is very robust and easy to use, its relatively low excitation and emission spectrum (*ex* 380/*em* 460 nm) may coincide with many small molecules, and hence the Rh110Gly substrates are preferred (*exc* 485/*emi* 520 nm). A recent improvement of the Rh110Gly substrates is based on the introduction of Rh110-morpholinecarbonyl (Rh110MP) [27], which has a similar excitation and emission spectrum as Rh110Gly but exhibits a much higher fluorescence intensity.

Since both AMC- and Rh110-based substrates do not represent true substrate mimetics due to the absence of the native isopeptide linkage, a TAMRA-K(Ub)G and TAMRA-K(ISG15)G substrate were developed [28] (Fig. 1). Here Ub or ISG15 is linked via an isopeptide bond to a tetramethylrhodamine (TAMRA) labeled Lys-Gly dipeptide. Protease activity is measured through fluorescence polarimetry (FP). This chapter describes step-by-step instructions on how to use the previously described suite of reagents to profile PLpro activity with necessary recommendations such that one can get reproducible results.

2 Materials

2.1 Protein Preparation

1. Plasmids expressing glutathione S-transferase (GST)-tagged versions of SARS-CoV-1 and SARS-CoV-2 PLpro proteins with codon usage optimized for expression in *Escherichia coli*. The plasmids carry the synthetic genes in the pGEX6P-1 vector (GE Healthcare, UK) at the BamHI and NotI sites and are synthesized by Gene Universal (USA) for expression as a PreScission protease-cleavable N-terminally GST-tagged protein[28].
2. Bacterial strain *E. coli* BL21(DE3) RIL Codon Plus (Stratagene).
3. 100-mg ampicillin per mL in water, filter sterilized.

Fluorogenic substrates used in this study	
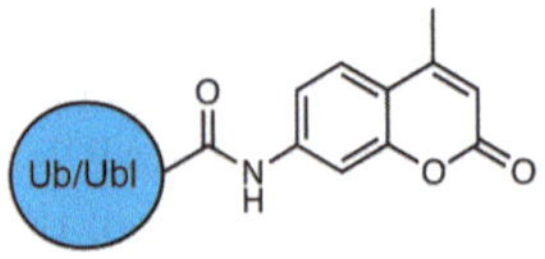 **AMC** **excitation / emission: 380 nm / 460 nm**	**Pros:** **1:** easy to use and amenable for high-throughput assay. **Cons:** **1:** does not mimic true substrate. **2:** excitation wavelength is in the UV range, hence it is not optimal for drug discovery as it can excite many screening compounds and cause false positives
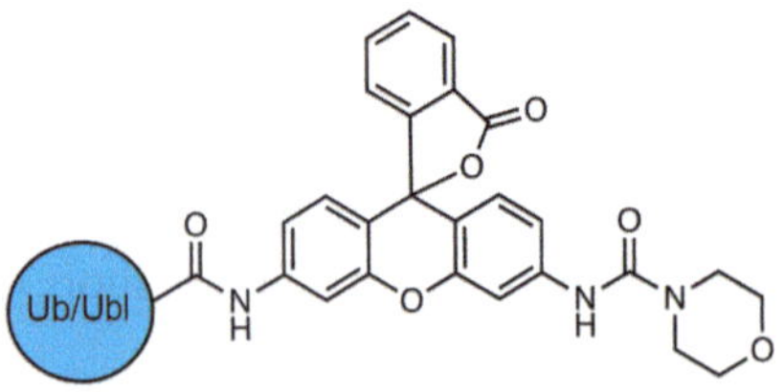 **Rh110MP** **excitation / emission: 485 nm / 520 nm**	**Pros:** **1:** easy to use and amenable for high-throughput assay. **2:** excitation wavelength is outside UV range, making it very suitable for screening small molecules. **Cons:** **1:** does not mimic true substrate.
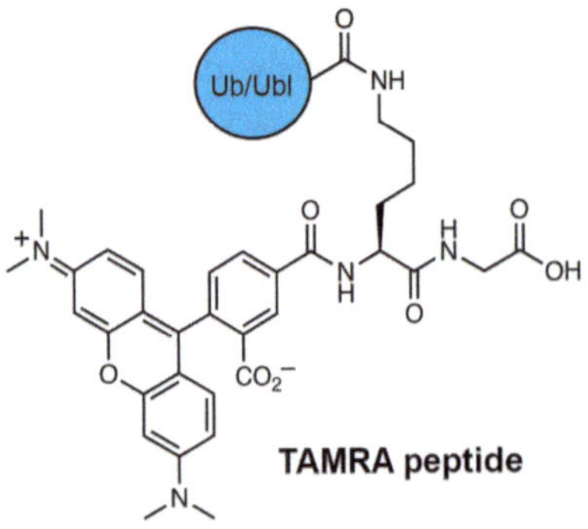 **TAMRA peptide** **excitation / emission: 540 nm / 590 nm**	**Pros:** **1:** easy to use and amenable for high-throughput assay. **2:** excitation wavelength is outside UV range, making it very suitable for screening small molecules. **3:** mimics a true substrate as Ub/Ubl is linked via an isopeptide linkage

Fig. 1 Fluorophore substrates used in this chapter. The chemical structures of fluorophores with their excitation/emission wavelengths and other properties are shown. AMC 7-amido-4-methylcoumarin, Rh110MP Rhodamine110-morpholinecarbonyl, TAMRA tetramethylrhodamine, nm nanometers

4. 1-M isopropyl-b-d-thiogalactopyranoside (IPTG) in water, filter sterilized.
5. 1-M $ZnSO_4$ in water, filter sterilized.
6. LB broth.
7. Lysis buffer: 50-mM Tris–Cl pH 8.0, 300-mM NaCl, 5% glycerol, 1-mM phenylmethylsulfonyl fluoride (PMSF), and 5-mM β-mercaptoethanol.
8. Equilibration buffer: 20-mM HEPES pH 7.5, 150-mM NaCl, 0.5-mM tris(2-carboxyethyl)phosphine (TCEP).
9. Refrigerated shaker (Innova 44R; New Brunswick Scientific).
10. Sonicator (Branson 450 Digital Sonifier).
11. Glutathione Sepharose 4B affinity chromatography resin (GE Healthcare).
12. PreScission protease (Cytiva).
13. A stock solution of 0.5-M reduced glutathione, pH 8.0 in water.
14. Chromatography systems: NGC purification system (Bio-Rad), equipped with gel filtration (Superdex-200 pg 26/60) and desalting (HiPrep 26/10) columns (GE Healthcare).
15. Glass Econo-Columns, 2.5 × 20 cm (Bio-Rad).
16. Centrifugal filter units: Amicon Ultra-4 and Ultra-15 (10-kDa molecular weight cutoff, MWCO) (Millipore).
17. Liquid nitrogen.

2.2 ABPs and Gel-Based Assay Preparation

1. Ub-VPS (UbiQ Bio B.V., product code UbiQ-193).
2. Ac-mISG15-VPS (C-terminal domain of mouse ISG15) (UbiQ Bio B.V., product code UbiQ-262).
3. Ac-hISG15-VPS (C-terminal domain of human ISG15) (UbiQ Bio B.V., product code UbiQ-311).
4. 100% DMSO.
5. 2.5-M sodium acetate buffer, pH 4.5.
6. 1-M HEPES buffer, pH 7.5.
7. 10× assay buffer: 500-mM HEPES pH 7.5, 1500-mM NaCl, 5-mM TCEP.
8. 4× SDS reducing loading buffer.
9. 4–20% gradient gels (Bio-Rad) or homemade gels.
10. Coomassie Brilliant blue stain (Bio-Rad, 1610436).
11. Standard electrophoretic equipment.

2.3 Fluorescent Substrates and Plate-Based Assay Preparation

1. Ub-AMC (R&D systems, Catalog# U-550-050).
2. Ub-Rh110MP (UbiQ Bio B.V., product code UbiQ-126).
3. TAMRA-Lys(Ub)-Gly (UbiQ Bio B.V., product code UbiQ-012).
4. ISG15-AMC (R&D systems, Catalog# U-553-050).
5. Ac-mISG15-Rh110MP (C-terminal domain of mouse ISG15) (UbiQ Bio B.V., product code UbiQ-127).
6. Ac-hISG15-Rh110MP (C-terminal domain of human ISG15) (UbiQ Bio B.V., product code UbiQ-309).
7. 100% DMSO.
8. Black 384-well assay plate, flat bottom, low flange (Corning, REF 3575).
9. PHERAstar plate reader equipped with suitable optic modules (BMG Labtech).
10. 10× assay buffer: 500-mM HEPES pH 7.5, 1500-mM NaCl, 5-mM TCEP.
11. Multichannel pipette.

2.4 Data Analysis

1. MARS software (version 3.20 R2).
2. Microsoft Excel (version 14.0.0 (100825)).
3. GraphPad (version 7.0a).

3 Methods

3.1 Preparing the PLpro for the Assay

1. Transform the pGEX6P-1-based plasmids designed to express the GST-tagged versions of SARS-CoV-1 and CoV-2 PLpro into *E. coli* strain BL21 (DE3) Codon Plus.
2. Grow the cells in LB broth supplemented with 100-μg/mL ampicillin at 37 °C with shaking until the optical density at 600 nm reaches 1.5.
3. Add 0.2-mM isopropyl-β-D-thiogalactopyranoside and 0.1-mM $ZnSO_4$ to induce protein expression followed by shaking overnight at 18 °C.
4. Harvest cells by centrifugation (6000 *g* for 15 min at 4°). Resuspend the cell pellets in lysis buffer.
5. In a beaker placed on an ice water bath, disrupt the cell pellets by sonication (5 min with 50% amplitude for 2 s *on* and 5 s *off*). Clear the lysate by centrifugation at 35,000 *g* for 30 min at 4 °C.

6. Meanwhile, pour 5–10 mL of Glutathione Sepharose 4B resin into the Econo-Column, and wash the beads with MilliQ water to remove the ethanol solution from the beads. Again wash the beads with the lysis buffer to equilibrate them.
7. Load the cleared lysis supernatant onto Glutathione Sepharose 4B resin, placed in the Econo-Column, and allow the supernatant to pass slowly. Wash the proteins bound to the resin with lysis buffer.
8. Elute the GST-tagged proteins in lysis buffer supplemented with 30-mM reduced glutathione pH 8.0.
9. Concentrate the fusion proteins in Amicon Ultra-15 centrifugal filters (10 kDa, MWCO) until 10 mL.
10. Add GST–PreScission protease at 1:1000 (w/w) to the protein samples and incubate overnight at 4 °C to cleave the GST tag.
11. Inject 10 mL of the target protein with cleaved GST tag and PreScission protease into a desalting column (HiPrep 26/10 desalting) to remove the glutathione.
12. Passage the fractions collected from the desalting column through fresh Glutathione Sepharose 4B resin to remove the cleaved GST and PreScission protease.
13. Concentrate the cleaved proteins in centrifugal filters 10 kDa (Amicon Ultra-15) until 10 mL, and further purify using Superdex-200 pg 26/60 size-exclusion columns with equilibration buffer.
14. After purification, concentrate the proteins to 5–10 mg/mL, using Amicon Ultra-15 or Ultra-4, aliquot for single use, and snap freeze in liquid nitrogen (*see* **Note 1**).

3.2 Gel-Based Assay Using Activity-Based Probes (ABPs)

3.2.1 Refolding and Stock Solution Preparation for the ABPs

All UbiQ probes are supplied as 50-μg lyophilized powder in tubes.

1. Dissolve the CTD mISG15-VPS and CTD hISG15-VPS probes in 2-μL DMSO. Add this solution to 91-μL MilliQ water followed by the addition of 2 μL of 2.5-M sodium acetate pH 4.5 (final concentration of 50-mM sodium acetate). Buffer these stock solutions with assay buffer by adding 5 μL of 1-M HEPES pH 7.5, resulting in final probe concentrations of 52 μM (0.5 mg/mL) and 54 μM (0.5 mg/mL), respectively, and a final pH of pH 7.5 (*see* **Note 2**).
2. Dissolve the Ub-VPS probe in 2-μL DMSO, and add this stock to 93-μL MilliQ water followed by the addition of 5 μL of 1-M HEPES pH 7.5 to generate a final probe concentration of 50 μM (0.5 mg/mL) (*see* **Note 3**).

3.2.2 Setting Up the Assay

Perform the entire experiment at 4 °C (*see* **Note 4**).

1. Dilute all protein samples in 1× assay buffer, and incubate the CTD mISG15-VPS probe stock (prepared in previous section) and proteins at 4 °C before starting the experiment.
2. Spin both the probe and PLpros at high speed on a micro-centrifuge to remove any precipitate.
3. Keep the final reaction volume at 120 μL in 1× assay buffer. Maintain the PLpro and probe concentration at 1 μM and 3 μM, respectively (*see* **Note 5**). Add the PLpro first, and start the reaction by adding the CTD mISG15-VPS probe stock (*see* **Note 6**).
4. Collect samples (15 μL) at 2, 5, 10, 30, 60, and 120 s, respectively (*see* **Note 7**), and add to 1.5-mL micro-centrifuge tubes pre-dispensed with 5-μL 4× SDS reducing loading buffer to quench the reaction.
5. Prepare a negative control by incubating a sample with only 1-μM PLpro for 120 s.
6. Spin the above loading dye-treated samples, including negative control, and load 5–7 uL of the supernatant on a 4–20% gradient gel. Stain the gel using Coomassie blue (*see* **Note 8**) (Fig. 2).
7. Repeat the above **steps (1–6)** for CTD hISG15-VPS and Ub-VS probes, respectively.

3.3 Kinetic Assay Using Fluorescent Substrates

3.3.1 Preparation of Stock Solutions of Fluorescent Substrates

1. Reconstitute UbiQ-126, UbiQ-127, UbiQ-309, and UbiQ-012 as discussed before for ABPs. In short, dissolve the reagent in DMSO, add to MilliQ water, and buffer with 50-mM sodium acetate pH 4.5. Finally, buffer this stock using 1-M HEPES pH 7.5, affording a final stock of 50 μM (*see* **Notes 2** and **3**).
2. Dilute UbiQ-126 and UbiQ-012 to 20-μM stocks, and dilute UbiQ-127 and UbiQ-309 to 10 μM in 1× assay buffer (2× working stock).
3. Directly dilute Ub-AMC and ISG15-AMC (R&D systems) in 1× assay buffer to a final concentration of 20 μM and 10 μM (2× working stocks), respectively.

3.3.2 Preparation of Kinetic Assay with the Fluorescent Probes

Perform all kinetic assays at room temperature, i.e., 25 °C (*see* **Note 9**).

1. Perform the assay in a black 384-well format assay plate (*see* **Note 10**), keeping the final reaction volume at 20 μL.
2. Dilute PLpro to 40 nM (2× working stocks) in 1× assay buffer.

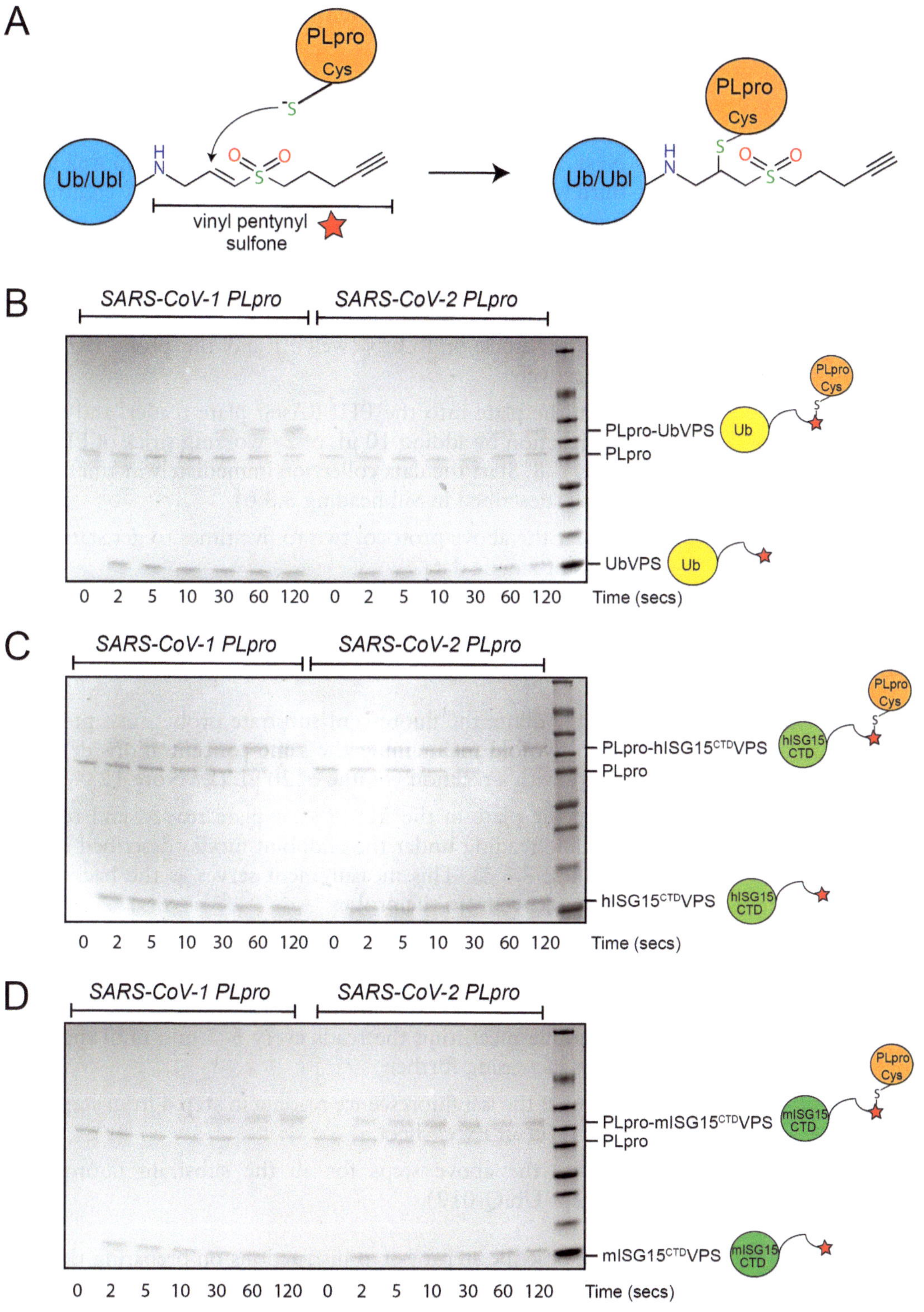

Fig. 2 Summary of results for gel-based assays using activity-based probes (ABPs). (**a**) Chemical structure of vinyl pentynyl sulfone (VPS) and its mode of action when it comes in contact with PLpro active cysteine. (**b**). Gel-based assay to show qualitative labeling of PLpro from SARS-CoV-1 (left part of gel) and SARS-CoV-2 (right part of gel) using Ub-VPS. Similar labeling uses the C-terminal domain of human ISG15-VPS (**c**) and the *C*-terminal domain of murine ISG15-VPS (**d**)

3. In the first well of the assay plate, add 20 μL of 2× working stock of fluorescent probe (Ub-AMC). Using a multichannel pipette, add 10 μL of 1× assay buffer to the subsequent seven wells. Then using a single-channel pipette, transfer 10 μL of the probe from the first well to the second well and mix properly. Next, transfer 10 μL of the probe from the second well to the third well and mix. Repeat this procedure until the eighth well, such that each well has 10 μL of solution. Hence the substrate is serially diluted twofold to achieve eight different concentrations (*see* **Note 11**).
4. Prepare two 20-μL samples as negative controls, including 1× working stock of PLpro (well 9) and the probe (well 10), respectively.
5. Place the plate into the PHERAstar plate reader, and initiate the reaction by adding 10 μL of 2× working stock of PLpro to wells 1–8. Start the data collection immediately in kinetic plate mode (described in Subheading 3.3.6).
6. Repeat the above protocol two to five times to get statistically relevant kinetic data.
7. Repeat the above **steps (1–6)** to generate data for Ub-Rh110MP, ISG15-AMC, CTD mISG15-Rh110MP, and CTD hISG15-Rh110MP, respectively.

3.3.3 Preparation of the Standard Curve

1. Serially dilute the fluorescent substrate probes (except UbiQ-012) twofold maintaining the same concentrations described above with a reaction volume of 20 μL (*see* **Note 12**).
2. Place the plate in the PHERAstar plate reader, and obtain a baseline reading under the endpoint mode (described in Subheading 3.3.5). This measurement serves as the background fluorescence from the probes.
3. Add a minimal volume (preferably 0.5–1 μL) of concentrated PLpro (CoV-1-PLpro) to those eight wells. After 5 min of incubation, take a reading in the plate mode (*see* **Note 13**).
4. Continue measuring the reads every 5–7 min until the signal stops increasing further.
5. Subtract the last fluorescence reading in **step 4** from **step 2** and save it on an Excel sheet.
6. Repeat the above steps for all the substrate fluorophores (except UbiQ-012).

3.3.4 Setting Up the PHERAstar Plate Reader

Here we describe step-by-step instructions on preparing the plate reader to measure endpoint and plate mode readings for standard curve and enzyme kinetics, respectively.

1. After opening the PHERAstar application, look for the "manage protocol" tab and click on the "New" tab.
2. Under "measurement method," choose "Fluorescence Intensity" for UbiQ-126, UbiQ-127, UbiQ-309, Ub-AMC, and ISG15-AMC or "Fluorescence Polarimetry" for UbiQ-012.
3. Next, under "reading mode," choose either "Endpoint" for making the standard curve or "Plate mode" for kinetic analysis. When "Endpoint" mode is executed, only one read is collected. But in the "Plate mode," continuous reads are collected depending on the "no. of cycles" entered (explained in Subheading 3.3.6).

3.3.5 Using Endpoint Mode (for Generating Standard Curve)

1. Under the Basic parameter tab, for the "Microplate" drop-down menu, select Corning 384-Black low flange (*see* **Note 14**).
2. Next, choose the correct optic module based on the probe used. For AMC, Rhodamine110, and TAMRA, select FI 355460, FI 485520, and FP 540-20 590-20, respectively (*see* **Note 15**).
3. Choose "top" for the optic option. Keep all other settings as default.
4. Under the layout tab, choose the wells for which data needs to be collected. Here use the option to label certain wells as a positive or negative control.
5. Click on "Start measurement."
6. On the new page, select "optical focal and gain adjustment." For gain adjustment, set the target value as 60–80% and click "start adjustment" (*see* **Note 16**).
7. Click "measure one plate" to take the reading (*see* **Note 17**).

3.3.6 Using Kinetic Plate Mode (for Kinetic Assay)

1. The parameters that appear under this option remain almost the same as endpoint mode. Thus, choose similar options as discussed above. There is an additional window called "Kinetic Window 1," which gives the option for entering "no. of cycles" and "cycle time."
2. Enter 1000 for a number of cycles and 9 s as cycle time (*see* **Note 18**).
3. As discussed before, under the "layout window," choose the samples for which the data will be collected and click "start measurement."
4. On the new page, choose appropriate values of focal height and gain adjustment (*see* **Note 19**).
5. Click "Measure one plate" to collect data (*see* **Note 17**).

3.4 Data Analysis

3.4.1 Generation of Standard Graph

1. Open the Excel file generated in Subheading 3.3.3.
2. Plot a curve using the chart option keeping the concentration (μM) on *X*-axis and fluorescence reading on *Y*-axis.
3. Under the "format trendline" tab, select "display equation" and "R-squared value" on the chart (*see* **Note 20**).
4. Similarly, generate equations and R-squared values for all the probes.

3.4.2 Analysis of the Raw Kinetic Data

1. Copy the data generated in the kinetic mode from the PHER-Astar page and paste it into the Excel sheet. These data includes the fluorescence reading and time intervals for each substrate concentration used.
2. Normalize the data such that at time point = 0, there was no fluorescence for each substrate concentration (*see* **Note 21**).
3. Then using the equations generated in the "Standard curve" section above, convert the fluorescence reading to concentration (μM) for each substrate concentration.
4. Based on the data generated in **step 3**, plot a graph using *Concentration* on the *Y*-axis and *Time* on the *X*-axis.
5. Calculate the slopes for each substrate concentration used from **step 4**, and save this information on an Excel sheet. These values serve as the rate of reactions or enzyme activities for each fluorophore substrate (*see* **Note 22**).

3.4.3 Generation of Kinetic Parameters

1. Open GraphPad Prism and under "Use tutorial data," choose the last option for more tutorials.
2. Under the drop-down menu, choose "Michaelis-Menten" equation and click "Create."
3. On the new page, remove the prefilled data under "Substrate" and "Enzyme activity" tab; enter the values for the used fluorescent substrate concentration (Subheading 3.3.2) and rate of reaction generated from the previous section ("Polishing Raw kinetic data").
4. Click on "Analyze" and choose *Non-linear regression (curve fit)*.
5. In the next page, choose "Michaelis-Menten" for fitting the data. By default, the fitting method used is *least squares (ordinary) fit*.
6. Calculate the kinetic parameters like V_{max} and K_{m}, including the Michaelis–Menten graph. Calculate K_{cat} based on the standard formula: $K_{cat} = V_{max}/E$ where E is the PLpro concentration (Figs. 3 and 4) (*see* **Note 23**).

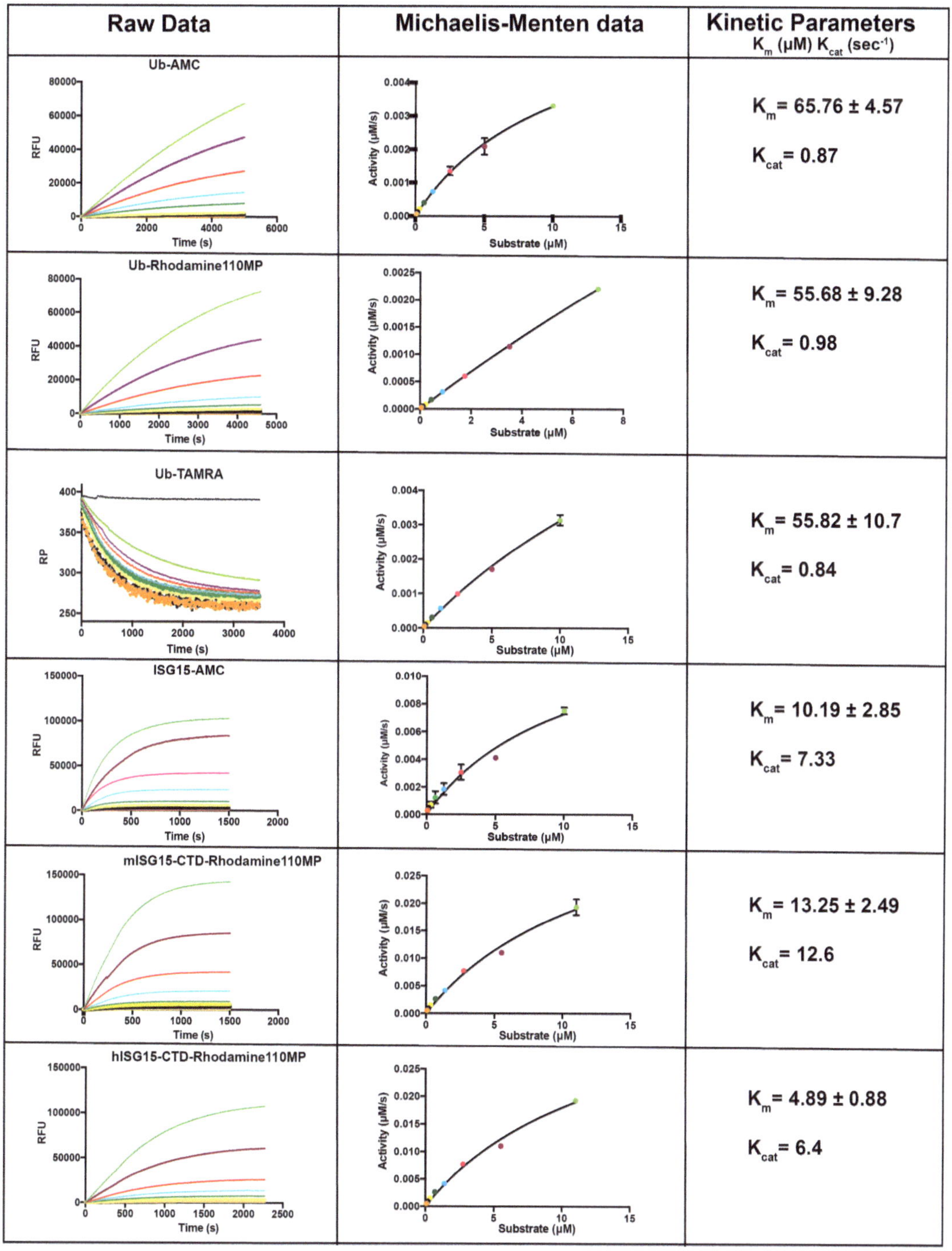

Fig. 3 Summary of results showing Michaelis–Menten kinetic parameters for SARS-CoV-1 PLpro using various fluorescent-labeled substrates. Shown here is the plot between relative fluorescence unit (RFU) and relative polarimetry (RP) with time in seconds, termed as "raw data." Each color represents substrate concentration. Using the raw data, the Michaelis–Menten graph was plotted to calculate kinetic parameters using GraphPad software

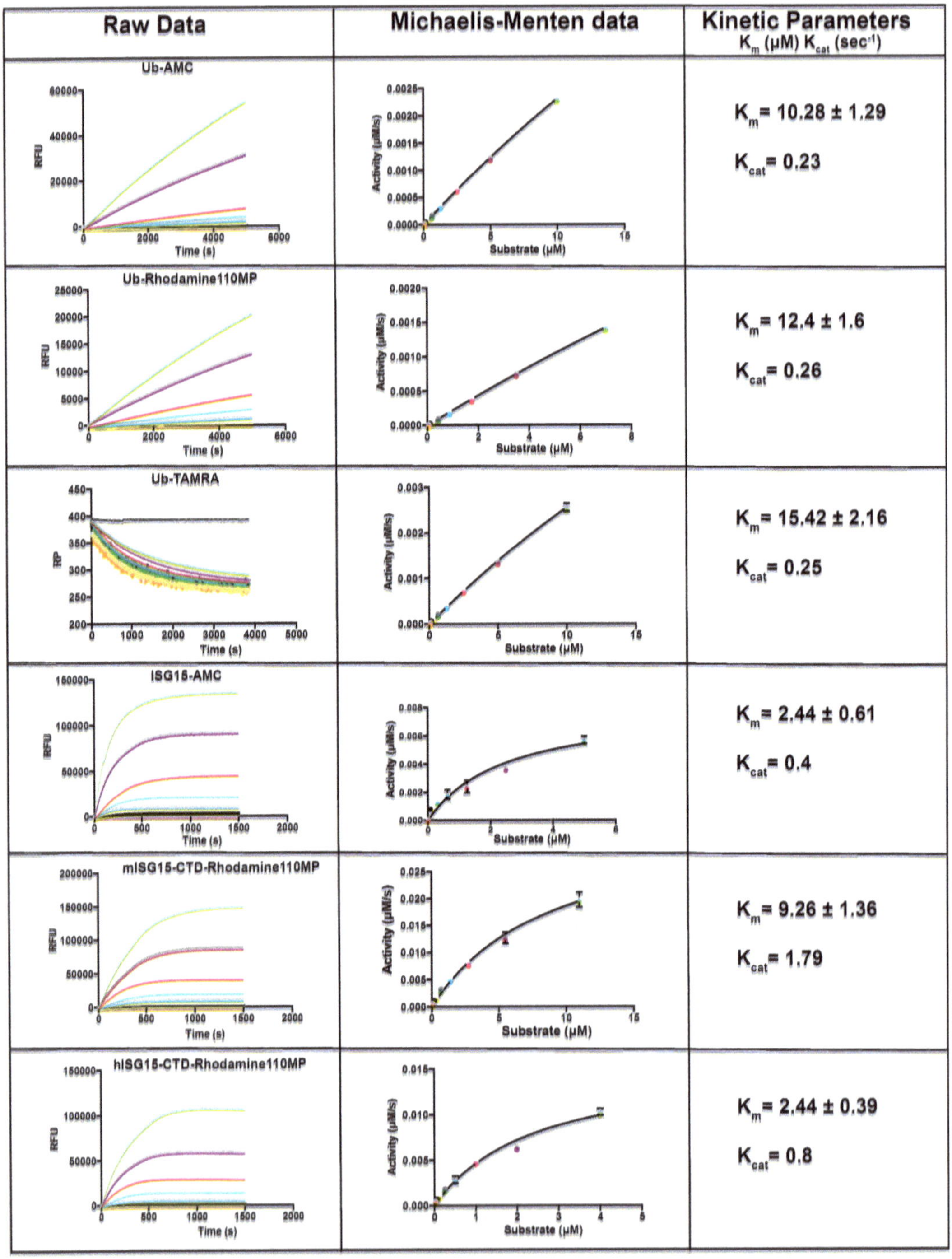

Fig. 4 Summary of results showing Michaelis–Menten kinetic parameters for SARS-CoV-2 PLpro using various fluorescent-labeled substrates. Shown here is the plot between relative fluorescence unit (RFU) and relative polarimetry (RP) with time in seconds, termed as "raw data." Each color represents substrate concentration. Using the raw data, the Michaelis–Menten graph was plotted to calculate kinetic parameters using GraphPad software

4 Notes

1. When using sonication to isolate proteins, optimizing the sonication cycles/amplitude is recommended since over-sonication can irreversibly oxidize the active cysteine of PLpros and affect the ABPs-mediated conjugation. Moreover, the use of untagged PLpro from S-CoV-1 and S-CoV-2 is advised to maintain consistency when comparing activity.
2. It is crucial to reconstitute the probes in the lyophilized powder form according to the manufacturer's protocol, especially the order of solvent added to make the stock solutions. If this reconstitution is not done properly, the probes may precipitate.
3. If the probes precipitate (e.g., due to a high concentration), spin the solution at high speed and collect the supernatant. Next, run a SDS-PAGE gel with a standard protein (such as ubiquitin) to estimate the concentration of the remaining probe solution.
4. The reaction can be performed on ice, but it is recommended to do the assay in a cold room to avoid temperature fluctuations. The assay can be performed at room temperature, but shorter time points are advised (*see* **Note 7**).
5. It is recommended to keep the molar ratio between PLpro and probe at 1:3. If there is no limitation of the probe, a higher ratio like 1:5 can be tested.
6. After the addition of the probe to initiate the reaction, it is advised to flick the tube and then start the timer to ensure the probe is evenly mixed in the micro-centrifuge tube.
7. These time points were chosen based on how fast all the PLpro protein was modified by the probe. If one needs to go for shorter time points for sensitive experiments like kinetic analysis, it is recommended to use a quench-flow or stop-flow instrument, which is an automated instrument used to quench enzymatic reactions at a millisecond timescale.
8. The gel can be stained using SYPRO Ruby protein gel stain for a more sensitive reading depending on the need.
9. The kinetic assays can be performed at higher temperatures (e.g., 30–37 °C). However, this will increase the evaporation rate, which might affect the data if the incubation time is long. Maintaining the proteins and performing the experiment at room temperature makes it straightforward to replicate. Hence it is recommended to carry out the assay at room temperature.
10. *Important*: Please use a black deep well and not a white assay plate for doing the experiment as it has low background

fluorescence, less light scattering, and reduced signal crosstalk. It is also recommended to use alternate wells of the plate for the assay as it prevents accidental spillage of the probe and also works well with a typical 8 or 12 multichannel pipette (for 384-well plate).

11. Care should be taken that the probes are mixed thoroughly before being dispensed to the next well, and air bubbles are avoided.
12. Another way to plot a standard curve is to use free fluorophore, i.e., not attached to a protein. Specific dilutions of the fluorophore can be prepared in the 1× assay buffer and readings taken on a plate reader. The data can be processed as explained in the *Data Analysis* section.
13. It is necessary to keep the focal height for all experiments identical, including standard curve calculation, for consistent data analysis and comparison. Therefore, adding the minimum volume of the concentrated enzyme would not change the height considerably.
14. Choose the exact plate if possible because all the focal and gain adjustment calculations depend on this information.
15. Under the drop-down menu, only those filter modules appear which are already placed in the machine. If you do not see the correct optic module, you have to physically put it into the machine.
16. *Important*: The focal height and gain adjustment values are crucial and should be kept constant as much as possible to get relevant and consistent data. These values depend on the fluorescent probe type, the volume of the sample, the selected optic module, and the plate type. Therefore, keeping a consistent set of values for focal height and gain for a specific fluorescent probe is recommended.
17. Before clicking on "measure one plate," check one last time on the top right of the page whether the right optic module is selected.
18. The "no. of cycles" is a crucial parameter and decides how long the data collection will run for the kinetic assay. One can try to optimize the reaction time, by varying the number of cycles, and click on "check timing" tab on the same page to see the exact duration. If the number of probes is limited and you want to save time on optimization, it is recommended to run the collection for a longer time period, and once the signal reaches saturation, stop the run. The data will be saved until that time. The "cycle time" takes on a default value and depends on how many wells you select on the "layout window."

19. For a specific probe, the values for focal height and gain adjustment should be kept the same for both "Endpoint mode" and "Plate mode." This detail is very important since only then the standard graph generated from "Endpoint mode" will correlate with the kinetic data from "Plate mode."
20. An R-square value of close to 1 is expected if the fluorescence readings are in the linear range.
21. Data normalization can also be done using the *division method*, but this leads to loss of signal, and the data generated from the standard curve cannot be correlated to it. Hence it is recommended to use the *subtraction method*.
22. It is recommended to use the linear range of the graph to calculate the slope. All calculations of the kinetic parameters depend on this.
23. If the graph plotted between the rate of reaction and substrate concentration does not follow Michaelis–Menten equation and instead forms a straight line, then the slope of the graph gives K_{cat}/K_m value. It's difficult to get the K_m and K_{cat} individually.

Acknowledgments

This work was supported by NIH R01 GM115568, R01 GM128731, and CPRIT RR200030 (S.K.O.). The Drag laboratory is supported by "TEAM/2017-4/32" project, which is carried out within the TEAM program of the Foundation for Polish Science, co-financed by the European Union under the European Regional Development Fund. The content of this study is solely the responsibility of the authors and does not necessarily represent the official views of the NIH.

Conflict of Interest F.E.O. and A.N.N. declare competing financial interests as co-founders and shareholders of UbiQ Bio B.V. All other authors have no competing interests.

References

1. Huang C, Wang Y, Li X, Ren L, Zhao J, Hu Y et al (2020) Clinical features of patients infected with 2019 novel coronavirus in Wuhan, China. Lancet 395(10223):497–506
2. Wang C, Horby PW, Hayden FG, Gao GF (2020) A novel coronavirus outbreak of global health concern. Lancet 395(10223):470–473
3. Patchett S, Lv Z, Rut W, Békés M, Drag M, Olsen SK et al (2021) A molecular sensor determines the ubiquitin substrate specificity of SARS-CoV-2 papain-like protease. Cell Rep 36(13):109754
4. Li C, He Q, Qian H, Liu J (2021) Overview of the pathogenesis of COVID-19. Exp Ther Med 22(3):1–10
5. Lu R, Zhao X, Li J, Niu P, Yang B, Wu H et al (2020) Genomic characterisation and epidemiology of 2019 novel coronavirus: implications for virus origins and receptor binding. Lancet 395(10224):565–574
6. Zhou P, Yang X-L, Wang X-G, Hu B, Zhang L, Zhang W et al (2020) A pneumonia outbreak associated with a new coronavirus of probable bat origin. Nature 579(7798):270–273

7. Báez-Santos YM, John SES, Mesecar AD (2015) The SARS-coronavirus papain-like protease: structure, function and inhibition by designed antiviral compounds. Antivir Res 115:21–38
8. Harcourt BH, Jukneliene D, Kanjanahaluethai A, Bechill J, Severson KM, Smith CM et al (2004) Identification of severe acute respiratory syndrome coronavirus replicase products and characterization of papain-like protease activity. J Virol 78(24): 13600–13612
9. Ziebuhr J (2005) The coronavirus replicase. In: Coronavirus replication and reverse genetics, pp 57–94
10. Barretto N, Jukneliene D, Ratia K, Chen Z, Mesecar AD, Baker SC (2005) The papain-like protease of severe acute respiratory syndrome coronavirus has deubiquitinating activity. J Virol 79(24):15189–15198
11. Barretto N, Jukneliene D, Ratia K, Chen Z, Mesecar AD, Baker SC (2006) Deubiquitinating activity of the SARS-CoV papain-like protease. Springer, The Nidoviruses, pp 37–41
12. Lindner HA, Lytvyn V, Qi H, Lachance P, Ziomek E, Ménard R (2007) Selectivity in ISG15 and ubiquitin recognition by the SARS coronavirus papain-like protease. Arch Biochem Biophys 466(1):8–14
13. Clementz MA, Chen Z, Banach BS, Wang Y, Sun L, Ratia K et al (2010) Deubiquitinating and interferon antagonism activities of coronavirus papain-like proteases. J Virol 84(9): 4619–4629
14. Devaraj SG, Wang N, Chen Z, Chen Z, Tseng M, Barretto N et al (2007) Regulation of IRF-3-dependent innate immunity by the papain-like protease domain of the severe acute respiratory syndrome coronavirus. J Biol Chem 282(44):32208–32221
15. Frieman M, Ratia K, Johnston RE, Mesecar AD, Baric RS (2009) Severe acute respiratory syndrome coronavirus papain-like protease ubiquitin-like domain and catalytic domain regulate antagonism of IRF3 and NF-κB signaling. J Virol 83(13):6689–6705
16. Fu Z, Huang B, Tang J, Liu S, Liu M, Ye Y et al (2021) The complex structure of GRL0617 and SARS-CoV-2 PLpro reveals a hot spot for antiviral drug discovery. Nat Commun 12(1): 1–12
17. Osipiuk J, Azizi S-A, Dvorkin S, Endres M, Jedrzejczak R, Jones KA et al (2021) Structure of papain-like protease from SARS-CoV-2 and its complexes with non-covalent inhibitors. Nat Commun 12(1):1–9
18. Ekkebus R, Flierman D, Geurink PP, Ovaa H (2014) Catching a DUB in the act: novel ubiquitin-based active site directed probes. Curr Opin Chem Biol 23:63–70
19. van Tilburg GB, Elhebieshy AF, Ovaa H (2016) Synthetic and semi-synthetic strategies to study ubiquitin signaling. Curr Opin Struct Biol 38:92–101
20. Xin B-T, Gan J, Fernandez DJ, Knobeloch K-P, Geurink PP, Ovaa H (2019) Total chemical synthesis of murine ISG15 and an activity-based probe with physiological binding properties. Org Biomol Chem 17(48):10148–10152
21. Hewings DS, Flygare JA, Bogyo M, Wertz IE (2017) Activity-based probes for the ubiquitin conjugation–deconjugation machinery: new chemistries, new tools, and new insights. FEBS J 284(10):1555–1576
22. Daczkowski CM, Dzimianski JV, Clasman JR, Goodwin O, Mesecar AD, Pegan SD (2017) Structural insights into the interaction of coronavirus papain-like proteases and interferon-stimulated gene product 15 from different species. J Mol Biol 429(11):1661–1683
23. Klemm T, Ebert G, Calleja DJ, Allison CC, Richardson LW, Bernardini JP et al (2020) Mechanism and inhibition of the papain-like protease, PLpro, of SARS-CoV-2. EMBO J 39(18):e106275
24. Rut W, Lv Z, Zmudzinski M, Patchett S, Nayak D, Snipas SJ et al (2020) Activity profiling and crystal structures of inhibitor-bound SARS-CoV-2 papain-like protease: a framework for anti-COVID-19 drug design. Sci Adv 6(42):eabd4596
25. Shin D, Mukherjee R, Grewe D, Bojkova D, Baek K, Bhattacharya A et al (2020) Papain-like protease regulates SARS-CoV-2 viral spread and innate immunity. Nature 587(7835): 657–662
26. Hewings DS, Heideker J, Ma TP, AhYoung AP, El Oualid F, Amore A et al (2018) Reactive-site-centric chemoproteomics identifies a distinct class of deubiquitinase enzymes. Nat Commun 9(1):1–17
27. Terentyeva TG, Van Rossom W, Van der Auweraer M, Blank K, Hofkens J (2011) Morpholinecarbonyl-Rhodamine 110 based substrates for the determination of protease activity with accurate kinetic parameters. Bioconjug Chem 22(10):1932–1938
28. Geurink PP, El Farid Oualid AJ, Hameed DS, Ovaa H (2012) A general chemical ligation approach towards isopeptide-linked ubiquitin and ubiquitin-like assay reagents. Chembiochem 13(2):293

Chapter 12

Next-Generation Phage Display to Identify Peptide Ligands of Deubiquitinases

Anastasios Spiliotopoulos, Sigrun K. Maurer, Maria T. Tsoumpeli, Juan A. F. Bonfante, Jonathan P. Owen, Kevin C. Gough, and Ingrid Dreveny

Abstract

Phage display (PD) is a powerful method and has been extensively used to generate monoclonal antibodies and identify epitopes, mimotopes, and protein interactions. More recently, the combination of next-generation sequencing (NGS) with PD (NGPD) has revolutionized the capabilities of the method by creating large data sets of sequences from affinity selection-based approaches (biopanning) otherwise challenging to obtain. NGPD can monitor motif enrichment, allow tracking of the selection process over consecutive rounds, and highlight unspecific binders. To tackle the wealth of data obtained, bioinformatics tools have been developed that allow for identifying specific binding sequences (binders) that can then be validated. Here, we provide a detailed account of the use of NGPD experiments to identify ubiquitin-specific protease peptide ligands.

Key words Phage display, Next-generation sequencing, Biopanning, Bioinformatical analysis, Deubiquitinase, Peptides

1 Introduction

The human genome encodes 56 ubiquitin-specific proteases (USPs), the largest family of deubiquitinating enzymes (DUBs) that are crucial regulators of ubiquitin-mediated proteasomal degradation and virtually all cellular pathways, including processes as diverse as signaling, immune responses to pathogens, DNA damage repair, receptor trafficking, and gene regulation [1]. Critical to USP function is a network of protein–protein interactions that determines USP specificity, regulation of the catalytic activity, and recruitment to target substrates. Large-scale proteomics and more focused studies have identified numerously confirmed and putative USP binding partners [2]. Linear and short-sequence motif recognition is emerging as a theme for transient interactions central to

Julie A. Maupin-Furlow and Mariola J. Edelmann (eds.), *Deubiquitinases: Methods and Protocols*,
Methods in Molecular Biology, vol. 2591, https://doi.org/10.1007/978-1-0716-2803-4_12,

these signaling pathways [3]. For example, USP7's non-catalytic regions, including the TRAF-like and UBL (ubiquitin-like) domains, recognize binding motifs in p53, MDM2, viral proteins [4], and the epigenetic regulators UHRF1 [5] and DNMT1 [6]. Phage display (PD) in combination with next-generation sequencing (NGS) is a powerful method, here termed NGPD, that has significantly eased the process of identifying ligands to any given target [7] and minimized the need for repetitive selection rounds [8]. We adapted an NGPD protocol followed by the evaluation of obtained sequences using *z*-scores to identify peptide-binding motifs for USP11 [9].

Over the last decade, a significant advance has been made in developing peptide phage display libraries due to their versatility for various purposes such as antibody epitope mapping, mimotope discovery, and protein hot spot identification. Peptides have been displayed mainly on the major coat protein (pVIII) or the pIII protein and include standard linear and cyclic libraries of various lengths [10, 11] as well as more complex ones such as bicyclic [12] and noncanonical amino acid libraries [13] for drug discovery.

Identifying linear antibody epitopes (e.g., for Western blot antibody reagents) is a relatively straightforward approach, including recent examples such as the simultaneous epitope mapping of 200 COVID antibodies in a single run using a linear commercial library and NGS [14]. The same approach has also been applied to identify more complex, structural/discontinuous epitopes for antibodies or protein–protein interactions, with peptide libraries being used for mimotope discovery, ultimately leading to hot spot identification and/or the discovery of binding partners [9, 15–17]. Recent computational advances will further facilitate this approach [18].

Here, we provide a detailed protocol for discovering peptide binders (schematically depicted in Fig. 1), which will be applicable to USPs and other targets. The chapter covers the production of His-tagged target proteins for biopanning, phage display peptide library construction and propagation for pVIII display, biopanning with His-tagged target protein, and screening phage-peptide-binding events by phage ELISA. Moreover, this chapter overviews screening for peptide binders by next-generation sequencing, associated data analysis methods, and initial evaluation using synthetic peptides in ELISA assays.

2 Materials

2.1 General

All solutions and buffers should be autoclaved or filter-sterilized as appropriate. Reagents should be analytical reagent grade. Suitable equipment to use includes shaking incubator, sonicator, FPLC, electrophoresis equipment, Bio-Rad Gene Pulser Xcell

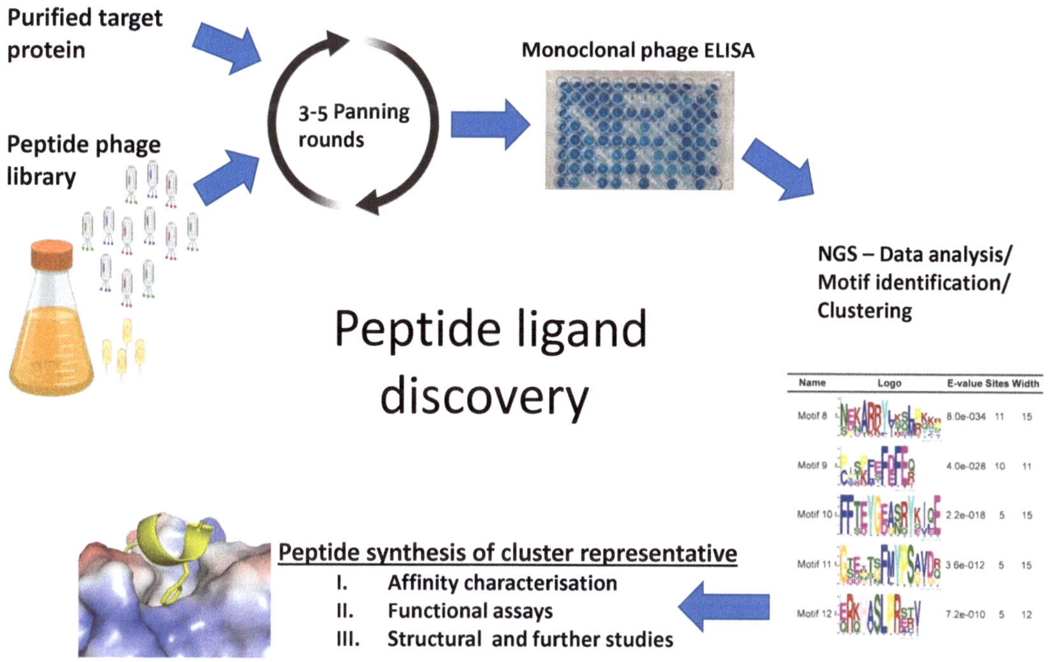

Fig. 1 Overview of the NGPD method. A schematic overview of the technique detailing the different steps involved in discovering peptide binders to a given target by combining phage display biopanning with next-generation sequencing

electroporator, thermal cyclers, Qubit 2.0 Fluorometer (Invitrogen, UK), NanoDrop spectrophotometer, Ion Torrent sequencing machine, or service provider.

2.2 Recombinant Protein Expression and Purification

1. Overexpression plasmid of His-tagged target protein (e.g., in a pET vector).
2. Heat-shock competent cells BL21 (DE3) Codon Plus-RP strain (*E. coli* B F$^-$ *ompT hsdS*(r_B^- m_B^-) *dcm*$^+$ Tet^r *gal endA* Hte [*argU proL* Cam^r]).
3. Antibiotics as required for the vector and *E. coli* strain (e.g., kanamycin 50-µg/µL and chloramphenicol 35-µg/µL working concentrations).
4. Lysogeny broth (LB) agar (per liter): 20-g agar, 10-g NaCl, 10-g tryptone, 5-g yeast extract.
5. LB microbial growth media (per liter): 10-g NaCl, 10-g tryptone, 5-g yeast extract.
6. 2xYT microbial growth media (per liter): 5-g NaCl, 16-g tryptone, 10-g yeast extract.
7. Isopropyl ß-D-1-thiogalactopyranoside (IPTG).
8. Bacterial Protein Extraction Reagent (BPER, Thermo Scientific).

9. Broad range unstained protein ladder.
10. SDS-PAGE gels.
11. Reducing SDS-PAGE sample loading dye 3×: 0.135-M Tris–Cl pH 6.8, 30% (v/v) glycerol, 3% (w/v) SDS, 0.03% bromophenol blue, 0.15-M DTT.
12. Tris–glycine SDS-PAGE running buffer 1×: 25-mM Tris base, 192-mM glycine, 0.1% SDS, pH 8.3.
13. Coomassie staining solution: 0.25% (w/v) Brilliant Blue R-250, 25% (v/v) isopropanol, 25% (v/v) glacial acetic acid.
14. Coomassie de-staining solution: 15% (v/v) isopropanol, 15% (v/v) glacial acetic acid.
15. Immobilized metal affinity chromatography (IMAC) buffer A, filtered and degassed: 50-mM Tris–Cl, pH 8, 300-mM NaCl, 20-mM imidazole, 10% glycerol.
16. IMAC buffer B, degassed: 50-mM Tris–Cl, pH 8, 300-mM NaCl, 500-mM imidazole.
17. Size exclusion chromatography (SEC) buffer, filtered and degassed: 50-mM Tris–Cl, pH 7.5, 150-mM NaCl, 1% (v/v) glycerol.
18. FPLC protein purification system.
19. HiTrap chelating HP column 5 mL (Cytiva) charged with nickel ions or similar.
20. Superdex S75 or S200 16/60 prep grade (PG) chromatography column (Cytiva).
21. Centrifugal concentrators at suitable molecular weight cutoff.

2.3 Phagemid Library Construction, Panning, and Clone Screening

1. Phagemid template (in this protocol, pc89 is used) [19].
2. Primers suitable for site-directed mutagenesis (SDM) (forward, 5′-GCGGAGGAGCGCCCAAT-3′; reverse, 5′-TTCCTCGC TCACTGACTC-3′), phosphorylated at their 5′ ends.
3. Suitable sequencing primer: 5′-CTTTATGCTTCCGGCTCG TATG-3′.
4. UltraPure DNase/RNase-Free Distilled Water.
5. Q5 High-Fidelity DNA Polymerase with 5× concentrated buffer.
6. Q5 GC enhancer reagent 5× concentrated.
7. Deoxynucleotide (dNTP) solution mix (10 mM).
8. DEPC-treated, molecular-grade water.
9. Agarose.
10. Deep-well plates.
11. Nancy 520 or similar DNA gel stain.

12. 6× DNA loading dye.
13. TAE buffer (agarose gel running buffer): 2-M Tris–HCl, 1-M acetic acid, 0.05-M EDTA, pH 8 (50×).
14. 1-kb DNA molecular weight marker.
15. Gel and PCR Clean-up kit (Macherey-Nagel™ NucleoSpin™ Gel and PCR).
16. DpnI.
17. T4 DNA ligase (4000 U).
18. TG1 *Escherichia coli*: F′ traD36 proAB lacIqZ ΔM15] supE thi-1 Δ(lac-proAB) Δ(mcrB-hsdSM)5(rK - mK -) (Lucigen).
19. Gene Pulser cuvette (0.1 cm).
20. SOC outgrowth medium: 2% (w/v) tryptone, 0.5% (w/v) yeast extract, 10-mM NaCl, 2.5-mM KCl, 10-mM $MgCl_2$, 10-mM $MgSO_4$, and 20-mM glucose.
21. 2YT: 16% (w/v) tryptone, 10% (w/v) yeast extract, 5% (w/v) NaCl.
22. 2YT agar: 15% (w/v) tryptone, 10% (w/v) yeast extract, 5% (w/v) NaCl, 15-g/L agar.
23. Ampicillin: 150 mg/mL.
24. DNA isolation kit.
25. Phagemid (in this protocol pc89_SapI_free).
26. NNK-randomized primers: pc89_forward 5′-GATTGCTCTT CGGATCCCGCAAAAGCGGCCTTTG −3′and pc89_reverse 5′-GGTAGCTCTTCGATC(MNNX16)GAATTCA CCCTCAGCAGCGA)-3′.
27. Qubit™ dsDNA BR Assay Kit (Thermo Fisher Scientific).
28. Bioassay dish (L × W × H: 245 mm × 245 mm × 25 mm).
29. His-tagged target protein and His-tagged non-related control protein.
30. Blocking solution: 6 or 3% (w/v) skimmed milk powder in PBS (mPBS).
31. Phage-peptide library.
32. Dynabeads™ His-Tag Isolation & Pulldown.
33. Magnetic rack (DynaMag™-2 Magnet).
34. Phage elution buffer: 100-mM triethylamine.
35. Neutralizing buffer: 1-M Tris, pH 7.4.
36. Helper phage: MK13KO7 used at 2×10^{12} pfu/mL.
37. PEG-8000 20% (w/v)/2.5-M NaCl: 200-g PEG-8000 and 146.1-g NaCl per liter.

38. D-(+)-Glucose solution: 25% (w/v) D-(+)-glucose filter-sterilized.
39. QIAprep Spin M13 Kit.
40. Agencourt AMPure XP kit.
41. P8-Forward1NGS primer: 5′-**GTAATCCTTGTGGTATCG**GATGCTGTCTTTCGCTGC-3′.
42. P8 Reverse1NGS primer: 5′-CTAGAACATTTCACTTACGGTTTTCCCAGTCACG-3′.
43. Akey-BC-linker1 (Forward): 5-CCATCTCATCCCTGCGTGTCTCCGACTCAGxxxxxxxxxx**GTAATCCTTGTGGTATCG**-3′. Underlined is the 10-bp barcoded region, with many variants available to use. This region varies by five bases from every other primer. Codons in bold are the regions that are identical in primers in both first- and second-step PCRs.
44. P1prim-linker2 (Reverse): CCTCTCTATGGGCAGTCGGT-GAT**CTAGAACATTTCACTTAC**-3′. Codons in bold are the homologous linker regions, identical for first- and second-step PCRs.
45. Sterile 96-deep-well clear polypropylene plate.
46. Nunc MaxiSorp flat-bottom 96-well plates.
47. Anti-M13 bacteriophage antibody HRP (Sigma-Aldrich).
48. 3,3′,5,5′-Tetramethylbenzidine (TMB) blotting substrate solution.
49. 0.45-μm pore size Millipore filter.
50. Minimal media (M9) plates:
    - (a) Make sterile 5× M9 salts; mix 32-g $Na_2HPO_4.7H_2O$ [*or* 21.35-g $Na_2HPO_4.2H_2O$ *or* 15-g Na_2HPO_4 (anhydrous)], 7.5-g KH_2PO_4, 2.5-g NH_4Cl, 1.25-g NaCl, and distilled water to a total volume of 500 mL.
    - (b) Prepare sterile 20% (w/v) $MgCl_2$ solution.
    - (c) Prepare sterile 100-mg/mL thiamine hydrochloride solution (filter-sterilized).
    - (d) On the day of use, mix 1.5-g agar powder in 75-mL H_2O, autoclave, and allow to cool to 50 °C; then add 25-mL 5× M9 salts, 1.6-mL 25% glucose, 100-μL $MgCl_2$, and 50-μL thiamine hydrochloride. Pour 20 mL per 9-cm-diameter plate; allow to set.

3 Methods

3.1 Recombinant His-Tagged Target Protein Production for Panning

A one- or two-step protocol comprising of affinity and optional size exclusion chromatography depending on the purity of the sample after the first step can be used. The sample purity requirements (>90%) are not as strict as for many biophysical experiments.

3.1.1 His-Tagged Target Protein Overexpression

1. Transform 1 μL of the plasmid DNA expression construct (~100 ng/μL) into 50 μL of BL21(DE3) CP competent cells by heat shock or electroporation, add 500-μL LB broth media, and place into shaker incubator at 37 °C, 180 rpm for 1 h.
2. Plate 100 μL of the cell culture onto an LB agar plate with suitable antibiotics and incubate at 37 °C for ~16 h (static).
3. Use a single colony to inoculate 100 mL of LB broth containing chloramphenicol and the construct-specific antibiotic; grow cells at 37 °C with 180 rpm shaking overnight.
4. On the following day, inoculate 3 × 800 mL (depending on expression levels) 2xYT growth medium containing antibiotics 1:100 with the overnight culture, and continue growing at 37 °C to an OD_{600} of ~0.6. Make a note of the OD_{600} before induction and take a 1-mL sample (pre-induction sample) in a microcentrifuge tube, centrifuge for 3 min at 5000 g, and freeze small pellet at −20 °C.
5. Induce recombinant protein expression by adding IPTG to a final concentration of 0.5 mM. Lower the temperature to 18 °C and grow for about 16 h (*see* **Note 1**).
6. After this time, take a 1-mL post-induction sample in the same way as the pre-induction sample. Harvest the culture in a suitable high-volume centrifuge for 30 min at 4500 g at 4 °C and collect the pellet.
7. Resuspend the cell pellet in 25-mL LB and transfer to a 50-mL centrifugation tube, centrifuge again for 15 min at 4500 g, discard the supernatant, and make a note of the weight of the pellet. The cell pellet can be used immediately or frozen down prior to IMAC purification after checking for successful protein overexpression.

3.1.2 Checking for Successful Overexpression

1. A simple method to evaluate soluble and insoluble protein overexpression fractions, using B-PER solution (chemical lysis; *see* **Note 2**): Pre- and post-induction 1-mL sample pellets are defrosted and resuspended in B-PER by adjusting their respective OD_{600} using 50-μL B-PER per OD_{600} = 0.6 units. Vortex samples for 1 min before centrifugation for 5 min at 13,000 g.

2. Transfer the supernatant to a fresh microcentrifuge tube (soluble fraction). Resuspend the remaining pellet in the same volume of IMAC buffer A as B-PER used in the previous step (insoluble fraction; this requires rigorous resuspending; do not recentrifuge).
3. Mix samples to 1× with 3× SDS sample loading buffer. Heat the mixtures on a heat block for 10 min at 97 °C, and then apply 5-μL sample per lane on SDS-PAGE alongside a protein ladder to establish the correct size of the protein bands.
4. Wash gels in dH_2O and add Coomassie staining solution. Leave for 1 h on a gel shaker; then wash again with dH_2O. Add de-staining solution and leave for 15 min; then repeat. Once the gel is fully destained, wash with dH_2O.

3.1.3 IMAC Nickel Affinity Chromatography Purification of His-Tagged DUB

1. Resuspend protein pellets in IMAC buffer A (use 10-mL buffer per gram pellet weight) on ice.
2. Sonicate the cell suspension for 8 min at an amplitude of 18 kHz, switching between 10 s on and 5 s off (*see* **Note 3**) on ice. Transfer the solution to centrifugation tubes and clear the lysate by centrifugation for 1 h at 24,000 g at 4 °C.
3. During the centrifugation step, prepare the HiTrap chelating HP column (*see* **Note 4**) with 5-column volume (CV) dH_2O, 1-CV 50-mM $NiCl_2$, 5-CV dH_2O, 1-CV IMAC buffer B (500-mM imidazole), and 5-CV IMAC buffer A (20-mM imidazole) using a peristaltic pump.
4. After centrifugation, filter the lysate using a 0.45-μm filter, and load the lysate on the pre-equilibrated HiTrap column using a sample loop or peristaltic pump. Wash the column with an additional 5 CV of buffer A.
5. Prepare the FPLC system with IMAC buffers A and B using the appropriate buffer lead. Attach the column and wash with an additional 5-CV buffer A to remove further contaminations.
6. Elute the protein using a linear gradient of imidazole over 20 CV from 0 to 100% IMAC buffer B using a set flow rate of 2 mL/min, and collect ~2.5-mL fractions. Add 1-mM DTT (*see* **Note 5**) to peak fractions to reduce cysteines and keep fractions at 4 °C.
7. Analyze the peak fractions by SDS-PAGE (include lysate, pellet, and flow-through), and measure protein concentration using a Nano spectrophotometer.
8. Collect samples that contain the target protein, and concentrate the fractions to ~2.5 mL (depending on concentration) using a centrifugal concentrator at a suitable MW cutoff. This sample can be used straight away or further purified using SEC if required (*see* **Note 6**).

3.1.4 SEC Purification of His6-Tagged Target Protein

1. Equilibrate a Superdex S75, S200 16/60 PG, or alternative column in a suitable MW range with 1.5-CV SEC buffer.
2. Centrifuge the ~2.5-mL protein sample from the IMAC step for 10 min at 13,000 g prior to loading the sample onto the column via a sample loop.
3. Use a constant flow rate suitable for the column and collect fractions.
4. Analyze the peak fractions by SDS-PAGE and determine the concentration.
5. Using a centrifugal protein concentrator, concentrate pure protein fractions to the desired concentration, flash-freeze aliquots in liquid N_2, and store them at −80 °C until further use.

3.2 Inverse PCR Removal of a SapI Restriction Enzyme Site Within the Phagemid Vector

The library production method uses the SapI enzyme to cleave inverse PCR products and facilitate high-efficiency ligation. If your phagemid vector carries a SapI site, it can be removed using inverse PCR as follows. The site-directed mutagenesis (SDM) primers, as listed in Materials, are designed for blunt-end ligation, and the example used here is for the pc89 phagemid [19]. *See* Fig. 2 for the site-directed mutagenesis (SDM) strategy.

1. Isolate the pc89 phagemid vector from TG1 *E. coli* using a plasmid DNA isolation kit. Determine DNA concentration using Qubit. The Qubit working solution is prepared by

SDM forward primer 5'-GCGGAGGAGCGCCCAAT-3'
SDM reverse primer 5'-TTCCTCGCTCACTGACTC-3'

SapI restriction site 5'-NNNNGAAGAGC-3'
3' -NCTTCTCG-5'

Vector sequence and alignment of primers

5'-...CGCAGCGAGTCAGTGAGCGAGGAAGCGGAAGAGCGCCCAATACGCAAA...-3'
3'-...GCGTCGCTCAGTCACTCGCTCCTTCGCCTTCTCGCGGGTTATGCGTTT...-5'

↓

Mutated vector sequence

5'-...CGCAGCGAGTCAGTGAGCGAGGAAGCGGAGGAGCGCCCAATACGCAAA...-3'
3'-...GCGTCGCTCAGTCACTCGCTCCTTCGCCTCCTCGCGGGTTATGCGTTT...-5'

Fig. 2 SDM strategy to remove SapI site from the phagemid vector pc89. The forward primer introduces a site-directed change at a single nucleotide that removes the SapI recognition sequence. After inverse PCR the vector is blunt ligated and transformed. Underlined bases are complementary between the vector and primers. The shaded bases are the SapI recognition site. The SapI cleavage site is also shown

diluting the BR Reagent 1 in 200 in the BR buffer. 10 μL of the standards (0 ng and 100 ng) is added to 190 μL of working solution. 2 μL of the samples is added to 198 μL of working solution. Allow the incubation with the samples and the BR buffer for a minimum of 5 min at RT.

2. Dilute pc89 phagemid template to 10 ng/μL with UltraPure DNase/RNase-Free Distilled Water.
3. Set up PCR reaction in 50 μL as follows using the SDM primers: 10-μL Q5 buffer (5×), 10-μL GC enhancer, 1-μL dNTPs (10 mM), 1 μL of forward primer (10 μM), 1 μL of reverse primer (10 μM), 0.5 μL of Q5 enzyme, 1 μL of 10-ng/μL template, and 25.5 μL of UltraPure water.
4. Carry out PCR amplification for 95 °C for 5 min and then 30 cycles of 95 °C for 30 s, 55 °C for 50 s, and 72 °C for 5 min, followed by a single incubation at 72 °C for 5 min.
5. Add 12 μL of 6× loading to the PCR reaction. Resolve the PCR amplicon on a 1% agarose (w/v) gel, containing 6 μL of Nancy 520 alongside a 1-kb DNA molecular weight marker.
6. Excise and purify the band of the correct size (around 3400 bp for pc89) using a gel cleanup kit. Elute in 20-μL prewarmed (37 °C) UltraPure DNase/RNase-Free Distilled Water.
7. Determine DNA concentration using Qubit, as described above.
8. Set up a digestion reaction using DpnI restriction enzyme (to remove template DNA). 1 μg of DNA is digested for 2 h at 37 °C, followed by 20-min enzyme heat inactivation at 80 °C.
9. Purify DNA using a NucleoSpin kit and measure the DNA concentration with Qubit.
10. Set up a ligation reaction as follows: 1 μg of DNA ligated with 0.4 μL of T4 DNA ligase and 4 μL of T4 buffer (10×) in 40-μL total reaction (using UltraPure DNase/RNase-Free Distilled Water) for 18 h at 4 °C, followed by 20 min at 80 °C incubation for enzyme heat inactivation.
11. Again, purify DNA using a NucleoSpin kit and measure the DNA concentration.
12. Mix 0.5 μg of purified and ligated DNA with 50 μL of TG1 electrocompetent cells, transfer to an electroporator cuvette and pulse using an electroporator at 1.8 kV, and recover for 1 h at 37 °C, 220 rpm in 1 mL of SOC outgrowth medium. Include a transformation of the pre-ligated DNA to estimate the presence of the template DNA.
13. Plate 100 μL of the transformation volume onto 2YT agar plates with 150-μg/mL ampicillin and 1% (w/v) glucose, and incubate at 37 °C, static for ~16 h.

14. Pick single colonies and inoculate 5 mL of 2YT medium with 150-μg/mL ampicillin and 1% (w/v) glucose each; incubate at 37 °C, 220 rpm for ~16 h.
15. Isolate DNA according to manufacturer's instructions.
16. Sequence DNA using the listed sequencing primer to ensure accurate site-directed mutagenesis (in this protocol, the product is designated pc89_SapI_free phagemid).

3.3 Peptide Library Construction in a Phagemid for pVIII Display

The method is based on library construction within the pc89 phagemid vector that displays the peptide on pVIII. Alternative phagemid vectors are available, and the method can be adapted for pIII display. The example given here is for a 16mer peptide display library but can be adapted for other sized peptides, linear or constrained.

1. Isolate the pc89_SapI_free phagemid vector from *E. coli* using a plasmid DNA isolation kit. Determine DNA concentration using Qubit.
2. Dilute pc89_SapI_free phagemid template to 10 ng/μL using UltraPure DNase/RNase-Free Distilled Water.
3. Set up PCR reaction as follows using the NNK randomized primers (the given volumes are for 1× reaction; 40× reactions are carried out): 10-μL Q5 buffer, 10-μL GC enhancer, 1-μL dNTPs (10 mM), 1 μL of forward primer (10 μM), 1 μL of reverse primer (10 μM), 0.5 μL of Q5 enzyme, 1 μL of 10-ng/μL template, and 25.5 μL of UltraPure water.
4. Carry out PCR amplification for 95 °C for 3 min and then 30 cycles of 95 °C for 30 s, 57.5 °C for 30 s, and 72 °C for 2 min and 20 s, followed by a single incubation at 72 °C for 5 min.
5. Add 12 μL of 6× loading dye to every PCR reaction. Resolve the PCR amplicons on a 1% agarose (w/v) gel, containing 6 μL of Nancy 520 and TAE buffer to confirm amplification; include 1-kb DNA molecular weight marker.
6. Excise the correctly sized band (around 3400 bp in this example) and purify using the NucleoSpin™ Gel and PCR Clean-up kit. Elute in 20-μL prewarmed at 37 °C UltraPure™ DNase/RNase-Free Distilled Water.
7. Pool elution fractions and determine DNA concentration using Qubit.
8. Set up a digestion reaction using SapI and DpnI restriction enzymes. 1 ug of DNA per reaction was digested for 2 h at 37 °C, followed by 20 min enzyme heat inactivation at 80 °C.
9. Purify DNA using a NucleoSpin kit and measure the concentration using Qubit.

10. Set up ligation reaction as follows: 1–2 ug of DNA ligated with 2 μL of T4 DNA ligase and 20 ul of T4 buffer (10×) in 200-μL total reaction volume incubated for 6 h at 4 °C, 18 h at 16 °C, and 6 h at RT, and then followed by 20 min at 80 °C incubation for enzyme heat inactivation. Keep aside 1 μL of DNA after DpnI/SapI digestion and before the ligation step to carry out a separate ligation to estimate the presence of the remaining DNA template.
11. Purify DNA using a NucleoSpin kit and measure DNA concentration using Qubit.
12. Mix 1 μg of purified and ligated DNA with 50 μL of TG1 electrocompetent cells, transfer to an electroporator cuvette, and pulse using the electroporator at 1.8 kV and then recover for 1 h at 37 °C, 220 rpm in 1 mL of SOC outgrowth medium. Ten to 20 transformations should be carried out in total. Include transformation of the pre-ligated DNA to estimate the presence of the template DNA.
13. Prepare tenfold serial dilutions from 10 μL of transformation reaction, each at a final volume of 100 μL in 2YT with 150-μg/mL ampicillin (w/v). Spread each dilution (containing 10 μL of transformation reaction and a further eight further tenfold serial dilutions) onto 2YT agar plates with 150-μg/mL ampicillin and 1% (w/v) glucose.
14. Spread the remainder of the transformations onto square bioassay plates containing 2YT agar with 150-μg/mL ampicillin and 1% (w/v) glucose. Incubate all Petri and bioassay dishes at 37 °C, static for 16 h.
15. Count the colonies on the serial dilution plates to estimate the library size. Library size should be consistent across at least two consecutive serial dilutions.
16. Add 1–3 mL of 2YT medium with 150-μg/mL ampicillin and 1% (w/v) glucose onto each bioassay plate, and carefully scrape all colonies using a cell scraper (*see* **Note** 7).
17. Collect all the medium-containing bacterial cells, mix them from all bioassay plates by rotation for 5 min, add 20% (w/v) glycerol, prepare 500-μL aliquots, and store at −80 °C.

3.4 Phage-Peptide Library Propagation

The method described uses a 16mer peptide library within the pc89_SapI_free phagemid but is suitable for use with other phage-display peptide libraries using a phagemid system. *See* Fig. 3 for the cloning strategy.

1. Streak minimal media plates with TG1 from glycerol stocks. Incubate at 37 °C overnight. As a control, streak the TG1 onto a 2YT agar plate containing ampicillin and glucose and grow overnight at 37 °C. Any growth observed indicates contamination of the TG1 stock.

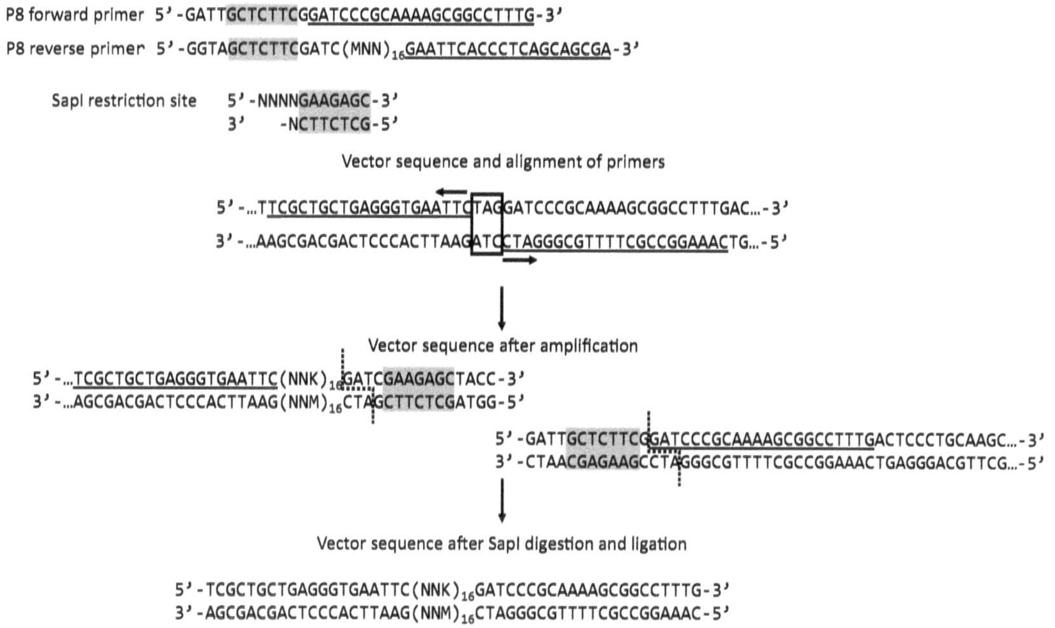

Fig. 3 Library cloning strategy for p8 display in phagemid vector pc89. The P8 forward primer introduces a SapI restriction site that, once cleaved, produces a GAT 5′ overhang. The p8 reverse primer introduces 16 NNK codons and a SapI cleavage site that produces a CTA 5′ overhang once digested. Upon ligation, the vector is reformed, and the 16 NNK codons replace the original TAG stop codon (boxed). Underlined bases are complementary between the vector and primers. The shaded bases are the SapI recognition site. The SapI cleavage site is also shown. Dotted lines indicate the SapI cleavage of the amplified vector

2. Grow the TG1-16mer phagemid library until mid-log phase to a 2-L flask, add 500 μL of glycerol library stock to 500 mL of 2YT containing 150-μg/mL ampicillin and 1% (w/v) glucose, and grow to OD_{600} = 0.4 to 0.6 in a shaking incubator at 180 rpm.
3. Add M13KO7 helper phage to a multiplicity of infection (MOI) of 10 and incubate for 45 min at 37 °C static.
4. Centrifuge culture at 3000 g for 10 min and discard the supernatant. Remove all traces of glucose-containing supernatant with a second quick spin (30 s) to collect supernatant adhering to the walls of the centrifuge vessel. Discard any supernatant.
5. Resuspend cell pellet in double the initial volume of 2YT containing 150-μg/mL ampicillin and 50-μg/mL kanamycin (no glucose), shaking at 200 rpm overnight at 30 °C.
6. Pick a single TG1 colony from a minimal agar plate and inoculate 10 mL of 2YT medium. Incubate at 37 °C with shaking for approximately 4 h to reach OD_{600} = 0.5–1.0. These cells are for determining phage titer and will also be used to confirm that no contamination is present within this culture.

7. Plate 100 μL of the TG1 culture on a 2YT agar plate containing 150-μg/mL ampicillin and 1% (w/v) glucose to check for contamination.
8. Centrifuge the phage library culture for 20 min at 8000 g. Collect the supernatant containing phage. The supernatant can be used directly in phage display experiments or can be further purified/concentrated with PEG. PEG precipitated phage is usually used in the first round of panning and phage present in culture supernatant used for subsequent rounds.
9. Precipitate the phage by adding 250-mL 20% (w/v) PEG-8000/2.5-M NaCl for every liter of phage supernatant.
10. Place on ice for 30–60 min; then centrifuge at 8000 g for 30 min at 4 °C. Discard the supernatant and remove residual PEG/NaCl by centrifuging two more times at 8000 g for 30 s, making sure to remove any liquid at the bottom of the centrifuge vessel after each spin.
11. Resuspend the phage pellet by adding PBS (typically 10 mL total for each 1 L of initial supernatant), and allow the PBS resuspension to soak 5–10 min on ice.
12. Pipette gently to resuspend the purified phage pellet. Then mix by rotation for 30 min at RT, spin at 13,000 g for 10 min, and collect the supernatant.
13. Titer phage immediately. Dilute 10 μL of phage to 1 mL in PBA; then carry out four further 100-fold serial dilutions into PBS. Add 20 μL of the dilutions to 180 μL of the TG1 culture. Also, include diluent PBS buffer-only control. Leave at 37 °C static for 30 min for the phage to infect bacteria. Spread 100 μL of the infected TG1 onto 2YT agar plates containing 150-μg/mL ampicillin and 1% (w/v) glucose. Incubate plates at 37 °C overnight. Count colonies for each dilution, and calculate the original number of phages in the stock, taking into account the dilution of the phage used. Phage present in culture supernatant should be at ~1×10^{12} per mL and PEG precipitated phage at least 1×10^{13} per mL.
14. Phage should be stored short-term 7–10 days at 4 °C or long-term at −80 °C as a glycerol stock.

3.5 Biopanning with Phage Display Targeting His-Tagged Target Protein

Biopanning is performed in solution against a poly-His-tagged protein. Magnetic beads are used to pull the antibody-phage complexes from the solution (*see* **Note 8**).

1. Block the PEG precipitated phage library in a 1.5-mL microcentrifuge tube with a 1:1 dilution in 6% (w/v) mPBS, and incubate for 1 h at RT on a roller mixer.

2. Target protein (100 μg) is bound to Dynabeads His-Tag Isolation & Pulldown in 1.5-mL microcentrifuge tubes, following the manufacturer's instructions. Block target-coated beads with 3% (w/v) mPBS for 1 h at RT using a roller mixer. Following blocking, discard the supernatant using a magnetic rack, and wash beads four times with 1× binding/wash buffer (50-mM sodium phosphate, 300-mM NaCl, 0.01% (v/v) Tween 20, pH 8).
3. Add 1 mL of the blocked library [containing around 10^{12} phage] to the beads coated with the target, and incubate for 1 h at RT on a roller mixer.
4. Beads are then washed ten times with PBS containing 0.1% (v/v) Tween and ten times with PBS. Bound phage is eluted from the beads using 1 mL of a 100-mM triethylamine solution (incubated for 10 min at RT), and the eluted phage is neutralized by adding 500 μL of 1-M Tris–Cl pH 7.4.
5. Mid-log phase TG1 cells (10 mL) are infected by adding half of the eluted phage library (store the other half at 4 °C) directly to the culture in suspension and incubating statically for 30 min at 37 °C.
6. Pellet bacteria by spinning at 3000 g for 10 min and then resuspend pellet in 1-mL 2YT. Remove 10 μL of cells, and plate the remaining cells onto 2YT agar bioassay dishes with 150-μg/mL ampicillin and 1% (w/v) glucose to select for infected cells: Incubate overnight at 37 °C.
7. The 10 μL of cells is used to titrate the eluted phage. Carry out a tenfold serial dilution of the bacteria in 1% (v/v) glucose in 2YT medium. Plate out onto 9-cm 2YT agar plates with 150-μg/mL ampicillin and 1% (w/v) glucose. Incubate overnight at 37 °C. Between 10^4 and 10^7 phage eluted from the panning rounds is expected.
8. Make glycerol stocks of each selection by scraping the lawn of bacteria from the bioassay dishes using a cell spreader and 4.5 mL of 2YT media with 150-μg/mL ampicillin and 1% (w/v) glucose. Add glycerol to 20% (v/v), mix on a rotating mixer for 15 min, and then divide into 1-mL aliquots, and store at −80 °C.
9. For further panning rounds, phage particles can be prepared from glycerol stocks as described above for the original library. Panning experiments usually consist of between three and five rounds of panning (*see* **Notes 8** and **9**).

3.6 Screening Phage-Peptide-Binding Events

Next-generation sequencing (NGS) to detect binding events can reveal additional binders compared to the traditional random picking of clones [7, 8]. The following describes both random picking and then NGS analysis methods.

3.6.1 Random Clone Picking to Screen for Peptide Binders

After three to five rounds of panning, random colonies from the titration plates can be picked to assess peptide binding/enrichment against targets. This step is achieved by screening through a phage ELISA.

(a) Colony Picking

1. To a sterile 96-deep-well plate, add 100 μL of 2YT media containing 150-μg/mL ampicillin and 1% (w/v) glucose per well.
2. Pick colonies from the titration plate and inoculate the deep-well plate. One well should be left empty as a blank control. Usually, one to four plates are screened per panning.
3. Incubate the culture plate with shaking (300 rpm) at 37 °C for around 2–4 h, or until all inoculated wells have reached an estimated OD_{600} of 1.

   Once this stage is reached, the experiment can proceed, or the plates can be stored at 4 °C overnight; alternatively, 50 μL of 2YT media with 50% glycerol can be added to each well and flash-frozen with dry ice for long-term storage at −80 °C.
4. Take 5 μL of culture at OD_{600} of ~1, and inoculate a new culture in a 96-deep-well plate containing 200 μL per well of 2YT media with 150-μg/mL ampicillin and 1% (v/v) glucose using a multichannel pipette.
5. Incubate the deep-well plate with shaking (300 rpm) at 37 °C for around 2.5 h, or until the cultures have reached the mid-log phase ($OD_{600} = 0.5$).
6. Next, mix 4 mL of 2YT media with 350 μL of helper phage stock (at around 10^{12} cfu/mL). Add 10 μL of the diluted helper phage solution to each well of the 96-deep-well plate. Incubate statically for 30–60 min at 37 °C.
7. Spin the plates at 3000 g for 10 min and remove the supernatant (around 200 μL).
8. Resuspend infected bacteria pellets with 400 μL of 2YT media with 150-μg/mL ampicillin and 50-μg/mL kanamycin (no glucose). Incubate overnight with shaking (~300 rpm) at 30 °C.
9. On the next day, harvest the phage-containing supernatant (around 400 μL) through 10-min centrifugation at 3000 g.

10. Block the phage solutions in mPBS, by making a 1:1 dilution of the supernatant with 6% (w/v) mPBS, and incubate for at least 1 h at RT.

    The blocked phage solution will be used as primary antibodies for ELISA screening.

(b) Monoclonal ELISA

1. His-tagged target proteins and controls (100 μL) are coated on MaxiSorp plates at a 10–50-μg/mL concentration, through an overnight incubation at 4 °C.

    MaxiSorp well coating should be performed the same day as overnight helper phage incubation (point a8) to have blocked antibody-phage and coated plates on the same day. As a control, use a protein of approximately the same molecular weight (MW) and affinity tag as the target protein.

2. On the next day, remove excess solution from the wells and wash the plate twice in 300 μL of PBS.
3. Block the coated and washed MaxiSorp wells for 1 h at RT, with a 3% (w/v) solution of mPBS, using 350 μL per well.
4. Remove mPBS from the wells and wash the plate 3× in 300 μL of PBS containing 0.1% (v/v) Tween 20 and then 3× in 300 μL of PBS.
5. The blocked phage solution is then added to the MaxiSorp wells (100 μL) and incubated for 1 h at RT. It is recommended to test each phage clone within the same plate (i.e., by coating wells with the target protein and control adjacent to each other) in order to minimize effects arising from plate-to-plate variability.
6. Blocked phage is then removed and the plate washed four times in 300 μL of PBS containing 0.1% (v/v) Tween 20 and then 3× in 300 μL of PBS.
7. Next, 100 μL of anti-M13 bacteriophage antibody (HRP) (1/2000) in 3% (w/v) mPBS is added to each well and incubated for 1 h at RT.
8. Unbound antiphage antibody is removed and the plate washed 4× in 300 μL of PBS containing 0.1% (v/v) Tween 20 and then 3× in 300 μL of PBS and once with water.
9. Finally, add 100 μL of TMB substrate per well and incubate for at least 5 min until a color change is noted.
10. Absorbance is measured at 650 nm using a plate reader.

3.6.2 Screen for Peptide Binders by Next-Generation Sequencing

This methodology requires additional steps within the panning process to maximize true peptide binders' selection to the target. At the final selection round, the panning is carried out against the target protein and, in parallel, against a control protein (similar molecular weight with the same affinity tag). In addition, panning against the target and control proteins is carried out in multiple replicates (e.g., five each). Each panning step produces a phage eluate and a concomitant TG1 glycerol stock of the eluted phagemid pool (as described for round 1 panning above).

(a) Preparation of Samples for NGS

1. For each phage sample for the final round of panning (selected against both target and control proteins), single-stranded DNA (ssDNA) is extracted and purified using the QIAprep Spin M13 Kit (*see* **Note 10**). The ssDNA is ethanol precipitated and stored as a dry pellet to be used as a PCR template.
2. Carry out PCR amplification in a two-step process, with primers designed to give a final product of around 335 bp and containing the required tags for the NGS step (here described for the Ion Torrent platform; Fig. 4).
3. For the first step, a primer set specific for phagemid vector is used (here, the pc89 vector). The primers should be designed to amplify the 16mer peptide sequence flanked by two small linker regions to a size of 275 bp (P8-ForwardNGS and P8-ReverseNGS primers). Primer regions homologous to the primers used in the second-step PCR are also included. PCR reactions are 1 μL of the template (10 ng) mixed with 25.5-μL DEPC water, 10-μL Q5 buffer, 10-μL GC enhancer, 1-μL dNTPs (10 mM), 1 μL of forward primer (10 μM), 1 μL of reverse primer

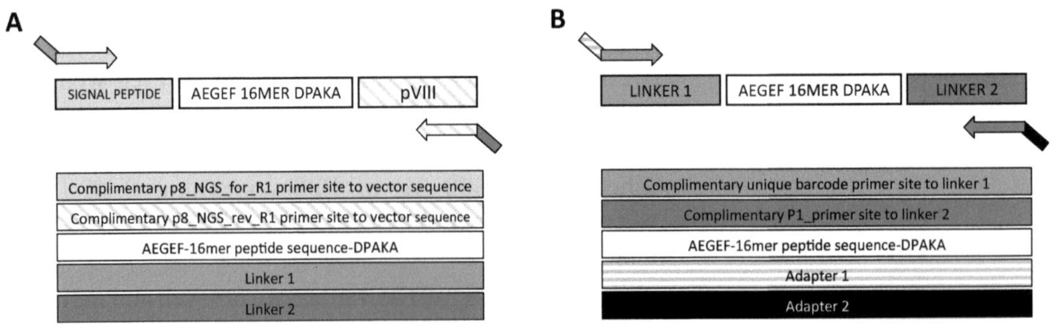

Fig. 4 Overview of the amplification of the 16mer peptide gene region for NGS analysis. The region AEGEF-16mer peptide-DPAKA is amplified in a primary PCR that introduces linker regions upstream and downstream (**a**). In the secondary PCR, a primer homologous to the upstream linker 1 introduces a distinct barcode, and the "A" adapter for Ion Torrent sequencing (Akey-BC-linker1), the second primer, homologous to linker 2, introduces a "P1" adapter for Ion Torrent sequencing (P1prim-linker2)

(10 μM), and 0.5 μL of Q5 enzyme. The PCR conditions are 95 °C for 3 min [95 °C for 30s, 60 °C for 30s, 68 °C for 30s] ×30 and last 72 °C for 5 min.

4. Purify amplicons using a PCR Clean-up kit after resolution on a 3% (w/v) agarose gel.
5. For the second-step PCR, purified amplicon (1 μL) from the first-step PCR is used as a template with a new primer set. One of these primers contains a distinct barcode (Akey-BC-linker1); the second contains a universal P1 primer compatible with Ion Torrent sequencing (P1prim-linker2). Differently barcoded primers are used to amplify each different sample. 1 μL of amplicon from the **step 1** PCR is amplified as described before, with the PCR conditions 95 °C for 3 min [95 °C for 30 s, 63 °C for 30 s, 68 °C for 30 s] × 12 and 72 °C for 5 min.
6. Analyze amplicons on a 3% (w/v) agarose gel to confirm amplification of the correct size and estimate amplicon concentration using Qubit. Pool all samples in equal amounts.
7. Run the pooled PCR amplicon on a 3% (w/v) agarose gel and purify the amplicon using a gel extraction kit. The expected size of the amplicon is 335 bp.
8. Further purify the amplicon with a PCR Clean-up kit and run on a 3% (w/v) agarose gel to confirm the presence of a single band.
9. Finally, the concentration of the pooled DNA is measured with Qubit and diluted as required by the commercial Ion Torrent service provider.

(b) Next-Generation Phage Display Data Analysis

1. Deep sequencing of all samples can be performed using an Ion Torrent commercial service on machines ranging from a PGM to an S5 and using chips ranging from 318 to 540 chips. Which combination is used depends on the depth of sequence required and how many samples are multiplexed. Manufacturer estimates of the read numbers expected from these chips ranged from 4–5.5 million reads for a 318 chip up to 60–80 million reads for a 540 chip.
2. 200-base read lengths are used for each chip. Sample requirements for the commercial service used were for amplicon DNA to be supplied in a 30–50-μL volume containing a total of 300–400-ng DNA if quantified using a Bioanalyzer or Qubit fluorometry.

3. Data are downloaded as ".gz"-suffixed files, compressed using the standard GNU zip (gzip) compression algorithm. The sequencing data are in FASTQ format and are processed on a PC running Ubuntu Linux. Decompression of the compressed FASTQ file is performed at the terminal using gunzip:

```
gunzip -dkf INPUT.fastq.gz
```

4. Conversion of the resultant FASTQ format sequence data file into FASTA format can be performed using several methods with the caveat that sequences must be maintained in single-line format, or if the data is supplied in multiline format, it is converted into single-line FASTQ format at this stage. The following "sed" or "awk" commands typed at the terminal in the directory containing the input file can be used to convert single-line FASTQ format sequencing data files into single-line FASTA format files. Both programs come preinstalled in a Linux environment without any further installation requirements. Alternatively, a suite of bioinformatics tools, such as those found in SeqKit [20], can be installed to provide this and further functionality for processing FASTQ and FASTA files.

```
sed -n '1~4s/^@/>/p;2~4p' INFILE.fastq > OUTFILE.fasta

awk 'NR % 4 == 1 || NR % 4 == 2' INFILE.fastq | sed -e 's/^@/
>/' > OUTFILE.fasta

seqkit fq2fa -w 0 -o OUTFILE.fasta INFILE.fastq
```

5. Demultiplexing is performed on the resultant FASTA file by searching the 5′ region for exact matches to the Ion Torrent barcode sequences used in the second round PCR amplification of the panning-derived DNA samples "binning" FASTA sequences into separate files dependent on which barcode has been identified. This demultiplexing step can be accomplished using freely available software such as FASTX-Toolkit; in this case, three files are needed, each containing Ion Torrent barcode sequences of identical length. Given that the standard set of Ion Torrent barcodes are either 10, 11, or 12 nucleotides in length, the first file will contain those of 10 nucleotides in length in tab-separated format with the first column containing the barcode identifier, which will be used to label the appropriate output FASTA file, and the second column containing the barcode sequence. The second and third files will contain those barcode sequences, which are 11 and

12 nucleotides long, respectively, along with their corresponding barcode identifiers. Lines starting with "#" are ignored, for example:

```
# ion barcodes length 10 nt
```

```
BC01  CTAAGGTAAC
BC02  TAAGGAGAAC
BC03  AAGAGGATTC
BC04  TACCAAGATC
BC05  CAGAAGGAAC
```

6. Invocation is carried out at the terminal from the folder in which the data is located, our preference being to set the number of allowed mismatches within the DNA sequence compared to the barcode sequence to zero to ensure only exact matches to the barcode sequence are considered a match:

```
cat INPUT.fasta | fastx_barcode_splitter.pl –mismatches 0 --
bc BARCODES10.txt --bol --prefix "fastx_" --suffix ".fasta"
>barcodes10.log
```

7. Using this method, an input file of "INPUT.fasta" generates output files of "fastx_BC01.fasta," "fastx_BC02.fasta," and so on, generating one demultiplexed FASTA file for each barcode identified. Barcode DNA sequences were removed from FASTA sequences at this stage by deleting any nucleotides up to and including the barcode DNA sequence. Demultiplexing software can be obtained online, such as SeqKit [20], FASTX-Toolkit [21], or Flexbar [22]. Care should be taken to ensure that output FASTA files from whichever software is used are in single-line format; i.e., line wrapping is avoided. If sequencing data is supplied in line-wrapped format or subsequently demultiplexed into such a format, conversion into single-line format must be performed.
8. FastX_barcode_splitter.pl only identifies barcodes at the exact start (or end) of a sequence (*see* **Note 11**) and does not remove the barcode from the sequence. An alternate method of identifying an exact match to a barcode sequence located within five nucleotides of the 5′ end of a DNA sequence and subsequently removing everything up to and including the barcode sequence CTAAGGTAAC when found can be performed using the following command at the terminal:

```
cat INPUT.fasta | grep --no-group-separator -E -B 1 '^.{0,5}
CTAAGGTAAC'

| sed 's/^.\{0,5\}CTAAGGTAAC//' > BC01sequence.fasta
```

9. A convenient method to process all Ion Torrent barcode sequences sequentially is to create a text file containing one such line similar to the above for each barcode to be processed, changing the barcode sequence at both positions in the line and redirecting the output to an appropriately labeled file. Saving the text file with the appendix ".sh," for example, as demultiplex.sh, and running by typing the following at the terminal:

```
bash demultiplex.sh
```

   Executing the above command will process all lines within the script, attempting to run each new line after the preceding line has finished processing (*see* **Note 12**).

10. For certain library constructs where sequencing was performed in the reverse direction, FASTA sequences were reverse complemented at this stage, and a new file was generated from each demultiplexed FASTA file generated in **step 8**. Reverse complementing sequences can be performed using SeqKit specifying the -r and -p options to reverse and complement, respectively, turning off linewidth control and specifying the sequence type, or alternatively using a Perl script to perform the process, for example:

```
seqkit seq -v -t DNA -w 0 -r -p -o OUTPUT.fasta INPUT.fasta
```

   or

```
perl reverse_complement.pl --infile INPUT.fasta --outfile
OUTPUT.fasta
```

11. Translation of DNA sequences into amino acid sequences in each of the three forward frames was performed using a custom Perl script in which amber stop codons were translated to "q" instead of the standard translation "*" symbol, to allow for identification of peptides where translational readthrough could occur in amber-suppressor strain (SupE) TG1 and hence produce viable phage displaying full-length peptide with glutamine (Q) encoded at the stop codon position. Similarly, translation of ochre and opal stop codons was custom coded to

characters "o" and "p," respectively (with the -opq option at the command line), although significant translational readthrough would not be expected at these stop codons.

For each DNA FASTA file, the script is run three times, specifying the required frame, e.g., –f 1 or –f 2 or –f 3 as an argument on each occasion. Output is redirected to a file using the standard Linux redirector ">."

```
perl translate.pl -i input.FASTA -s opq -f 1 > frame1.fasta
```

At this stage, each of the three translated frame files was concatenated together so that subsequent identification of P8 amino acid sequences N-terminally and C-terminally flanking the variable peptide region could be performed by reading from a single protein FASTA file irrespective of which frame the desired sequences were located. The "cat" command is used to join the three translated frame files together:

```
cat frame1.fasta frame2.fasta frame3.fasta >OUTPUT.fasta
```

12. Peptide sequences located between flanking N-terminal and C-terminal vector amino acid sequences (AEGEF and DPAKA, respectively, for the pc89 pVIII-derived libraries) were identified by a further Perl script which must be supplied with the FASTA file generated in the previous step as its input file, along with two files containing, respectively, a list of the N- and C-terminal conserved motifs to be identified, with the additional option to discard peptide sequences that contain matching amino acid flanking regions if the insert length is shorter than a specified cutoff (the default being to retain all sequences with at least one amino acid between the motifs).

    Output files are suffixed with ".LR.fa" to indicate the sequences within the file were identified as having matched both the N-terminal and C-terminal motif sequences with those which matched only the N-terminal (output file suffixed .L.fa) or C-terminal (suffixed .R.fa) sequences.

    The script is executed as follows and in this example specifies a minimum insert size of five amino acids:

```
perl iterate_motifs.pl --infile INPUT.fasta --outfile OUTPUT
--peptide 5 --left AEGEF.txt --right DPAKA.txt
```

An example input sequence and output sequence resulting from this processing step would be:

Input

```
>G2HKA:08736:10824
SLWYRMLSFAAEGEFRGSLSQPSPLMSRSNLDPAKAAFDSLQATEY......
```

Output

```
>G2HKA:08736:10824
RGSLSQPSPLMSRSNL
```

13. It should be noted that the files describing flanking sequences or motifs may contain multiple target sequences in each file, each of which will be paired with all target sequences in the opposing file when searching the protein FASTA file. As such, if 5 motifs are specified in the N-terminal file and 4 in the C-terminal file, all 20 combinations will be used as search patterns against the translated sequences. Particular amino acids can be excluded using the [^X] pattern, for example, AE[^G]EF will identify all motif sequences which do not contain a glycine in position 3 (*see* **Note 13**).
14. A subsequent Perl script counted and ranked identical peptide sequences in two FASTA files to tabulate frequencies on a per peptide basis, normalized frequencies relative to the total number of peptide sequences which passed the length and framework motif filtering in each file, and relative enrichment of any particular peptide from a panning against a target compared to its relative enrichment against a control panning (or library reads where no control panning was used). This was determined using a two proportion z-test to compare the sequence populations between two barcodes using the equation:

$$\left(z = \frac{p1 - p2}{\sqrt{\frac{p1(1-p1)}{n1} + \frac{p2(1-p2)}{n2}}} \right)$$

This reflects the ratio between the control and target proteins as well as the absolute frequencies and ranks them by relative statistical importance (n1 = sample 1 size, n2 = sample 2 size, p1 = proportion 1, p2 = proportion 2).

The script is executed by specifying two. LR.fa input files such as those generated in **step 12**, i.e., FASTA files where the flanking sequences surrounding the variable peptide sequence have been identified and removed. An

Table 1
Example output format from the Perl script used to analyze the NGS data

Sequence	INPUT1.fasta	INPUT2.fasta	Z-score	INPUT1.fasta_%	INPUT2.fasta_%
LTYNARKADPRLSAML	226640	177254	23.64	48.7150	46.1425
HSNLLRCLSSESKRSN	45315	32535	20.33	9.7402	8.4695
MKPAEPMKRSELTVTL	353	9	17.88	0.0759	0.0023
IKPSEPNKVTPTQDSM	294	0	17.15	0.0632	0.0000
TKTELASLSCARAGEW	127	0	11.27	0.0273	0.0000

optional "-p" switch will generate two further columns in the tabulated output consisting of the percentage abundance of each sequence within the two FASTA files:

```
perl compareZ_2.1.pl INPUT1.fasta INPUT2.fasta >output_table.
txt
```

Examples of the first few lines of a typical output are shown in Table 1 (also, *see* **Note 14**).

3.7 Initial Evaluation of Candidate Peptide Binders

1. Initial evaluation of peptide binders can be performed by coating ELISA MaxiSorp plates with synthetic peptides and controls (100 μL of 50 ug/mL) overnight at a concentration of 10–50 μg/mL at 4 °C. Alternatively, biotinylated peptides can be used either by immobilizing them on coated streptavidin on MaxiSorp plates or by coating plates with the protein and using the biotinylated peptides to probe with a streptavidin-poly-HRP antibody.
2. After washing and blocking as described above for the phage ELISA, incubate wells with the His-tagged target protein in blocking buffer.
3. After wash steps, immobilized peptides can be probed with the corresponding His-tagged target protein at 25 μg/mL in PBS. Detect bound protein using 100 μL of mouse anti-tetra His antibody (Qiagen; 1:1000) followed by anti-mouse horseradish peroxidase (HRP) conjugate (1:2000) and TMB.

4 Notes

1. Cultures can be scaled up as required. Protein expression conditions can be optimized with regard to parameters such as the use of the medium, OD_{600} at induction, IPTG concentration, expression temperature, and duration.
2. As an alternative to B-PER, micro-sonication (mechanical lysis) can be used to separate the soluble and insoluble fractions.

3. The optimal sonication conditions need to be optimized for the available sonicator. Do not use chemical lysis such as B-PER for the preparative protein purification, as detergents and other reagents contained in such solutions can influence downstream applications.
4. Alternatively, other nickel or cobalt purification columns can be used, or the protein purified in batch.
5. Some target proteins have exposed surface cysteines that require a reducing agent such as DTT or TCEP. Adding these agents to fractions after IMAC works well in our hands and avoids potential issues with a metal reduction during IMAC.
6. If protein is of sufficient purity after the IMAC step, perform buffer exchange into PBS or SEC buffer before further use to remove unwanted imidazole, which can lead to protein aggregation.
7. If the cells are too high a density to be scraped into 1 mL of media, add 5–10-mL media to the first plate, and use this media consecutively to resuspend cells from each remaining plate.
8. Panning in the solution can be performed with a wide array of variations. If the target protein is not fused to an affinity tag, it can be chemically coupled to one. The most common tag is biotin using a commercial kit such as the Lightning-Link® Biotin Conjugation Kit (Abcam). Beads such as Dynabeads™ M-280 Streptavidin (Thermo Fisher) and SpeedBeads™ Magnetic Neutravidin Coated particles (Merck) can be used to capture the tagged proteins. The biotinylated target can be added directly to the blocked phage library and incubated for 1 h at room temperature in a rolling machine to interact with peptide. Blocked streptavidin beads can then be used to pull out the target-bound phage through the use of a magnetic rack. Only 5 min are needed for the streptavidin-coated beads to bind the target. Subtraction steps with the control biotinylated protein can be carried out in the same stages as described above by direct addition to the blocked phage. Between rounds in the panning experiment, it is also possible to switch the type of magnetic bead (compared to the previous round) used to capture the biotinylated target from streptavidin-coated Dynabeads™ to SpeedBeads™ Magnetic Neutravidin Coated beads. This can help remove any nonspecific peptides that bind the beads used to capture the target. However, with NGS, filtering out the background should be possible even if these measures are not taken.
9. A complementary strategy for obtaining more specific peptides uses the subtraction of nonspecific binders. After the first round of biopanning, a subtraction step is performed. The subtraction step is achieved by adding a similar-sized protein possessing the same affinity tag as the target (e.g., poly-His-

tag) coupled to the beads and then added to the phage library. Nonspecific binders are then removed from the library through the use of magnetic beads, and the "subtracted" library is then used for further rounds of panning against the target protein. In summary, Dynabeads™ His-Tag Isolation & Pulldown are coated in the control protein and blocked exactly as described for the target protein. The blocked beads are then introduced to the blocked library, and the antibody phage is allowed to bind the unrelated protein for 1 h. Beads are then discarded using a magnetic rack and the subtracted supernatant used for a normal round of panning against the target protein.

10. Alternatively, the glycerol stocks of polyclonal phage produced after panning can be used as a source of dsDNA and DAN extracted using the QIAprep Spin Miniprep Kit and following the manufacturer's instructions.
11. Note that FastX_barcode_splitter.pl, which is part of the FASTX-Toolkit, does not remove the barcode sequence following the identification, and this must be performed as a separate step. Given that the default Ion Torrent barcodes can be 10, 11, or 12 nucleotides in length, the result is that if the barcode sequence is not removed, the frame in which the correct translation is found will vary depending on the length of the barcode sequence used in the PCR primer for that sample. Similarly, reverse complementing data before translation will result in the frame in which the correct in-frame translation appears being a function of the read length of that particular sequence.
12. Considerable improvements in run times can be obtained by running commands in parallel where suitable, using, for example, "GNU Parallel" [23]. Parallel processing can be applied to many processing steps; for example, demultiplexing a FASTA file containing barcoded sequences would rely on a single input file source, with multiple output files each of which is written independently and not reliant on the output from any other process to complete. In the given example of running "demultiplex.sh," where each of the 96 possible Ion Torrent barcodes is searched for and removed during demultiplexing of a DNA FASTA file using the "grep" and "sed" commands, the command to run the script can be adjusted to run on eight processors simultaneously by using the following:

```
parallel -j 8 < demultiplex.sh
```

13. While this method of processing Ion Torrent deep sequencing data initially proved very successful, as new Ion Torrent chips were developed, the size of the corresponding data files increased rapidly.

As such, pipeline scripts are written in a modular fashion such as this which rely on their input on text files stored on disk generated during a previous step of the process, soon led to the limiting factor being the rate at which data could be written to and read from storage drives and the corresponding size of files needing to be processed in such a manner at each step.

Running scripts in parallel where possible, for example, demultiplexing barcode sequences from a FASTA file using one processor per barcode, can result in a notable speed improvement only until the limiting factor becomes the rate at which data can be read from hard disk. As an alternative, the entire pipeline can be rewritten so that each sequence in the compressed FASTQ file is read just once from disk into memory and all processing steps performed sequentially on a per-sequence basis. All the processing steps such as barcode identification, translation, and insert identification are performed on that sequence before the resulting data is written out to disk, proved to considerably reduce run times as well as providing a simpler user experience for inexperienced users of the analysis pipeline.

This enhanced pipeline combines all previous scripts for the data processing pipeline into a single program, written in Perl, which carries out all steps from reading in either a compressed or uncompressed FASTQ file, demultiplexing, reverse complementing, translating, identifying N- and C-terminal flanking motifs, and optionally ranking sequences by frequency.

Z-test comparisons are still performed separately due to the range of potential target versus control comparisons, which may be considered on a per-experiment basis, and the possibility that normalized frequencies may be the only ranking score an end user may require for their analysis.

Intermediate files, such as the demultiplexed FASTA files and translations of each of the three frames, are not written to disk unless that option is selected, saving on storage requirements. The user must specify a FASTQ input file and the output file to be created, a file of barcodes in a tab-separated format consisting of a barcode ID and barcode sequence (one barcode per line), and two files containing the N-terminal and C-terminal amino acid sequences flanking the randomized region of interest, for example:

```
perl pipeline.pl --infile INFILE.fastq.gz --outfile EXAMPLE
--left N_list.txt --right C_list.txt --barcodes BCLIST.txt
```

The option to reverse complement each DNA sequence prior to translation is given by the "--revcomp" option; the minimum number of amino acids located between the N-terminal and C-terminal motifs to be considered a hit is specified by the "--minimum <integer>" option, with the default being a requirement for at least one amino acid to be present between the N- and C-terminal motifs; and generating a text file for each barcode containing ranked frequencies of all identified inserts is done using the "--rank" option.

Full output of all intermediate files, specified using the "--fulloutput" option, is the equivalent to specifying "--fasta," "--translate," and "--rank" and will generate demultiplexed DNA FASTA files, translations of each sequence in all three forward frames, and a text file with numerically ranked peptide sequences for each FASTA file.

14. Tab-separated tables generated in this manner are amenable to subsequent analysis, for example, filtering the z-test ranked peptide sequences based on which sequences are above an arbitrary cutoff value across replicate panning barcoded samples. These analyses were typically performed using inbuilt Linux tools such as awk, grep, sort, uniq, wc, and nl with additional Perl scripts written on an ad hoc basis for more complicated filtering and collating operations. For example, given a FASTA file containing many peptide sequences such as that generated by the output shown in **step 12**, a ranked listing of the peptides in descending numerical order can be produced using the Linux pipeline to pass data from one program to the next using the commands grep, sort, and uniq:

```
grep -v '>' INPUT.fasta | sort | uniq -c | sort -nr > OUTPUT.
txt
```

In this string of commands, the program "grep" is used to identify non-matches (using the -v switch) to lines containing the character ">" which forms the start of a FASTA sequence identifier. This generates a list consisting solely of the corresponding peptide sequences, which are then piped to the "sort" command to be alphabetically sorted. The data is now in the required input format for "uniq" to count adjacent repeated sequence lines (using the -c switch). The resulting data are processed again by sort, this time sorting in reverse numerical order (specified by combining the -r and -n switches), before being redirected to a text file.

Chaining simple inbuilt Linux commands together in this manner can provide highly effective and rapid methods for processing and previewing relevant information from large data sets.

Acknowledgments

This work was supported by the University of Nottingham and the Biotechnology and Biological Sciences Research Council (BBSRC) doctoral training studentships [grant number BB/M008770/1].

References

1. Mevissen TET, Komander D (2017) Mechanisms of deubiquitinase specificity and regulation. Annu Rev Biochem 86:159–192
2. Sowa ME, Bennett EJ, Gygi SP, Harper JW (2009) Defining the human deubiquitinating enzyme interaction landscape. Cell 138:389–403
3. Remenyi A, Good MC, Lim WA (2006) Docking interactions in protein kinase and phosphatase networks. Curr Opin Struct Biol 16:676–685
4. Pozhidaeva AK et al (2015) Structural characterization of interaction between human ubiquitin-specific protease 7 and immediate-early protein ICP0 of herpes simplex Virus-1. J Biol Chem 290:22907–22918
5. Zhang ZM et al (2015) An allosteric interaction links USP7 to Deubiquitination and chromatin targeting of UHRF1. Cell Rep 12:1400–1406
6. Cheng J et al (2015) Molecular mechanism for the substrate recognition of USP7. Protein Cell 6:849–852
7. Ngubane NA et al (2013) High-throughput sequencing enhanced phage display identifies peptides that bind mycobacteria. PLoS One 8: e77844
8. AC't Hoen P et al (2012) Phage display screening without repetitious selection rounds. Anal Biochem 421:622–631
9. Spiliotopoulos A et al (2019) Discovery of peptide ligands targeting a specific ubiquitin-like domain-binding site in the deubiquitinase USP11. J Biol Chem 294:424–436
10. Fagerlund A, Myrset AH, Kulseth MA (2008) Construction and characterization of a 9-mer phage display pVIII-library with regulated peptide density. Appl Microbiol Biotechnol 80: 925–936
11. Noren KA, Noren CJ (2001) Construction of high-complexity combinatorial phage display peptide libraries. Methods 23:169–178
12. Heinis C, Rutherford T, Freund S, Winter G (2009) Phage-encoded combinatorial chemical libraries based on bicyclic peptides. Nat Chem Biol 5:502–507
13. Tharp JM et al (2020) An amber obligate active site-directed ligand evolution technique for phage display. Nat Commun 11:1392
14. Qi H et al (2021) Antibody binding epitope mapping (AbMap) of hundred antibodies in a single run. Mol Cell Proteomics 20:100059
15. Ting JP et al (2018) Utilization of peptide phage display to investigate hotspots on IL-17A and what it means for drug discovery. PLoS One 13:e0190850
16. Cao J et al (2021) Phage-display based discovery and characterization of peptide ligands against WDR5. Molecules 26:1225
17. Gubeli RJ et al (2021) In vitro-evolved peptides bind monomeric actin and mimic actin-binding protein thymosin-beta4. ACS Chem Biol 16:820–828
18. Ashkenazy H et al (2021) Motifier: an IgOme profiler based on peptide motifs using machine learning. J Mol Biol 433:167071
19. Felici F, Castagnoli L, Musacchio A, Jappelli R, Cesareni G (1991) Selection of antibody ligands from a large library of oligopeptides expressed on a multivalent exposition vector. J Mol Biol 222:301–310
20. Shen W, Le S, Li Y, Hu F (2016) SeqKit: a cross-platform and ultrafast toolkit for FASTA/Q file manipulation. PLoS One 11: e0163962
21. FastX-Toolkit. http://hannonlab.cshl.edu/fastx_toolkit/
22. Dodt M, Roehr JT, Ahmed R, Dieterich C (2012) FLEXBAR-flexible barcode and adapter processing for next-generation sequencing platforms. Biology (Basel) 1:895–905
23. Tange O (2011) GNU parallel - the command-line power tool. USENIX Mag 2011:32–47

Chapter 13

Identification of Deubiquitinase Substrates in *Xenopus* Egg Extract

Valentina Rossio, Joao A. Paulo, and Randall W. King

Abstract

Deubiquitinases (DUBs) antagonize protein ubiquitination by removing ubiquitin from substrates. Identifying the physiological substrates of each DUB is critical for understanding DUB function and the principles that govern the specificity of this class of enzymes. Since multiple DUBs can act on the same substrate, it can be challenging to identify substrates using inactivating a single enzyme. Here, we outline a method that enables the identification of proteins whose stability depends on DUB activity and an approach to profile DUB specificity in *Xenopus* egg extract. By coupling broad DUB inhibition with quantitative proteomics, we circumvent DUB redundancy to identify DUB substrates. By adding back recombinant DUBs individually to the extract, we pinpoint DUBs sufficient to counteract proteasomal degradation of these newly identified substrates. We apply this method to *Xenopus* egg extract but suggest that it can also be adapted to other cell lysates.

Key words Deubiquitinases, Substrates, UPS, TMT-proteomics, Ubiquitin

1 Introduction

Reversible posttranslational modifications of proteins regulate virtually all essential functions of the cell. An example is a covalent attachment of ubiquitin, mediated by a cascade of enzymes in the ubiquitin–proteasome system (UPS), in which ubiquitin is initially activated by E1 and then attached to substrates by specific E2 and E3 enzymes [1, 2]. Ubiquitination targets some proteins to proteasomal degradation but can also regulate protein function independently of degradation [1, 3]. Deubiquitinases (DUBs) catalyze ubiquitin removal from proteins and process ubiquitin precursors, ensuring the availability of a free ubiquitin pool [4, 5].

DUBs comprise a large group of enzymes that fall into two main families: the zinc metalloproteases and the cysteine proteases [6, 7]. Although a number of DUBs have been well characterized, most DUBs still do not have any known substrates, making it

Julie A. Maupin-Furlow and Mariola J. Edelmann (eds.), *Deubiquitinases: Methods and Protocols*,
Methods in Molecular Biology, vol. 2591, https://doi.org/10.1007/978-1-0716-2803-4_13,

difficult to understand their functional roles or the molecular basis of substrate specificity. A major challenge in identifying DUB substrates is that multiple DUBs may function redundantly on the same substrates [8, 9]. Thus, the inactivation of a single DUB may not be sufficient to produce changes in substrate ubiquitination or stability.

Here, we describe a method that circumvents DUB redundancy to enable substrate identification in *Xenopus* egg extract. By chemically inhibiting many cysteine protease DUBs simultaneously, with the highly specific DUB inhibitor ubiquitin vinyl sulfone (UbVS) [10], we applied tandem mass tag (TMT)-based quantitative proteomics to identify proteins whose stability depends on DUB activity. By restoring the activity of specific DUBs in UbVS-treated extract, we illustrate how these substrates can be used to define enzyme–substrate relationships, overcoming hurdles associated with redundant enzyme function.

2 Materials

2.1 *Xenopus* Egg Extract Preparation

1. Extract buffer (XB): 100-mM KCl, 0.1-mM $CaCl_2$, 1-mM $MgCl_2$, 10-mM potassium HEPES (pH 7.7 with KOH), 50-mM sucrose.
2. MMR (1×): 100-mM NaCl, 2-mM KCl, 1-mM $MgCl_2$, 2-mM $CaCl_2$, 0.1-mM EDTA, 5-mM HEPES (pH 7.8 with NaOH).
3. Energy mix (20×): 150-mM creatine phosphate, 20-mM ATP, 2-mM EGTA, 20-mM $MgCl_2$ (pH 7.7).
4. Dejellying solution (prepare fresh within 1 hour of use): dissolve 5% w/v cysteine HCl in water (pH 7.8 with NaOH).
5. Protease inhibitors (1000×): mixture of leupeptin, chymostatin, pepstatin (final concentration of 10 mg/mL each) in DMSO.
6. Cytochalasin B (1000×): 10 mg/mL in DMSO.
7. Calcium ionophore A23187 (5000×): 10 mg/mL in DMSO.
8. Cycloheximide (100×): 10 mg/mL in water.
9. Liquid nitrogen.
10. Equipment: Sorvall centrifuge with HB-6 rotor, vortex, pH meter.

2.2 Reagents for Extract Treatment

1. UbVS (stock concentration can range from 150 to 250 mM) in 50-mM MES, pH 6.
2. MES buffer: 50-mM MES, pH 6.
3. Recombinant human ubiquitin, dissolved to 1–1.5 mM in XB.
4. TNT Quick Coupled Transcription/Translation System, rabbit reticulocyte lysate (Promega).

5. Standard PCR reagents and PCR purification kit.
6. ^{35}S-methionine 500-mCi, stabilized aqueous solution (PerkinElmer).
7. Human recombinant DUBs.
8. Equipment: thermomixer, nanodrop, PCR thermocycler.

2.3 SDS-Polyacrylamide Gel Electrophoresis and Immunoblotting

1. MES SDS running buffer (20×): 50-mM MES, 50-mM Tris–HCl, 0.1% SDS, 1-mM EDTA (pH 7.3).
2. Tris-acetate SDS running buffer (20×): 50-mM tricine, 50-mM Tris–HCl, 0.1% SDS (pH 8.24).
3. 4× LDS sample buffer: 988-mM Tris–HCl, 2.04-mM EDTA, 8% LDS, 40% glycerol, 0.88% Coomassie Brilliant Blue G250, 0.7-mM phenol red.
4. Polyacrylamide protein gels 4–12% and 3–8% and PVDF transfer membrane (0.45-micron pore size).
5. Transfer buffer: 25-mM Tris–HCl, 192-mM glycine, 20% methanol.
6. Tris-buffered saline (TBS-T; 10×): 1.5-M NaCl, 500-mM Tris–HCl, pH 7.5, 1% Tween.
7. Blocking solution: 5% dry milk in TBS-T.
8. Primary antibody: anti-Ube2C antibodies (R & D systems).
9. Secondary antibody: anti-rabbit IgG conjugated with horseradish peroxidase (HRP).
10. SuperSignal West Femto Sensitivity Substrate.
11. Equipment: heat block, Novex mini-cell gel running apparatus, power supply, Mini PROTEAN Tetracell transfer apparatus, Amersham Imager 600 (Amersham).

2.4 Sample Preparation for TMT-Proteomic Analysis

All the following reagents used to prepare proteomic samples should be prepared with HPLC grade solvents:

1. Pierce BCA protein assay.
2. Quenching buffer: 50-mM Tris–HCl, pH 8.8, 1% SDS, 5-mM DTT, protease inhibitor tablet.
3. 200-mM 3-[4-(2-Hydroxyethyl)-1-piperazinyl] propanesulfonic acid (EPPS or HEPPS) buffer, pH 8.5 (NaOH).
4. Lysyl endopeptidase (LysC), trypsin, TMT10plex Isobaric Label Reagent Set (Thermo Fisher).
5. C18-based Sep-Pak prepacked cartridge 100 mg (Waters).
6. Sep-Pak washing buffer: 5% acetonitrile (ACN), 1% trifluoroacetic acid (TFA).
7. Sep-Pak elution buffer: 70% ACN, 1% TFA.

8. Peptide resuspension buffer (prior to HPLC fractionation): 10-mM ammonium bicarbonate, 5% ACN, pH 8.
9. 300 Extend C18 column, 2.1 × 150 mm, 3.5 micron for peptide fractionation (Agilent).
10. C18 solid-phase extraction disks (SPE) for creating homemade StageTips or commercially available StageTips.
11. StageTip washing buffer: 5% ACN, 1% formic acid.
12. StageTip elution buffer: 70% ACN, 1% formic acid.
13. Resuspension buffer: 5% ACN, 5% formic acid.
14. Equipment: speed vacuum concentrator, benchtop centrifuge, vortex, Agilent 1260 HPLC system, Orbitrap Fusion mass spectrometer coupled to a Proxeon EASY-nLC 1000 liquid chromatography (LC) (Thermo Fisher).

2.5 Coomassie Staining and Drying of SDS-Polyacrylamide Protein Gels

1. Coomassie blue solution: 0.2% Coomassie blue, 7.5% acetic acid, 50% methanol.
2. Destaining solution: 40% methanol, 10% acetic acid.
3. Drying gel solution: 50% methanol, 10% acetic acid.
4. Phosphor screen (Fuji).
5. Equipment: gel dryer (Bio-Rad), Typhoon Biomolecular Imager (Amersham).

3 Methods

3.1 Preparation of Interphase Xenopus Egg Extract

Interphase *Xenopus* egg extract is prepared based on Murray's protocol [11] and further adapted with a few modifications from Salic and King [12]. Extracts are prepared from eggs laid overnight by frogs placed separately in each container with 2 L of MMR solution (18 °C). Each single batch of eggs is examined, and if they contain a considerable number of abnormal eggs (such as white color, abnormal morphology, or connected), they are omitted. All the other eggs are pooled together and processed to prepare the egg extract. All the steps are carried out at room temperature, if not otherwise indicated.

1. Wash the collected eggs in three volumes of MMR (prechilled to 16 °C) and allow them to settle. Repeat two additional times with fresh MMR.
2. Incubate the eggs with three volumes of 5% cysteine to remove the egg jelly coat (dejellying). This procedure is complete when all the eggs are tightly packed together (5–10 minutes).
3. Wash the eggs gently with MMR, as you did in **step 1**, until the buffer is transparent (typically three to five times).

4. Activate the eggs with two volumes of MMR containing 2-μg/mL calcium ionophore to drive the eggs into interphase. Incubate until the pigmented area of the eggs contracts, indicating that egg activation has occurred, typically about 5 minutes.
5. Wash eggs three times with two volumes of extract buffer (XB) (prechilled at 16 °C). Wash one last time with one volume of XB containing 10-μg/mL protease inhibitors. Incubate eggs in this solution for 20 minutes. Remove the XB.
6. Carefully pour the eggs into a centrifuge tube (*see* **Note 1**) containing a small volume of XB with 100-μg/mL cytochalasin B and 10-mg/mL protease inhibitors (*see* **Note 2**). Cytochalasin B addition is important to prevent actin polymerization. For a 50-mL centrifuge tube, we use 2 mL of XB. Spin tubes at low speed (Sorvall HB-6 rotor at 860 rpm (127 g) at 4 °C) for 1 minute to pack the eggs, and aspirate excess buffer (*see* **Note 3**).
7. Crush the eggs by spinning for 15 minutes at 4 °C in an HB-6 rotor at 12,000 rpm (23,000 g). The cytoplasmic layer (the middle tan layer between the yellow lipid layer on the top and the dark yolk at the bottom) is removed by needle puncture (*see* **Note 4**) and collected in a clean falcon tube placed in ice.
8. Add to the extract cytochalasin B and protease inhibitors at final concentrations of 10 μg/mL and cycloheximide at 100 μg/mL. Mix well by pipetting. Repeat **step 7** once again.
9. Supplement the extracts with 1× energy mix and 4% glycerol. Mix well by pipetting to dissolve the glycerol (*see* **Note 5**). The final extract should be clear and of a tan-yellowish color. Aliquot the extract, snap freeze in liquid nitrogen, and store at −80 °C (*see* **Note 6**).

3.2 Evaluation of Egg Extract Quality

After extract preparation, we suggest verifying the activity of the extract by checking the degradation of known proteasome substrates. Typically, degradation of the known APC substrates, cyclin B1 or securin, is measured after driving the extract into mitosis by adding nondegradable cyclin B1 [13].

3.3 Monitoring Level of Ubiquitin Depletion by Analysis of E2 Charging

Since protein synthesis is inhibited in the extract by cycloheximide, the only source of free ubiquitin is generated by the action of DUBs on their substrates. Inhibiting multiple DUBs simultaneously with UbVS depletes free monomeric ubiquitin [13, 14]. Indeed, treatment of extract with UbVS must be coupled with the addition of exogenous ubiquitin to the extract to sustain UPS activity. As a readout of ubiquitin availability, the charging status of the E2s enzymes is monitored by using nonreducing gels to access the fraction of Ube2C that is present in the form of a ubiquitin thioester (charged Ube2C). Ube2C is fully charged in unperturbed extracts but is rapidly discharged in UbVS-treated extract in a manner that can be fully rescued with exogenous ubiquitin [13, 14]. Therefore, monitoring Ube2C charging is a convenient measure of functional ubiquitin availability in the extract (*see* **Note 7**). We observed that 10-μM UbVS is sufficient to inhibit multiple DUBs ($n = 35$) and rapidly deplete free ubiquitin [14]. At this concentration of UbVS, ubiquitin depletion can be fully restored by the addition of 50-μM exogenous ubiquitin [14]. Although higher UbVS concentrations may inhibit more DUBs, we have found that it is also more challenging to sustain Ube2C charging with exogenously added ubiquitin. We observed that 50-μM ubiquitin is not sufficient to restore ubiquitin availability in extract treated with 30-μM UbVS [14]. Since ubiquitin dynamics can vary in different batches of extract, we suggest determining the UbVS concentration sufficient to induce ubiquitin depletion while allowing for the rescue of ubiquitin availability after adding exogenous ubiquitin.

1. Thaw the *Xenopus* egg extract rapidly in hand and then incubate on ice.
2. Split the extract into four tubes (*see* **Note 8**). Add to each tube MES buffer (control) or UbVS at the final concentrations of 5, 10, and 20 μM, respectively (*see* **Notes 9** and **10**). Mix by pipetting gently multiple times to be sure that the UbVS is well mixed. Collect time 0 from all the conditions by pipetting 1 mL of extract into 10 mL of nonreducing LDS sample buffer and briefly vortex it. Keep the sample on ice until the end of the experiment. Incubate the control- and UBVS-treated extracts at 24 °C with gentle shaking in a thermomixer (1250 rpm) for 10 minutes. After 10 minutes, collect samples from the different conditions as described for the time 0 sample.
3. Perform nonreducing SDS-PAGE electrophoresis (*see* **Note 11**) followed by immunoblotting with antibodies against Ube2C (*see* **Note 12**). After 10 minutes, 10-μM UbVS should be sufficient to induce rapid ubiquitin depletion, visualized as a complete loss of Ube2C thioester.
4. Next, check if ubiquitin availability can be rescued in UbVS-treated extract by adding exogenous ubiquitin to the extract. Thaw an aliquot of the *Xenopus* egg extract, as before. Treat the extract with 10-μM UbVS for 10 minutes, as previously. Split the extract into four clean tubes. Add XB (control) or human recombinant ubiquitin to each tube at the final concentrations of 10, 20, and 50 μM, respectively. Mix by pipetting a couple of times gently. Collect time 0 and treat it as in **step 2**. Incubate at 24 °C with gentle shaking, as before. Collect time points at 5, 30, and 60 minutes and quench them with nonreducing LDS sample buffer.

5. Analyze samples by nonreducing SDS-PAGE and immunoblotting with Ube2C antibodies.
6. In our experimental conditions, 50-μM ubiquitin is typically sufficient to sustain Ube2C charging for at least 1 hour in extract pre-treated with 10-μM UbVS (*see* **Note 13**).

3.4 Identification of Proteins Protected from Degradation by DUBs with TMT-Proteomics

To identify proteins protected from degradation by DUBs, TMT-based quantitative proteomics can be used [15] to compare protein levels in different conditions in a quantitative manner. Proteins that decrease specifically in the presence of UbVS or UbVS/ubiquitin but are stable in extract treated with XB or ubiquitin represent potential DUB substrates. In this experiment, E2 charging is monitored in parallel with performing the proteomic analysis in the same experiment as a control for DUB inhibition by UbVS and restoration of ubiquitin charging by exogenous ubiquitin. At each time point of the experiment described below, you will collect samples for proteomic analysis (*see* **Note 14**) and one sample for analysis by nonreducing SDS-PAGE, followed by immunoblotting to measure E2 charging. The logic and expected findings from this experiment are illustrated in Fig. 1.

1. Determine the protein concentration of the extract. We use a BCA assay kit according to the manufacturer's instructions. Each sample collected and processed for TMT-proteomics should contain at least 100 μg of protein. The *Xenopus* egg extract that we prepare typically has a concentration of ~50 mg/mL, and thus we usually use 2 μL of extract for each sample.
2. Depending on the number of samples you want to collect, calculate the amount of extract required, based on 2 μL per sample, and add a further 5–10 μL. Distribute this volume of extract to each of two tubes. Treat one sample with UbVS to achieve a final concentration of 10-μM UbVS and the other with an equivalent volume of MES buffer (control). Mix by pipetting gently multiple times to ensure that the UbVS is well mixed.
3. Incubate samples at 24 °C with gentle shaking (1250 rpm).
4. After 10 minutes, split both samples in half. For each pair of tubes, add ubiquitin to achieve a final concentration of 50 μM to one tube, and add an equivalent volume of XB to the other. Mix by pipetting gently multiple times.
5. Collect 2 μL for proteomic analysis as time 0 from the extract treated with only XB (control) by pipetting it in a clean 1.5-mL tube and adding 100 μL of quenching buffer for mass spectrometry. After buffer addition, mix by pipetting, vortex for at least 30 seconds, and flash freeze the samples and store at −80 °C. Samples can be stored for years. In parallel, collect time 0 from all the conditions by pipetting 1 μL of extract into 10 μL of nonreducing LDS sample buffer and vortex briefly.

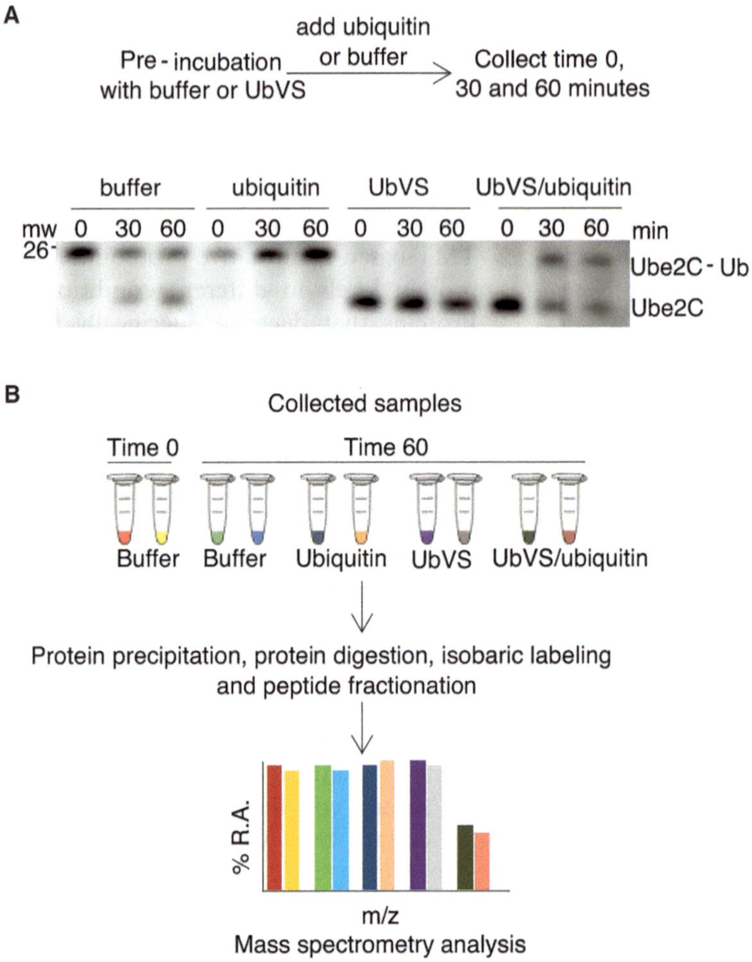

Fig. 1 Identifying proteins protected from degradation by DUBs with TMT-proteomics. (**a**) Top: Experimental scheme. Treat the extract with 10-μM UbVS or buffer for 10 minutes. After 10 minutes, split both samples in two and add 50-μM ubiquitin to one sample and buffer to the other. Collect samples for proteomic analysis at time 0 from extract treated with only buffer. Collect time 0 from all the conditions for Ube2C immunoblot analysis and quench with nonreducing LDS sample buffer. Incubate the remaining extract at 24 °C, and collect samples for proteomic analysis after 1 hour and samples for E2 charging analysis after 30 minutes and 1 hour. Bottom: Immunoblot analysis of Ube2C is shown. (**b**) Summary of treatment and analysis of the samples for TMT mass spectrometry. R.A., relative amount. The graph represents the expected behavior of potential DUB substrates, whose abundance should decrease specifically in the presence of UbVS/ubiquitin but remain stable in all the other conditions

6. Incubate the remaining extract at 24 °C (1250 rpm) for 1 hour. Collect samples for proteomic analysis and E2 charging as described above.
7. Before treating the samples for mass spectrometry analysis, analyze the samples by nonreducing SDS-PAGE and immunoblotting for Ube2C. UbVS treatment should induce ubiquitin depletion visualized as the discharge of Ube2C. The addition of ubiquitin should be sufficient to recharge Ube2C for the experiment duration.

3.5 Processing Samples for TMT-Proteomic Analysis

3.5.1 Methanol–Chloroform Precipitation, Protein Digestion, and TMT Isobaric Labeling

1. Thaw frozen samples rapidly. Add 400 μL of 100% methanol to each sample and vortex for 5–10 seconds. Add 100 μL of 100% chloroform and vortex for 5–10 seconds. Lastly, add 300 μL of water and vortex for 5–10 seconds (*see* **Note 15**).
2. Centrifuge the samples for 5 minutes at top speed (23,000 g). After centrifugation, you will see the formation of two layers (one aqueous and one organic). A thin white disk will form at the interface between the two layers, the proteins. Aspirate both layers using a P1000 pipette, starting from the top layer, leaving the disk intact (*see* **Note 16**).
3. Wash the protein pellet twice by adding 800-μL methanol. Vortex briefly and centrifuge at 23,000 g for 2 minutes. Remove the supernatant without drying the pellet completely (usually, around 10 μL of methanol remains).
4. Resuspend the samples in 100 μL of 200-mM EPPS (pH 8.5). Add LysC at a 1:50 enzyme/protein ratio and incubate samples overnight at room temperature with gentle agitation (*see* **Note 17**). Add trypsin at 1:100 protease/protein ratio and digest the samples for 6 hours at 37 °C (*see* **Note 18**).
5. Halt digestion by adding 25-μL anhydrous acetonitrile (ACN) to each sample.
6. Add 10 μL of TMT labeling reagent (the stock is at 20 μg/μL) to the digested peptides. Mix with a gentle vortex (2 seconds). Incubate for 1 hour at room temperature, briefly vortexing the samples every 10 minutes.

3.5.2 Quality Control Analysis

In TMT-based quantitative proteomics, each sample is separately labeled with a different isobaric TMT. The samples are subsequently combined, fractionated, and analyzed by mass spectrometry to identify and quantify each sample's relative abundance of peptides. Prior to proceeding with the final pooling of the samples and their fractionation, it is important to perform a quality control step in which a small amount of each sample is pooled together and analyzed by mass spectrometry to confirm that peptide digestion is successful, whether each of the labeled samples contains roughly the same amount of total peptides and if the degree of labeling is sufficient [16]. The labeling efficiency should be greater than 95%

before the analysis. If the labeling efficiency is <95%, we suggest repeating the labeling (**step 6** of Subheading 3.5.1). For example, if you have a labeling efficiency of 80%, we suggest adding 20% more labels and incubating for another hour. We suggest repeating the label check after adding more label. If the labeling efficiency is sufficient, we pool the final sample. We sum the TMT signal for each channel for all peptides and adjust the amount to mix for each sample using the sample with the lowest TMT signal as the baseline. Samples should vary by less than 1.5-fold. If the ratio between samples is greater than 1.5-fold, it is better to perform the experiment again and collect new samples. Some possible causes of large disparities include inaccuracy in concentration estimate with the BCA assay, improper digestion conditions (e.g., pH or protein/protease ratio), and/or inconsistent protein precipitation.

1. Pool together 2 μL of each labeled sample and add 100 μL of 1% formic acid to the combined samples. Freeze the remaining samples for analysis without pooling them by placing at −80 °C. These samples will be processed following the quality control analysis described in this section.
2. Desalt and clean up the combined sample by solid-phase extraction (SPE), using commercially available StageTips (*see* Subheading 2.4) or a homemade equivalent. C18 disks 0.4 mm in diameter and 0.5 mm in length have a binding capacity of 2–4 μg of digested proteins [17]. We use the equivalent of six disks per StageTip. StageTips are made manually by inserting these small disks of C18 beads embedded in a soft mesh of PTFE (polytetrafluoroethylene) into P200 pipette tips [17].
3. Place a StageTip adaptor in a 2-mL microcentrifuge tube and insert the StageTip in the adaptor. Add 100-μL methanol, centrifuge for 3 minutes at 1500 g, and discard the eluate. The time is an estimate as it is important that all of the added buffer flows out of the tip before the new buffer is added and that the tip does not dry entirely from excessive centrifugation time.
4. Add 100 μL of 1% formic acid and 70% ACN, centrifuge as before, and discard the eluate.
5. Equilibrate the StageTip with 100 μL of 1% formic acid and 5% ACN, centrifuge, and discard the eluate.
6. Load the sample from **step 1** on the StageTip, centrifuge, and discard the eluate.
7. Wash StageTip with 100 μL of 1% formic acid and 5% ACN, centrifuge, and discard the eluate.
8. Elute with 100 μL of 1% formic acid and 70% ACN, centrifuge, and dry the sample by vacuum centrifugation.
9. Add 100 μL of 5% ACN and 5% formic acid. The sample is now ready for mass spectrometry analysis.
10. Perform the ratio and label checks as described [16].

3.5.3 Peptide Desalting and Fractionation Via Basic pH HPLC

1. If the results show an acceptable ratio and the sample is adequately labeled, thaw the labeled samples from **step 1** on ice. Add hydroxylamine at a final concentration of ~0.3%, and incubate for 15 minutes at room temperature to quench the labeling reaction (*see* **Note 19**).
2. Combine the samples based on the results of the ratio check. If all the samples contain the same amounts of peptides, you can combine them 1:1. If there are slight differences among the samples (less than 1.5-fold), adjust the relative amounts. Freeze the sample in dry ice and evaporate all the ACN by vacuum centrifugation (*see* **Note 20**).
3. To clean up and desalt the sample, use one C18-based Sep-Pak cartridge under vacuum (pull through at about 1 mL/min). Sep-Pak units are solid-phase extraction (SPE) cartridges. The principle of operation is identical to that of StageTips, but Sep-Paks can accommodate a larger capacity of peptides. The binding capacity of the column is estimated to be 1% of the column matrix. In this case, for 1 mg of peptides (ten samples, 100 μg each), a 100-mg Sep-Pak cartridge (column volume 3 mL) can be used.
4. Activate the Sep-Pak with 1-column volume (CV) of methanol. Wash with 1 CV of a solution containing 70% ACN, 1% *trifluoroacetic acid* (TFA), and equilibrate with 1 CV of 1% TFA, 5% ACN.
5. Acidify the sample by adding a few μL of TFA 100% (pH 2–3) and load the sample (typically 1–1.5 mL) to the column.
6. Wash with 1 CV of 1% formic acid (FA), 5% ACN, and elute with 500–750 μL of 70% ACN, 1% formic acid.
7. Freeze the sample and evaporate off the ACN by vacuum centrifugation.
8. Resuspend peptides in 500 μL of 10-mM ammonium bicarbonate and 5% ACN at pH 8.
9. Fractionate up to 300 mg of the peptide mixture (with a 60-min linear gradient from 13% to 42% ACN in 10-mM ammonium bicarbonate pH 8) into a total of 96 fractions of 200 μL each, and collect them in a 96-well plate. Use an Agilent 300 Extend C18 column (3.5-μm particles, 4.6-mm ID, and 220-mm long) on an Agilent 1260 HPLC (flow rate: 0.6 mL/min) or equivalent column and LC.
10. Next, we suggest concatenating these fractions to improve proteome coverage and reduce mass spectrometry acquisition time [18]. We combine and consolidate 96 fractions into 24. Specifically, we combine four alternating wells vertically into one fraction (i.e., wells A1, C1, E1, and G1 will constitute fraction 1; B1, D1, F1, and H1, fraction 2; etc.), resulting in 24 super-fractions. Note that each super-fraction has on

average 12.5 mg of peptides. Because peptides elute horizontally in the 96-well plate during HPLC fractionation, these four wells combined have minimal overlapping peptides. On the other hand, because adjacent wells in the plate contain overlapping peptides, we analyze only 12 super-fractions from non-adjacent columns [18]. We usually analyze the columns that have even numbers. The remaining 12 super-fractions (collected from the columns with odd numbers) can be stored at −80 °C as a backup if needed or if specific peptides of interest are sought.

11. Desalt each fraction via StageTip as described previously (**step 2**, Subheading 3.5.2). Lyophilize to dryness by vacuum centrifugation, and reconstitute in 10 μL of 5% ACN and 5% formic acid for LC–MS/MS.
12. Inject 4 μL onto the mass spectrometer. This volume should represent 3–4 μg of peptides. We analyzed our samples on an Orbitrap Fusion mass spectrometer coupled with Proxeon EASY-nLC 1000 liquid chromatography (LC). For peptide detection, identification, and quantification, see Rossio et al. [14].

3.6 Validation of Candidate Substrates Identified by TMT-Based Proteomic Analysis

We suggest independently validating the findings of the proteomics experiment. We usually in vitro translate and label the candidate DUB substrates with ^{35}S-methionine [14]. Substrates are added to *Xenopus* egg extract, and stability is monitored by autoradiography. The addition of UbVS together with ubiquitin should induce their specific degradation. As an alternative to in vitro transcription/translation, an immunoblotting experiment with specific antibodies against the candidate substrates can be performed.

1. Amplify the cDNAs of potential DUB substrates by PCR, using a primer that includes a T7 promoter, to allow T7-dependent transcription. Clean up the reaction using a PCR purification kit.
2. Translate the PCR product using a coupled transcription/translation TNT system in the presence of ^{35}S-methionine. Use the amount of DNA and methionine suggested by the manufacturer's instructions of the coupled transcription/translation TNT system.
3. Incubate the reaction at 30 °C for 90 minutes.
4. Add cycloheximide to a final 100-μg/mL concentration to the translation mix and incubate for 10 minutes at room temperature (*see* **Note 21**).
5. Add the translation mix to the *Xenopus* extract (at 8% final volume) pre-treated with XB, ubiquitin, UbVS, or UbVS/ubiquitin, as in the proteomic experiment. Mix well, collect a sample immediately after substrate addition (time 0), and

quench it with sample buffer. Incubate the extract at 24 °C in the orbital mixer (1250 rpm) for 1 hour. Collect samples at 30 and 60 minutes for all the conditions and quench with sample buffer.

6. Boil the samples for at least 2 minutes, spin at 23,000 × *g* (30 seconds) in a bench centrifuge, and analyze by reducing SDS-PAGE followed by autoradiography on film or phosphorimaging.
7. The radioactive signal of the translated substrates should decrease over time, specifically in the extract treated with UbVS or UbVS/ubiquitin [14].

3.7 Identification of DUBs Sufficient to Rescue Substrate Degradation in UbVS-/Ubiquitin-Treated Extract

To identify DUBs that are sufficient to counteract proteasomal degradation of candidate substrates, we tested a panel of recombinant DUBs for the ability to stabilize ^{35}S-methionine-labeled substrates in extracts treated with UbVS/ubiquitin. In these conditions, the UbVS is fully consumed by reaction with endogenous DUBs, so that the activity of the added DUB will be not affected. This approach circumvents DUB redundancy to reveal the activity and specificity profile of single DUBs. The presence of the endogenous DUBs in *Xenopus* extract is in the concentration range from 5 nM to more than 2 mM [19]. We suggest starting with a high concentration of enzyme (500 nM–1 μM), followed by a dose–response analysis of active DUBs.

1. Before testing whether DUBs can rescue protein degradation, it is essential to confirm the activity of each recombinant, purified DUB. For cysteine protease DUBs, we do this by confirming that the DUB is UbVS-reactive.
2. To verify DUB enzyme activity, incubate every single DUB with a saturating amount of UbVS for 1 hour at 37 °C. We suggest incubating 1-μM DUB with 10 μM of UbVS (*see* **Note 22**). After 1 hour, add sample buffer to quench the reaction, boil for 2 minutes at 100 °C, and analyze by SDS-gel electrophoresis (*see* **Note 23**). If the DUB is active, it will covalently react with the UbVS and be visible as a gel shift of around 9 kDa.
3. Stain the gel with Coomassie blue solution for 1 hour. Incubate the gel in the destaining solution as long as needed.
4. After the activity of the DUB is verified, it can be added to the extract. It is important to demonstrate that the DUB added to the extract is not getting inactivated by unreacted UbVS present in the extract (*see* **Note 24**). Incubate the extract 30 minutes with UbVS or XB (control), and then add one of the recombinant DUBs of interest to the extract (*see* **Note 25**). Mix well, take a time point (time 0), and quench it with sample buffer. Continue to incubate the extract and take time points at 15 and 30 minutes and quench them with sample buffer.

5. Analyze the samples by reducing SDS-PAGE followed by immunoblotting with specific antibodies against the recombinant DUB. If the DUB is inactivated by residual UbVS present in the extract, it will be present as a gel shift of around 9 kDa, specifically in UbVS-treated extract (*see* **Note 26**).
6. After the DUBs are verified not to be inactivated after their addition to the UbVS-treated extract, you can proceed to identify the DUBs that can rescue degradation of the protein substrates revealed by the previous proteomic experiment (*see* **Note 27**).
7. Translate the PCR product of your selected substrates using the coupled transcription/translation system in the presence of ^{35}S-methionine, as done previously. Pre-treat the extract with UbVS for 30 minutes or with MES buffer as a control. Add the in vitro labeled substrates to both the reactions (*see* **Note 28**) and in addition ubiquitin to the UbVS-treated extract. Keep the reactions on ice.
8. Split the extract treated with UbVS/ubiquitin and contain the labeled substrates into clean tubes kept on ice. The number of tubes will depend on the number of DUBs tested in the experiment, as each DUB should be tested in an independent reaction (*see* **Note 29**).
9. Add XB (control) and the recombinant DUBs to the extract kept on ice and mix very well. As previously, collect time 0 from each reaction and incubate the rest at 24 °C, as previously. Take samples at 30 and 60 minutes.
10. Process and analyze the samples by reducing SDS-PAGE as before, followed by autoradiography on film or phosphorimaging.
11. Determine the fraction of substrate remaining at 60 minutes relative to time zero for each condition by quantifying the radioactive signal. We use the Quantity One software (Bio-Rad). If a DUB is sufficient to antagonize substrate degradation, you will see an increase in the fraction of substrate remaining at 60 minutes compared to the reaction lacking the DUB. Since multiple DUBs can act on the same substrates, more than one DUB may rescue substrate degradation.

4 Notes

1. The transfer of the eggs into the tube containing the XB must be done gently because the eggs are easily broken after they have been dejellied.
2. When preparing this buffer, vortex immediately after you add cytochalasin as it tends to precipitate.

3. For small preparations, a microcentrifuge can be used. In this case, pack the eggs with a brief spin at 160 g, remove the excess buffer, and then crush the eggs at 14,000 rpm (21,000 × *g*).
4. Heating the needle with a flame helps in puncturing the tube.
5. Undissolved glycerol is visible as a transparent gel at the bottom of the falcon tube.
6. The extract can be defrosted once. Thawing and refreezing the extract are not recommended due to potential activity loss.
7. Since ubiquitin antibodies cannot distinguish between free monomeric ubiquitin and unreacted UbVS in extract, this method is less helpful in monitoring free ubiquitin availability in UbVS-treated extract.
8. The extract is very viscous. Cutting the pipette tip helps to pipet the correct volume of extract.
9. We started the UbVS dose–response at 5-μM UbVS because cysteine protease DUBs in *Xenopus* egg extract have an estimated total concentration of around 7 μM [19].
10. The extract should constitute at least 75% of the reaction volume in all experiments since excessive dilution can decrease extract activity.
11. Nonreducing conditions are essential to preserving the thioester bond of Ube2C with ubiquitin. Because the thioester is sensitive to heat, it is important to keep collected samples on ice and not boil them. Performing gel electrophoresis in a cold room is preferable to avoid excessive gel heating.
12. We usually load 50 μg of protein per lane of the gel, as this amount is sufficient to detect most proteins by immunoblotting.
13. We observed that the UbVS concentration required to induce ubiquitin depletion decreases at longer incubation times. For example, 5-μM UbVS can induce the full discharge of Ube2C by 30 minutes (data not shown). This time and concentration dependence likely reflects the ongoing conjugation of ubiquitin to extract proteins by E2 and E3 enzymes in the extract.
14. All samples for mass spectrometry analysis should be collected at least in duplicate, with triplicate or quadruplicate being preferable for obtaining lower p-values during statistical analysis.
15. Both volumes and tubes can be scaled up, if necessary.
16. By gently tilting the tube, the protein pellet attaches to the side of the tube, making it easy to aspirate both layers.
17. The solution may turn cloudy due to the precipitation of undigested proteins. This precipitation is not a problem as the cloudiness disappears with digestion by trypsin.

18. We observed that the number of identified peptides increased significantly with Lys-C/trypsin sequential digestion compared to single-enzyme digestion.
19. We usually quench the labeling reaction after the label check. If the check reveals incomplete labeling, we add more labeling reagents. This check will not be possible if we have already quenched the reaction.
20. The samples do not need to be completely dry as overdrying pellets often makes them challenging to resuspend, and proteases remain active in up to 20% methanol.
21. The addition of cycloheximide to the translation mix ensures that there is no protein synthesis after adding the reticulocyte mix to the *Xenopus* egg extract. The extract has already been treated with cycloheximide (*see* Subheading 3.1).
22. Most DUBs should completely react with the UbVS [14]. However, some DUBs that require accessory proteins for activation may not react with UbVS when the individual enzyme is treated with the inhibitor, as is the case for USP14, which requires the proteasome [10].
23. Since most DUBs have a high molecular weight (more than 100 kDa), we suggest performing the SDS-gel electrophoresis with a gel containing a low polyacrylamide concentration (3–8%) in Tris-acetate SDS Running buffer. This concentration of polyacrylamide facilitates the detection of the mobility shift induced by the covalent addition of UbVS.
24. We observed that increasing the duration of UbVS incubation in an extract from 10 to 30 minutes is essential to consume all the unreacted UbVS present in the extract.
25. To establish that there is no residual unreacted UbVS in the extract, we usually add 800-nM USP7 because it is highly sensitive to UbVS.
26. If the DUB is inactivated by residual UbVS present in the extract, we suggest increasing the incubation time of the extract with UbVS or decreasing the concentration of UbVS.
27. We suggest starting with substrates that have been independently validated (*see* Subheading 3.4).
28. Multiple substrates can be examined together in the same experiment if the substrates have different molecular weights. In this way, you can test the effect of one DUB on the stability of multiple substrates simultaneously.
29. We suggest testing the DUB at a concentration of 800 nM to 1 μM.

Acknowledgments

We acknowledge W. Harper for the gift of plasmids (ORFeome collection). This work was supported by the Cell Biology Education and Fellowship Fund (V.R.), NIH grants R01 GM132129 (J.A.P.) and R35 GM127032 (R.W.K.).

References

1. Swatek KN, Komander D (2016) Ubiquitin modifications. Cell Res 26:399–422. https://doi.org/10.1038/cr.2016.39
2. Hershko A, Ciechanover A (1998) The ubiquitin system. Annu Rev Biochem 67:425–479. https://doi.org/10.1146/ANNUREV.BIOCHEM.67.1.425
3. Kulathu Y, Komander D (2012) Atypical ubiquitylation – the unexplored world of polyubiquitin beyond Lys48 and Lys63 linkages. Nat Rev Mol Cell Biol 13:508–523. https://doi.org/10.1038/NRM3394
4. Clague MJ, Barsukov I, Coulson JM, Liu H, Rigden DJ, Urbé S (2013) Deubiquitylases from genes to organism. Physiol Rev 93: 1289–1315. https://doi.org/10.1152/physrev.00002.2013
5. Komander D, Clague MJ, Urbé S (2009) Breaking the chains: structure and function of the deubiquitinases. Nat Rev Mol Cell Biol 10: 550–563. https://doi.org/10.1038/nrm2731
6. Clague MJ, Urbé S, Komander D (2019) Breaking the chains: deubiquitylating enzyme specificity begets function. Nat Rev Mol Cell Biol 20:338–352. https://doi.org/10.1038/s41580-019-0099-1
7. Mevissen TET, Komander D (2017) Mechanisms of deubiquitinase specificity and regulation. Annu Rev Biochem 86:159–192. https://doi.org/10.1146/annurev-biochem-061516-044916
8. Beckley JR, Chen JS, Yang Y, Peng J, Gould KL (2015) A degenerate cohort of yeast membrane trafficking DUBs mediates cell polarity and survival. Mol Cell Proteomics 14:3132–3141. https://doi.org/10.1074/MCP.M115.050039/ATTACHMENT/6D1E40C1-3C15-4C36-9FDF-39B39F544CF6/MMC1.ZIP
9. Kwon SK, Saindane M, Baek KH (2017) p53 stability is regulated by diverse deubiquitinating enzymes. Biochim Biophys Acta Rev Cancer 1868:404–411. https://doi.org/10.1016/J.BBCAN.2017.08.001
10. Borodovsky A, Kessler BM, Casagrande R, Overkleeft HS, Wilkinson KD, Ploegh HL (2001) A novel active site-directed probe specific for deubiquitylating enzymes reveals proteasome association of USP14. EMBO J 20: 5187–5196. https://doi.org/10.1093/emboj/20.18.5187
11. Murray AW (1991) Chapter 30 Cell cycle extracts. Methods Cell Biol 36:581–605. https://doi.org/10.1016/S0091-679X(08)60298-8
12. Salic A, King RW (2005) Identifying small molecule inhibitors of the ubiquitin-proteasome pathway in Xenopus egg extracts. Methods Enzymol 399:567–585. https://doi.org/10.1016/S0076-6879(05)99038-1
13. Dimova NV, Hathaway NA, Lee BH, Kirkpatrick DS, Berkowitz ML, Gygi SP, Finley D, King RW (2012) APC/C-mediated multiple monoubiquitylation provides an alternative degradation signal for cyclin B1. Nat Cell Biol 14:168–176. https://doi.org/10.1038/NCB2425
14. Rossio V, Paulo JA, Chick J, Brasher B, Gygi SP, King RW (2021) Proteomics of broad deubiquitylase inhibition unmasks redundant enzyme function to reveal substrates and assess enzyme specificity. Cell Chem Biol 28:487–502.e5. https://doi.org/10.1016/J.CHEMBIOL.2020.12.007
15. Navarrete-Perea J, Yu Q, Gygi SP, Paulo JA (2018) Streamlined Tandem Mass Tag (SL-TMT) Protocol: an efficient strategy for quantitative (phospho)proteome profiling using Tandem Mass Tag-synchronous precursor selection-MS3. J Proteome Res 17:2226–2236. https://doi.org/10.1021/acs.jproteome.8b00217
16. Gupta M, Sonnett M, Ryazanova L, Presler M, Wühr M (2018) Quantitative proteomics of Xenopus embryos I, sample preparation. Methods Mol Biol 1865:175–194. https://doi.org/10.1007/978-1-4939-8784-9_13
17. Rappsilber J, Ishihama Y, Mann M (2003) Stop And Go Extraction tips for matrix-assisted laser

desorption/ionization, nanoelectrospray, and LC/MS sample pretreatment in proteomics. Anal Chem 75:663–670. https://doi.org/10.1021/AC026117I

18. Yang F, Shen Y, Camp DG, Smith RD (2012) High-pH reversed-phase chromatography with fraction concatenation for 2D proteomic analysis. Expert Rev Proteomics 9:129–134. https://doi.org/10.1586/EPR.12.15

19. Wühr M, Freeman RM, Presler M, Horb ME, Peshkin L, Gygi SP, Kirschner MW (2014) Deep proteomics of the Xenopus laevis egg using an mRNA-derived reference database. Curr Biol 24:1467–1475. https://doi.org/10.1016/j.cub.2014.05.044

Chapter 14

Identification of Deubiquitinase Substrates in *Saccharomyces cerevisiae* by Systematic Overexpression

Heather E. Arsenault and Jennifer A. Benanti

Abstract

A significant hurdle to understanding the functions of deubiquitinases (DUBs) is the identification of their in vivo substrates. Substrate identification can be difficult for two reasons. First, many proteins that are degraded by the ubiquitin–proteasome system are expressed at relatively low levels in the cell, and second, redundancy between DUBs complicates loss of function screening approaches. Here, we describe a systematic overexpression approach that takes advantage of genome-wide resources available in *S. cerevisiae* to overcome these challenges and identify DUB substrates in cells.

Key words Deubiquitinase, Yeast, Overexpression, Western blotting, Cycloheximide chase

1 Introduction

Ubiquitination is a widespread, reversible protein modification that impacts almost all cellular processes. Although ubiquitination can control protein function in many ways, one of its primary roles is to stimulate regulated protein degradation via the ubiquitin–proteasome system (UPS) [1, 2]. To trigger proteasomal degradation, ubiquitin chains are added to protein substrates through the action of an enzymatic cascade consisting of an E1 ubiquitin-activating enzyme, an E2 ubiquitin-conjugating enzyme, and an E3 ubiquitin ligase, which recognizes specific protein substrates. Although most specificity in the ubiquitination system is thought to be conferred by the large family of E3s, increasing evidence suggests deubiquitinases (DUBs) also play critical roles in controlling protein stability by removing ubiquitin chains and protecting proteins from degradation in the UPS [3, 4]. Therefore, comprehensive identification of proteins whose stabilities are regulated by DUBs is necessary to understand the functions of DUBs in the cell.

Although the activities of many DUBs are well-characterized in vitro, identifying in vivo substrates has been challenging. For

Julie A. Maupin-Furlow and Mariola J. Edelmann (eds.), *Deubiquitinases: Methods and Protocols*,
Methods in Molecular Biology, vol. 2591, https://doi.org/10.1007/978-1-0716-2803-4_14,

proteins whose ubiquitination leads to degradation in the proteasome, deletion of a regulatory DUB should result in increased ubiquitination of the target protein, accelerated degradation, and a decrease in steady-state protein levels. This approach has been used successfully to identify DUBs that regulate individual substrates. However, DUB deletion approaches have been less successful in unbiased screens designed to identify substrates, since many ubiquitinated proteins are at low abundance in the cell. Proteins of low abundance are difficult to identify using standard proteomic approaches, and identifying a decrease in protein levels following the deletion of a regulatory DUB is even more challenging [5]. Another issue that complicates identifying substrates is potential redundancy between DUBs [6, 7]. When two or more DUBs can target a specific substrate, deletion or inhibition of a single DUB may not have an appreciable effect on substrate levels. Simultaneous deletion of multiple DUBs may be necessary to see change in target levels or half-life; however, screening all combinations of DUB deletions is not readily achievable.

In the budding yeast *Saccharomyces cerevisiae,* several genome-wide resources are available that can be leveraged to overcome these challenges. First, collections of yeast strains in which every open reading frame is tagged in the genome with a tandem-affinity purification (TAP) or green fluorescence protein (GFP) tag enable systematic quantification of most proteins in yeast [8, 9]. Second, plasmid libraries that contain all genes under the control of a galactose-inducible promoter enable systematic overexpression of all proteins [10]. These resources can be easily combined. For example, a collection of DUB overexpression plasmids can be transformed into a library of strains expressing tagged proteins of interest, and the effects of DUB overexpression on protein levels are determined systematically [11]. In this scenario, overexpression of a regulatory DUB stabilizes its substrates and increases protein levels. In addition, since overexpression is a gain of function approach, regulatory DUBs can be identified even if redundancy between DUBs occurs typically.

The design of DUB substrate screens is further facilitated by publicly available datasets that have quantified the half-lives of most proteins in yeast [12–15]. To identify DUBs that regulate protein stability, candidate substrates can be limited to unstable proteins likely to undergo proteasomal degradation. Here, we describe the screening of TAP-tagged proteasomal substrates by semi-high-throughput Western blotting (Fig. 1). Although the method detailed here focuses on identifying proteins whose stabilities are regulated by DUBs, we note that the same tools could be adapted to screen for non-proteolytic outcomes of deubiquitination. For instance, this could be achieved by measuring changes in protein localization utilizing the collection of GFP-tagged protein strains.

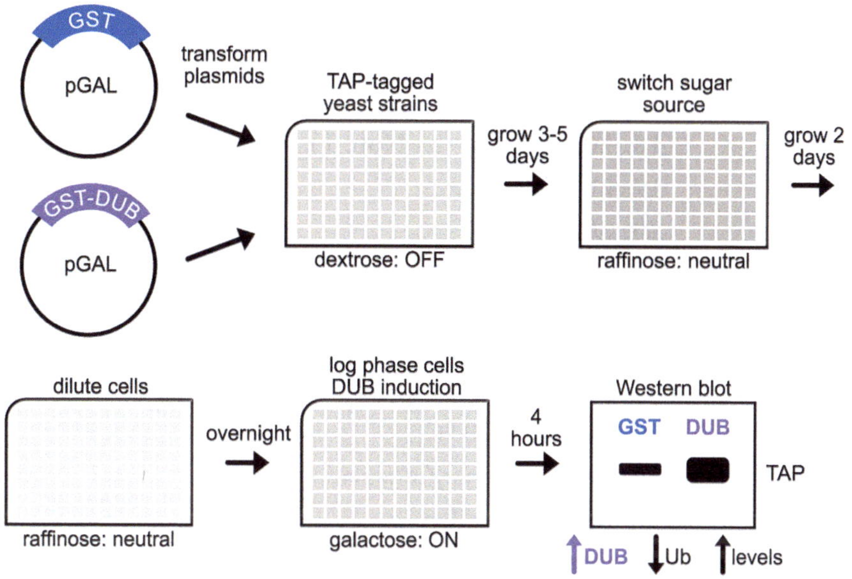

Fig. 1 Method for screening DUB substrates. Plasmids (pGAL-GST and pGAL-GST-DUB) are transformed into a selected subset of TAP-tagged yeast strains. Dextrose is present in the transformation media to keep expression of the *GAL1* promoter off. After transformants are grown, the carbon source is switched to raffinose to remove the active repression of the *GAL1* promoter that occurs in dextrose. To induce *GAL1* promoter plasmids, galactose is added to exponentially growing cells for 4 h. Protein levels of the TAP-tagged strains expressing GST and GST-DUB are run alongside each other to determine if the DUB increases substrate levels

2 Materials

Prepare all solutions using ultrapure water and store them at room temperature unless otherwise noted.

2.1 Equipment and Other Solutions

1. Water bath.
2. 30 °C microbiological incubator.
3. Aspirator.
4. Clear adhesive seals, sterile.
5. Aluminum adhesive seals, sterile.
6. Sterile plate lids.
7. Deep 96-well polypropylene plates, 2 mL.
8. Shallow 96-well polypropylene U-bottom plates, 0.3 mL.
9. 96-well PCR plates, 0.2 mL.
10. Multichannel pipette and tips (50–300 μL).
11. Standard pipettes and tips (10–1000 μL).
12. Sterile reagent reservoirs.
13. Gel loading tips.
14. Centrifuge with adapter for 96-well plates.

15. Thermocycler.
16. Spectrophotometer with cuvettes.
17. 20% dextrose.
18. 20% raffinose (*see* **Note 1**).
19. 20% galactose.
20. 30% glycerol.
21. 1.5-mL screw-cap tubes.
22. 15-mL conical centrifuge tubes.
23. 50-mL conical centrifuge tubes.
24. Cycloheximide (≥94%): prepare 50 mg/mL stock in ethanol. Store at −20 °C.
25. 96-well pinner/inoculating manifold.
26. Sterile test tubes.
27. Sterile 125 mL glass flasks.
28. Zip-top plastic bags.
29. Forceps.
30. Sheet protectors.

2.2 Yeast Media

1. YM-1 rich medium: 0.04 g of adenine and uracil, 5.5 g of ammonium sulfate, 11.0 g of peptone, 5.7 g of sodium hydroxide, 11.0 g of succinic acid, 5.5 g of yeast extract, 1.6 g of yeast nitrogen base without amino acids, ammonium sulfate, or carbohydrates. Adjust pH to 5.8 with 4 M NaOH and volume to 1 L. Sterilize by autoclaving and after cooled, supplement with 2% dextrose (dissolved in water and sterilized).
2. C-uracil synthetic medium (C-Ura): 5.0 g of ammonium sulfate, 1.4 g of C-uracil powder (all amino acids except uracil), 6.3 g of sodium hydroxide, 11.0 g of succinic acid, 1.4 g of yeast nitrogen base without amino acids, ammonium sulfate, or carbohydrates. Adjust pH to 5.8 with 4 M NaOH and volume to 1 L. Sterilize by autoclaving and after cooled, supplement with 2% dextrose or raffinose (dissolved in water and sterilized separately; *see* **Note 1**).

2.3 High-Throughput Yeast Plasmid Transformations

1. Sterile ultrapure water.
2. 1 M Lithium acetate.
3. 10 mg/mL salmon sperm DNA (ssDNA) (*see* **Note 2**). Store at −20 °C.
4. Plasmid DNA: pGAL-GST and pGAL-GST-DUB (*see* **Note 3**). Store at −20 °C.
5. 50% w/v PEG 3350 (*see* **Notes 4** and **5**).
6. TAP-tagged yeast strains (Horizon Discovery, *MATa his3Δ1 leu2Δ0 met15Δ0 ura3Δ0 XXX-TAP-HIS3MX*).

2.4 Protein Extraction with TCA

1. TCA buffer: 10% trichloroacetic acid (*see* **Note 6**), 25 mM ammonium acetate, 1 mM ethylenediaminetetraacetic acid (EDTA) disodium salt dihydrate, 10 mM Tris–HCl pH 8.0.
2. Resuspension solution: 3% sodium dodecyl sulfate (SDS), 0.1 M Tris–HCl pH 11.0.
3. 4× SDS-PAGE sample buffer: 8% SDS, 20% β-mercaptoethanol, 40% glycerol, 250 mM Tris–HCl pH 6.8, 0.2% bromophenol blue. Store at −20 °C.

2.5 SDS-PAGE and Western Blotting

1. Ultrapure water.
2. 30% acrylamide/bis solution (37.5:1). Store at 4 °C.
3. 1.5 M Tris–HCl pH 8.8.
4. 10% SDS.
5. 10% ammonium persulfate (APS) (*see* **Note 7**). Store at 4 °C.
6. Tetramethylethylenediamine (TEMED).
7. 1 M Tris–HCl pH 6.8.
8. Bio-Rad Criterion electrophoresis gel running cell.
9. Empty 26-well Criterion cassettes.
10. Bio-Rad Criterion blotter.
11. 5× running buffer (1 L): add 72 g of glycine, 15.1 g of Tris base base, and 5 g of SDS to ultrapure water and adjust volume to 1 L.
12. 1× running buffer: dilute from 5× stock to 1× immediately before use in ultrapure water.
13. 10× transfer buffer (1 L): add 24.21 g of Tris base, 112.56 g of glycine, and 18.75 mL of 20% SDS to ultrapure water and finalize volume to 1 L.
14. 1× transfer buffer (2 L): mix 200 mL of 10× transfer buffer, 400 mL of methanol, and 1.4 L of ultrapure water (*see* **Note 8**).
15. Use 1× transfer buffer for setting up transfer.
16. Nitrocellulose.
17. Whatman paper.
18. 10× PBS (4 L): add 8 g of KCl, 9.6 g of KH_2PO_4 (potassium phosphate monobasic), 320 g of NaCl, and 57.6 g of Na_2HPO_4 (sodium phosphate anhydrous dibasic), to ultrapure water. Adjust pH to 7.4. Adjust volume to 4 L and sterilize by autoclaving.
19. 1× PBST: 1× PBS diluted from 10× stock in ultrapure water with 0.05% Tween-20.
20. Blocking buffer: 4% instant nonfat dried milk in 1× PBST. Store at 4 °C.

21. Ponceau S (100 mL): add 1 g of Ponceau S and 1 mL of glacial acetic acid to 99 mL of ultrapure water.
22. Antibody buffer: 2% BSA, 0.02% NaN_3 in 1× PBST. Filter-sterilize and store at 4 °C.
23. Antibodies: rabbit anti-TAP (Invitrogen), mouse anti-GST (clone 4C10, BioLegend), mouse anti-PSTAIR (Sigma), goat anti-rabbit IgG-HRP, goat anti-mouse IgG-HRP.
24. Chemiluminescent substrate (*see* **Note 9**).
25. Bio-Rad ChemiDoc Touch Digital Chemiluminescent Imager (*see* **Note 10**).
26. Protein molecular weight marker.

3 Methods

3.1 High-Throughput Yeast Plasmid Transformations

The traditional high-efficiency lithium acetate method for yeast transformation can be scaled up from a single reaction to many reactions using 96 well plates [16]. Using a 96-well format allows for incorporating many unique plasmids into one strain of interest or a single plasmid into many strains of interest. In the example described here, the lithium acetate method is used to transform two plasmids into 1 set of 96 pre-selected TAP tag strains. First, a plasmid expressing glutathione-S-transferase from the *GAL1* promoter (pGAL-GST) is transformed into all strains in the 96-well plate. Second, the corresponding GST-DUB expressing plasmid (pGAL-GST-DUB) is transformed into these same strains. Following transformation and selection, GST and GST-DUB expression can be induced simultaneously to determine if TAP-tagged protein levels change following DUB overexpression.

1. From a 96-well plate of previously selected strains expressing TAP-tagged proteins, use a sterile pinner (*see* **Note 11**) to inoculate cells into a deep 96-well plate containing 300 μL YM-1/dextrose. This inoculation can be initiated from a frozen plate containing desired strains arrayed or from individual cultures that have been inoculated into a deep 96-well plate.
2. Cover the plate with a clear seal and incubate 1–2 days at 30 °C.
3. After cell pellets of equal size are observed in all wells, dilute 1:30 into a new deep 96-well plate containing 300 μL of YM-1/dextrose, and mix by pipetting. Incubate overnight 12–16 h at 30 °C.
4. Prepare transformation master mixes in 15 mL conical centrifuge tubes, vortex, and keep on ice. Make separate master mixes for pGAL-GST and pGAL-GST-DUB plasmids. An example of components can be found in Table 1 (*see* **Note 12**).

Table 1
Example recipe for plasmid transformation master mix

Component	One well	120 wells (96 + 24 extra, for one plate)
1 M lithium acetate	15 μL	1800 μL
10 mg/mL ssDNA (*see* **Note 2**)	4 μL (40 μg)	480 μL
Plasmid DNA (0.1–0.4 μg per well)	0.2 μL (~200 ng)	24 μL
Sterile water	30.8 μL	3696 μL
Total volume	50 μL	6 mL

5. Aliquot 100 μL of sterile water and 50 μL of cells from the overnight culture plate into two new deep 96-well plates.
6. Cover the plates with clear seals; vortex; then centrifuge at room temperature 1500 × g for 5 min. Pellets should be visible in all wells after spin.
7. Invert each plate and flick sharply over the sink twice to remove supernatants. Cover with sterile plate lid.
8. Pipette 50 μL of transformation master mixes into wells and mix by pipetting gently five to six times (*see* **Note 13**). Incubate at room temperature for 5 min.
9. Add 100 μL of 50% PEG with cut tips (*see* **Notes 4** and **5**) to wells and mix by pipetting or vortex for 2 s. Cover plates with aluminum seals and incubate at room temperature for 5 min.
10. Incubate in 42 °C water bath for 1 h (*see* **Note 14**).
11. Centrifuge plates at 1500 × g (room temperature) for 10 min.
12. Remove supernatant by inverting plates and flicking sharply over sink twice. Cells should be visible here but will not form clear pellets.
13. Resuspend cells in 300 μL of C-Ura/dextrose by gently pipetting.
14. Cover plates with clear seals and incubate in a 30 °C incubator for 4–6 days or until cell pellets are visible in each well (*see* **Note 15**).

3.2 Freezing and Thawing Cells

After transformants are grown up evenly in all wells, freeze aliquots for future use.

1. To freeze cells, aliquot 100 μL of 30% glycerol into a sterile, shallow 96-well polypropylene U-bottom plate.
2. Vortex transformation plate; then pipette 100 μL of saturated cells into a shallow-well plate containing glycerol. Mix cells and glycerol by pipetting.

3. Securely cover with aluminum adhesive seal and plate lid.
4. Immediately put the plate on dry ice for 10 min or until all wells are frozen.
5. Store plate at −80 °C.
6. To thaw cells, aliquot 300 μL of C-Ura/dextrose to a deep 96-well polypropylene plate (*see* **Note 16**).
7. Sterilize the pinner (*see* **Note 11**) by soaking in 100% ethanol.
8. Shake off excess ethanol from the pinner; then put it carefully through the flame.
9. Allow the pinner to cool slightly before putting onto frozen cells. Swirl to remove cells and use to inoculate a deep-well plate containing medium.
10. Cover the plate with a clear seal and incubate at 30 °C for 3–6 days.

3.3 Culturing Transformants and DUB Induction

1. Grow cells in C-Ura/dextrose for 4–6 days after plasmid transformation until cell pellets of equal size are visible in the bottom of each well (*see* **Note 17**).
2. Measure OD_{600} values for a few wells using standard single cuvettes and spectrophotometer. Saturated cultures in C-Ura medium should be between 3 and 5 OD_{600}. OD_{600} values may vary between wells.
3. Dilute saturated transformation plates 1:20 in 300 μL of C-Ura/raffinose and incubate at 30 °C for 2–3 days (*see* **Notes 18** and **19**).
4. After cells grow to saturation, dilute 1:300 into a new plate containing 1 mL of fresh C-Ura/raffinose, cover plates with clear seal, and incubate overnight (12–16 h) at 30 °C. This step aims to have cells growing in logarithmic phase the next day (*see* **Note 20**).
5. Measure OD_{600} values in a few wells. The ideal range before galactose induction is between 0.3 and 0.4 OD_{600} (*see* **Note 21**).
6. To induce GST and GST-DUB expression, add galactose to a final concentration of 2% for 4 h at 30 °C (*see* **Note 22**). Add clear seal and vortex plates gently to mix.
7. After incubation, use a multichannel pipette to add 200 μL of YM-1 (without sugar) to each well to aid in pelleting. Centrifuge the plate at 3,095 × g for 10 min.
8. Aspirate supernatant from each well using a gel loading tip attached to an aspirator, seal plates with a sterile aluminum adhesive seal, and freeze plate on dry ice. Store at −80 °C until protein extraction.

3.4 Protein Extraction with TCA

1. Add 240 μL of cold TCA buffer to pellets in 96-well plates with a multichannel pipette, but do not mix. Incubate at room temperature for 10 min (*see* **Note 23**).
2. When thawed, vortex plates in order to resuspend the pellets (*see* **Note 24**).
3. Centrifuge plates at 4 °C, 3,095 × g for 10 min.
4. Using a gel loading tip attached to an aspirator, aspirate TCA buffer from cell pellets (*see* **Note 25**).
5. Use a multichannel pipette to resuspend pellets in 60 μL of resuspension solution, transfer samples to 96-well PCR plates, and cover with an aluminum adhesive seal.
6. Heat samples at 95 °C in a thermocycler for 5 min. Remove plates and cool at room temperature for 5 min.
7. Centrifuge plates at 3,095 × g for 1 min.
8. Remove supernatant (60 μL) and transfer to new 96-well PCR plates with a multichannel pipette.
9. Add 20 μL of 4× SDS-PAGE sample buffer to supernatants and pipette up and down to mix.
10. Cover plates with an aluminum adhesive seal, vortex to mix, and then heat at 95 °C in a thermocycler for 5 min.
11. Allow samples to cool for a few minutes at room temperature if immediately proceeding to SDS-PAGE. Lysates can also be stored at −80 °C at this stage for up to several months.

3.5 SDS-PAGE, Western Blotting, and Data Quantification

3.5.1 SDS-PAGE and Western Blotting

This protocol uses the Bio-Rad Criterion gel system but can be modified for use with other electrophoresis systems. For an experiment with two 96-well plates (1 containing pGAL-GST and 1 containing pGAL-GST-DUB), all samples can be run on 8 Criterion gels with 26 wells each. The 8 gels can be run in 4 Criterion gel boxes or in 1 Criterion Dodeca cell, which can run up to 12 gels simultaneously.

1. For each 10% resolving gel (*see* **Notes 26** and **27**), make 10 mL solution in a 50 mL conical tube. This is done by combining 4 mL of ultrapure water, 3.3 mL of 30% acrylamide/bis solution (37.5:1), 2.5 mL of 1.5 M Tris–HCl pH 8.8, 0.1 mL of 10% SDS, 0.1 mL of 10% APS, and 0.004 mL of TEMED.
2. Invert solution gently five times and slowly pipette 9 mL into an empty Criterion cassette with a 10 mL pipette (*see* **Note 28**).
3. Immediately overlay resolving gel with ultrapure water and allow to polymerize for at least 1 h at room temperature.
4. After the gel has polymerized, add more ultrapure water to cover, place a wet paper towel on top, and wrap in plastic wrap. Store gel overnight at 4 °C (*see* **Note 29**).

5. The next day, unwrap and invert the Criterion cassette to pour off the water. Use a paper towel to remove any excess water. Let sit for 15–60 min before pouring the stacking gel.
6. For each gel, make a 5 mL stacking gel in a 15 mL conical tube by combining 3.4 mL of ultrapure water, 0.83 mL of 30% acrylamide/bis solution (37.5:1), 0.63 mL of 1 M Tris–HCl pH 6.8, 0.05 mL of 10% SDS, 0.05 mL of 10% APS, and 0.005 mL of TEMED.
7. Invert solution gently five times, and immediately add to the top of resolving gel. Slowly insert the comb. Allow the gel to polymerize at room temperature for 30 min.
8. Use freshly prepared lysates or thaw frozen lysates at room temperature. Vortex plate and do a quick spin.
9. Prepare a 1× running buffer working solution from the 5× running buffer stock. This dilution should be done the same day as running the gel.
10. After the stacking gel has polymerized, load the Criterion cassette into the gel running box, and add 1× running buffer according to the manufacturer's instructions.
11. Load 20 μL of GST and GST-DUB strains alongside each other into wells (*see* Fig. 2 and **Note 30**), and run at 140 V for 1.5 h or until the dye front is at the bottom of the gel.
12. When gel running is complete, presoak the nitrocellulose membrane and Whatman paper for 2 or more minutes in used 1× transfer buffer.
13. Add ice pack to the back of transfer cell and fill halfway with pre-chilled transfer buffer.
14. Break open Criterion cassette, remove the gel, and carefully assemble transfer sandwich in a cassette in the following order, with the black side of the transfer cassette facing down: sponge, Whatman paper, gel, nitrocellulose, Whatman paper, and sponge. Use a roller to squeeze out any excess bubbles between layers. Close transfer cassette and load in the proper orientation in the cell with the black side of cassette facing the ice pack. Add transfer buffer up to fill line.
15. Transfer at 0.45 A for 2 h (*see* **Note 31**).
16. Disassemble transfer sandwich, carefully remove nitrocellulose from cassette, and check for even protein levels by staining with Ponceau S for 2–5 min, then rinsing under ultrapure water. Trim edges of nitrocellulose as desired.
17. Add enough blocking buffer to cover gel and incubate for 15–30 min on a platform shaker at room temperature.
18. Pour off blocking buffer and briefly rinse two to three times in PBST.

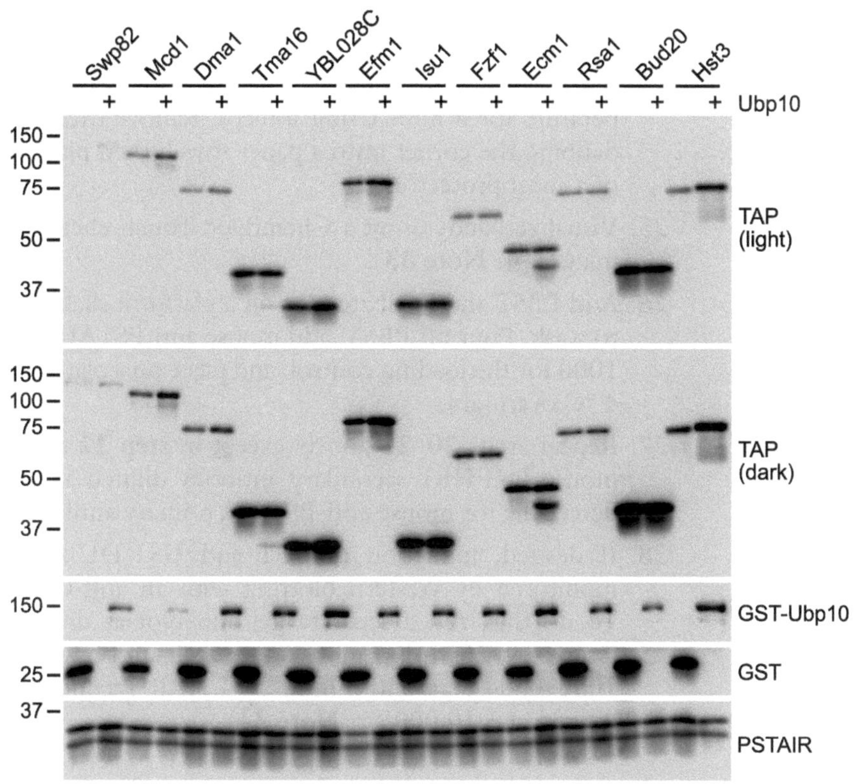

Fig. 2 Representative Western blot of TAP-tagged proteins following GST-Ubp10 overexpression. Exponentially growing TAP-tagged strains containing pGAL-GST or pGAL-GST-UBP10 were grown in C-Ura/raffinose. Galactose was added for 4 h to induce *GAL1* promoter expression. TAP-tagged proteins were visualized by Western blot; light and dark exposures are shown. GST blots show GST and GST-DUB protein levels. Note that GST and GST-Ubp10 panels have been cropped from the same blot due to the significant difference in their molecular weights. PSTAIR is shown as a loading control

19. Dilute rabbit anti-TAP primary antibody 1:1000 in antibody buffer and place on a platform shaker at 4 °C overnight.
20. The next day, remove TAP antibody and save the aliquot at 4 °C for future use (*see* **Note 32**).
21. Add enough PBST to cover the blot and incubate on a platform shaker for 15 min at room temperature. Repeat wash two additional times.
22. Remove PBST and add goat anti-rabbit IgG-HRP secondary antibody diluted at 1:10,000 in blocking buffer. Incubate on a platform shaker for 30 min at room temperature.
23. Remove and dispose of secondary antibody. Add PBST and incubate blot on a platform shaker for 15 min to wash. Repeat wash two additional times.

24. Remove blot from buffer with forceps and carefully dab corner on a paper towel to remove any excess buffer. Place blot flat onto plastic wrap; add ECL substrate; incubate at room temperature for 5 min. Using forceps, remove excess reagent by dabbing the corner onto a paper towel; then place blot inside of a sheet protector.
25. Visualize bands using a ChemiDoc Touch chemiluminescent imager (*see* **Note 33**).
26. Add PBST and incubate blot on a platform shaker for 15 min to wash. Pour off PBST, add mouse anti-PSTAIR diluted at 1: 1000 for the loading control, and place on a platform shaker at 4 °C overnight.
27. Repeat **steps 20–25** above, except in **step 22** add goat anti-mouse IgG-HRP secondary antibody diluted at 1:10,000 to detect the for mouse anti-PSTAIR primary antibody.
28. If desired, induction of GST and GST-DUB can also be monitored by Western blotting with an anti-GST antibody. To do this, run a separate gel and blot as described above (**steps 19–25**) using mouse anti-GST antibody diluted at 1: 1000 and goat anti-mouse IgG-HRP secondary antibody diluted at 1:10,000 (*see* **Note 34**).

3.5.2 Image Quantitation

1. Open images in Image Lab software. Determine lanes and bands. Measure volume (intensity) of each band for both TAP and PSTAIR. Export data to Excel.
2. To correct for loading, normalize TAP intensity values to PSTAIR values in each lane.
3. Determine fold change by dividing the normalized TAP tag value in the GST-DUB overexpression strain by the normalized TAP tag value in the control (GST) strain. Identify substrates that increase greater than twofold upon overexpression of the DUB of interest (*see* **Note 35**).

3.6 Cycloheximide-Chase Assay

DUBs deubiquitinate and stabilize target proteins by preventing their proteasomal degradation. The increased stability of proteins that increase following DUB overexpression can be confirmed using a cycloheximide-chase assay. Cycloheximide is a translation inhibitor used to determine protein degradation kinetics (*see* **Note 36**). Figure 3 shows an example cycloheximide-chase assay of Hst3-TAP following overexpression of GST-Ubp10, which was previously shown to stabilize Hst3 [11].

1. Inoculate strains expressing pGAL-GST and pGAL-GST-DUB into culture tubes containing 5 mL of C-Ura/raffinose and grow overnight at 30 °C. Cultures can also be started directly into sterile flasks containing 25 mL of C-Ura/raffinose and grown in a 30 °C water bath for 15–17 h overnight, aiming for log-phase cultures the following day.

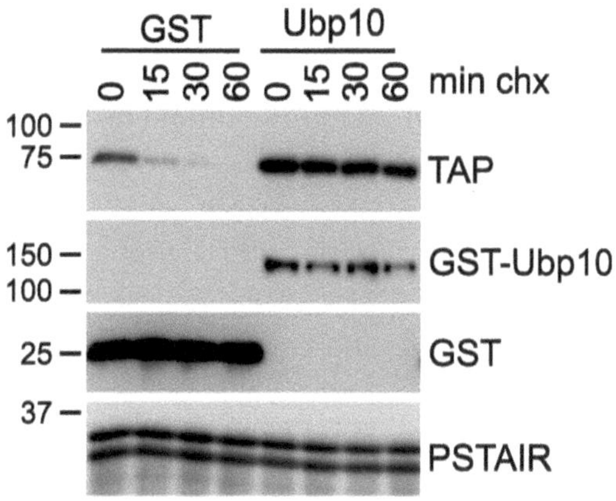

Fig. 3 Cycloheximide-chase assay follow-up on a representative DUB target. Exponentially growing Hst3-TAP strains containing pGAL-GST or pGAL-GST-UBP10 were grown in C-Ura/raffinose, and GST/GST-Ubp10 were induced by adding galactose for 4 h. Cycloheximide was added, and cells were collected at 15, 30, and 60 min. TAP, GST, and PSTAIR were visualized by Western blot. GST and GST-Ubp10 panels have been cropped from the same blot due to the significant difference in their molecular weights

2. If cultures are saturated the following day, dilute back to 0.2 OD_{600} in 20 mL of C-Ura/raffinose and allow cells to double at least once.
3. When cells are in log phase and at least 0.4 OD_{600}, add galactose for a final concentration of 2%. For a final volume of 20 mL, add 2 mL of 20% galactose to an 18 mL culture.
4. Incubate at 30 °C for 4 h to induce expression of GST and GST-DUB proteins.
5. Add 100 μL of YM-1 to 1.5 mL screw-cap tubes. After 4 h of galactose induction, collect 1 OD_{600} cell pellets at $t = 0$ min and freeze the pellet on dry ice (*see* **Note 37**).
6. Add cycloheximide for final concentration of 50 μg/mL. Collect time points as described in **step 5** at 15, 30, and 60 min. Time points may need to be adjusted depending on the half-life of the protein of interest.
7. Extract protein using TCA lysis (described above in Subheading 3.4 Protein Extraction with TCA), and run 15 μL on an SDS-PAGE gel (*see* **Note 38**).
8. Develop blot using a ChemiDoc Touch chemiluminescent imager, and quantify bands with the Image Lab software (as described in Subheading 3.5.2 Image Quantitation), if desired.

4 Notes

1. Raffinose should be filter-sterilized and not autoclaved because it is broken down during autoclaving.
2. Heat ssDNA at 95 °C for 5 min to denature; then keep on ice until ready to use.
3. Plasmids express GST/GST-DUB from the *GAL1* promoter and contain a 2-micron origin and *URA3* marker. Previously published plasmids [10] can be purchased from Horizon Discovery and should be verified by sequencing before use.
4. PEG is very viscous and is challenging to pipette accurately with standard beveled tips. Use sterile cut tips for aliquoting PEG by cutting off ends of tips and re-sterilizing.
5. PEG stocks should be used within 1 month after opening the sterilized bottle. Stocks older than 1 month can have increased PEG concentration due to water evaporation over time.
6. 100% w/v TCA stock can be made by dissolving 500 g of TCA in 227 mL of ultrapure water. Use caution as TCA is very corrosive.
7. Prepare 10% APS solution by adding 100 mg of APS to 900 μL of sterile, ultrapure water. This solution should be made fresh daily or stored at 4 °C and used within 4 days.
8. Make 1× transfer buffer the day before use and cool to 4 °C. The working stock of transfer buffer can be reused up to five times or within 2 months. Store at 4 °C.
9. 4 mL of chemiluminescent substrate is sufficient volume to cover a 7 × 13 cm blot. This substrate can be diluted to lower signal, depending on the intensity of the signal. A stronger chemiluminescent substrate (such as SuperSignal West Femto Maximum Sensitivity Substrate, Thermo Scientific) can also be used to increase signal.
10. Other equivalent digital chemiluminescent imagers can be used for data acquisition.
11. Sterilize pinner by soaking in 100% ethanol. Flick off excess ethanol; then carefully pass through a flame to sterilize.
12. For 1 full 96-well plate, prepare a master mix sufficient for 120 reactions.
13. Master mix can be added to a sterile reagent reservoir that matches the width of the multichannel pipette.
14. Place the plate in a zip-top plastic bag, seal with tape, and place on top of a rack inside a 42 °C water bath. Add a weight on top of the plate/bag to submerge the bottom half of the plate underwater. Take extra care to ensure water from the water

bath does not come in direct contact with the plate to maintain sterility.

15. The plate can be stored for up to 3 weeks at 4 °C for experiments.
16. Thawing cells into selective media with dextrose can help cells recover more quickly than in media containing raffinose.
17. If no clear cell pellets are observed, vortex the plate and return to the incubator. Cell pellets may be more visible the following day.
18. Raffinose is a neutral activator of the *GAL1* promoter. This step removes the active repression of the *GAL1* promoter that occurs in dextrose.
19. If cells grow poorly in media with raffinose, 1% dextrose may be supplemented. Subsequently, cells must be diluted and grown in media with only raffinose before galactose induction.
20. If cells grow unevenly, a fresh, saturated plate can be used to dilute cells the same day of the experiment. Let the cells grow at 30 °C until they have doubled once; then perform the galactose induction.
21. If OD_{600} readings are less than 0.3, continue growing at 30 °C until OD_{600} of 0.3–0.4 is reached. If OD_{600} values are too high, dilute cells back to 0.1–0.2 OD_{600} and allow cells to double once.
22. Example of volumes for DUB induction: 100 μL is removed from 1 mL overnight culture to measure the OD_{600} of the culture. 20% galactose is added to the remaining 900 μL of cells for a final concentration of 2% galactose.
23. Do not incubate the plate on ice, or the TCA buffer will freeze.
24. If pellets are not thawed after 10 min, the plate can be placed in a small amount of room temperature water for 2 min.
25. Aspiration of TCA buffer from cell pellets can be done by tilting the plate at a 45-degree angle, placing a gel loading tip attached to an aspirator in the corner of the well toward the bottom, and carefully removing the supernatant so as not to disturb the cell pellet. It is very important to completely remove the TCA buffer because any remaining TCA will alter the final pH of the sample.
26. 10% resolving gels are used to visualize proteins of interest between 25 and 250 kDa. This percentage is optimal because it works for most TAP-tagged proteins, each including a ~20 kDa tag, and separates the TAP-tagged proteins from the loading control (PSTAIR) on the gel. The gel percentage can be adjusted accordingly if desired.

27. Ninety-six TAP-tagged strains transformed with two plasmids (pGAL-GST and pGAL-GST-DUB) total 192 samples. Twenty-four samples each are then run on eight different gels.
28. Do not vortex. It is very important to mix gel gently so that bubbles are not introduced, which could interfere with polymerization or electrophoresis.
29. If possible, it is best to pour gels the day before instead of the same day.
30. GST and GST-DUB strains must be on the same gel to accurately calculate protein level changes upon DUB overexpression.
31. If two transfer boxes are used on the same power source, transfer at 0.90 A for 2 h.
32. Primary antibodies stored at 4 °C can be reused up to three to five times.
33. Capture a wide range of images to visualize proteins of low and high abundance (10 s to 20 min).
34. Western blots that confirm the induction and expression of GST and GST-DUB proteins are essential when setting up the screening system and testing the plasmids. However, it is not necessary to blot for GST in every experiment while performing a screen.
35. Any potential DUB substrates identified by high-throughput screening should be confirmed for strain accuracy. Here, the TAP tag is confirmed by designing oligos that anneal to the HIS3MX sequence (forward) and downstream ORF of interest (reverse) and performing PCR.
36. Candidate DUB substrates can also be validated by DUB gene deletion. Deleting a DUB gene may destabilize the substrate, leading to a decrease in protein levels and decreased stability in a cycloheximide-chase assay. Note that if redundant DUBs regulate a substrate, then deletion of a single DUB gene may be uninformative.
37. If desired, cells can be checked for ploidy by flow cytometry using standard methods [17].
38. The 96-well plate TCA lysis can be modified for a single-tube format. Collect 1 OD_{600} of cells (for instance, 2 mL of cells at an OD_{600} of 0.5), and adjust reagent volumes as follows: lyse in 300 μL of TCA buffer, resuspend in 75 μL of resuspension solution, and add 25 μL of 4× sample buffer before the final 95 °C incubation [18].

Acknowledgments

We thank members of the Benanti Lab for helpful discussions and suggestions on this chapter. This work was supported by National Institutes of Health grant R35GM136280 to J.A.B.

References

1. Finley D, Ulrich HD, Sommer T et al (2012) The ubiquitin-proteasome system of Saccharomyces cerevisiae. Genetics 192:319–360
2. Streich FC, Lima CD (2014) Structural and functional insights to ubiquitin-like protein conjugation. Biophysics 43:357–379
3. Clague MJ, Urbe S, Komander D (2019) Breaking the chains: deubiquitylating enzyme specificity begets function. Nat Rev Mol Cell Biol 81:1–15
4. Mevissen TET, Komander D (2017) Mechanisms of deubiquitinase specificity and regulation. Annu Rev Biochem 86:159–192
5. Isasa M, Rose CM, Elsasser S et al (2015) Multiplexed, proteome-wide protein expression profiling: yeast deubiquitylating enzyme knockout strains. J Proteome Res 14:5306–5317
6. Beckley JR, Chen J-S, Yang Y et al (2015) A degenerate cohort of yeast membrane trafficking DUBs mediates cell polarity and survival. Mol Cell Proteomics 14:3132–3141
7. Kwon S-K, Saindane M, Baek K-H (2017) p53 stability is regulated by diverse deubiquitinating enzymes. Biochim Biophys Acta 1868: 404–411
8. Ghaemmaghami S, Huh W-K, Bower K et al (2003) Global analysis of protein expression in yeast. Nature 425:737–741
9. Huh W-K, Falvo JV, Gerke LC et al (2003) Global analysis of protein localization in budding yeast. Nature 425:686–691
10. Sopko R, Huang D, Preston N et al (2006) Mapping pathways and phenotypes by systematic gene overexpression. Mol Cell 21: 319–330
11. Mapa CE, Arsenault HE, Conti MM et al (2018) A balance of deubiquitinating enzymes controls cell cycle entry. Mol Biol Cell 29: 2821–2834
12. Belle A, Tanay A, Bitincka L et al (2006) Quantification of protein half-lives in the budding yeast proteome. Proc Natl Acad Sci 103: 13004–13009
13. Martin-Perez M, Villén J (2017) Determinants and regulation of protein turnover in yeast. Cell Syst 5:283–294.e5
14. Kong K-YE, Fischer B, Meurer M et al (2021) Timer-based proteomic profiling of the ubiquitin-proteasome system reveals a substrate receptor of the GID ubiquitin ligase. Mol Cell 81:2460–2476
15. Christiano R, Nagaraj N, Fröhlich F et al (2014) Global proteome turnover analyses of the yeasts *S. cerevisiae* and *S pombe*. Cell Rep 9: 1959–1965
16. Gietz RD, Schiestl RH (2007) Microtiter plate transformation using the LiAc/SS carrier DNA/PEG method. Nat Protoc 2:5–8
17. Landry BD, Doyle JP, Toczyski DP et al (2012) F-box protein specificity for G1 cyclins is dictated by subcellular localization. PLoS Genet 8: e1002851
18. Landry BD, Mapa CE, Arsenault HE et al (2014) Regulation of a transcription factor network by Cdk1 coordinates late cell cycle gene expression. EMBO J 33:1044–1060

Chapter 15

Real-Time Deubiquitination Assays Using a Free Ubiquitin Sensor

Yun-Seok Choi and Robert E. Cohen

Abstract

Deubiquitinating enzymes cleave ubiquitin (Ub) from its attachment to another Ub, other proteins, peptides, or non-peptide adducts. In all cases, substrate hydrolysis by DUBs releases free Ub or polyubiquitin (polyUb) chains. Whereas most quantitative DUB assays depend on fluorescently labeled artificial substrates, employing a sensor able to detect Ub release in real time makes it possible to monitor DUB activity using virtually any Ub conjugate as a substrate. The protocols here describe the preparation of Atto532-tUI, a high-affinity sensor for free Ub, and its use in real-time deubiquitination assays.

Key words Ubiquitin, Deubiquitination, Deubiquitinase, DUB assay, tUI, Fluorescent ubiquitin sensor

1 Introduction

Deubiquitination is an essential component of many regulatory processes that depend on dynamic posttranslational modification of proteins by ubiquitin (Ub) [1–3]. Many different deubiquitination enzymes (also known as deubiquitinases, deubiquitylases, or DUBs) are responsible for the release of Ub from its covalent attachment to proteins (which can include other Ub(s)), peptides, or non-peptide adducts). To understand DUB specificities and functions and enable screens for DUB inhibitors, quantitative in vitro activity assays are essential. Most quantitative DUB assays employ artificial substrates such as Ub–fluorophore conjugates that, when cleaved by DUBs, generate products that are readily measured as changes in fluorescence [4]. Such assays can be quite sensitive, and the Ub derivatives commonly used as substrates (e.g., Ub–7-amido-4-methylcoumarin or Ub–rhodamine110-glycine) have the additional advantages of commercial availability and being acted on by many different DUBs. However, virtually all fluorogenic DUB substrates share some significant weaknesses:

Julie A. Maupin-Furlow and Mariola J. Edelmann (eds.), *Deubiquitinases: Methods and Protocols*,
Methods in Molecular Biology, vol. 2591, https://doi.org/10.1007/978-1-0716-2803-4_15,

their use in kinetic assays generally is uninformative regarding substrate specificity or enzyme rate and K_m with the physiological substrate(s), and use of artificial substrates can mask modes of regulation that are only manifested with physiological substrates [5–8]. For these reasons, we sought to establish an alternative DUB assay that would combine high sensitivity and the ability to monitor reactions in real time, but that also could be used with a large variety of Ub conjugates, including physiological substrates [9].

"Free" or unanchored Ub (or polyUb), i.e., Ub with a free C-terminal carboxylate, is a product common to all deubiquitination reactions. Our development of a general, quantitative DUB assay is based on the use of a fluorescently labeled sensor protein, Atto532-tUI, that shows increased fluorescence upon binding to the free Ub or polyUb released by deubiquitination enzymes [9]. Atto532-tUI was made by linking UBA^{UQ1} and $IsoT^{Buz}$ Ub-binding domains followed by site-specific derivatization with the fluorescent dye Atto532 (Fig. 1) [10, 11]. The $IsoT^{Buz}$ domain recognizes the free Ub C-terminus and UBA^{UQ1} binds to the canonical Ub surface hydrophobic patch. The combination of these two domains results in exceptionally high affinity for free Ub ($K_d = 66$ pM) and $>10^6$-fold discrimination against forms of Ub that lack a free C-terminus (e.g., Ub–protein conjugates). Thus, by including the Atto532-tUI sensor in a deubiquitination reaction, rapid and tight binding of free (poly)Ub released from Ub–protein conjugates can be detected as an increase in Atto532 fluorescence (Fig. 2).

In this chapter, we show how to prepare Atto532-tUI. Commercially available Atto532-maleimide is conjugated to the tUI protein, and then the unconjugated dye is removed. We describe how to set up a typical deubiquitination assay using Atto532-tUI to monitor DUB-catalyzed Ub release from a Ub–protein conjugate. A standard curve is generated to convert the increase in Atto532-tUI fluorescence to an absolute concentration of Ub product. These data are then used to determine the Michaelis–Menten parameters K_m and k_{cat}. Because there is no requirement for a label on the substrate in these deubiquitination assays, the experimental approach is extremely flexible and can be applied to a wide range of DUBs and DUB substrates.

2 Materials

For all buffers, it is recommended to avoid EDTA and decolorize the imidazole (*see* **Note 1**).

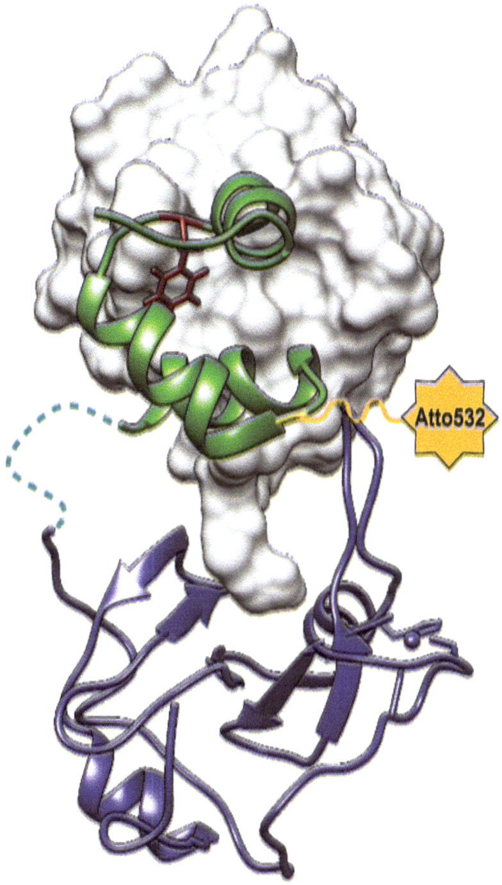

Fig. 1 Model of Atto532-tUI complex structure with Ub. Ub bound to Atto532-tUI is modeled as a composite of structures reported for Ub·UQ1^{UBA} (green; PDB 2JY6) and Ub·IsoT^{Buz} (purple; PDB 2G45) [10, 11]. The white surface structure is Ub. The dotted line represents a peptide linker between the IsoT^{Buz} and UQ1^{UBA} domains. Atto532-tUI binds Ub with $K_d = 0.066$ nM; a lower-affinity version called Atto532-tUI2 ($K_d = 1.2$ nM) has a F559V mutation in UQ1^{UBA}. The UQ1^{UBA} residues F559 (brown) and C573 (yellow) are highlighted; Atto532 fluorescent dye is conjugated to C573 of UQ1^{UBA}

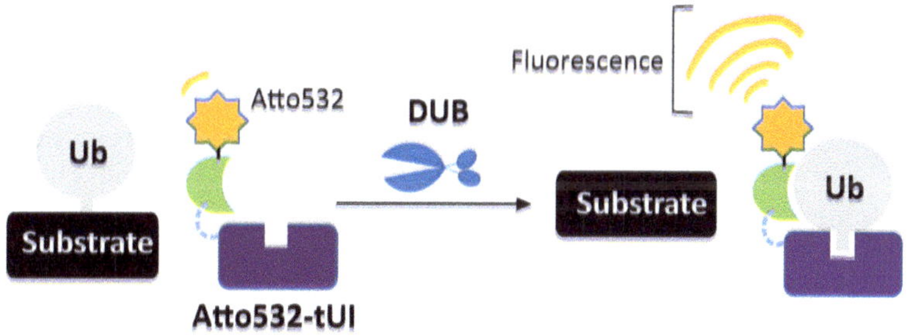

Fig. 2 Scheme of DUB assays monitored by product binding to Atto532-tUI. Ub binding to Atto532-tUI (or Atto532-tUI2) results in a 3.5- to 3.7-fold increase in fluorescence

2.1 Preparation of Atto532-tUI

1. Binding buffer: PBS, pH 7.4, 10-mM imidazole, and 10-mM β-mercaptoethanol.
2. Wash buffer: PBS, pH 7.4, 20-mM imidazole, and 10-mM β-mercaptoethanol.
3. Elution buffer: PBS, pH 7.4, 250-mM imidazole, and 10-mM β-mercaptoethanol.
4. Storage buffer for tUI protein: PBS, pH 7.4, with 1-mM tris (2-carboxyethyl)phosphine (TCEP).
5. Labeling buffer: 50-mM HEPES, pH 7.4, and 100-mM NaCl.
6. 6× SDS-PAGE sample loading buffer (important: omit the bromophenol blue dye if fluorescence detection near the dye front is required).
7. Activated charcoal (e.g., Norit CA).
8. Ni-NTA affinity resin.
9. Anhydrous dimethyl sulfoxide (DMSO) or *N*,*N*-dimethyl formamide (DMF).
10. Atto532-maleimide (ATTO-TEC, catalog no. AD 532-41).
11. β-Mercaptoethanol.
12. Tris(2-carboxyethyl)phosphine (TCEP).
13. 15% SDS-PAGE gels.
14. Benchtop microcentrifuge.
15. UV–visible spectrophotometer (e.g., Nanodrop).
16. Shaking incubator for 1.5-mL microcentrifuge tubes.
17. Fluorescence imager (e.g., Typhoon FLA 9500).
18. Spectrofluorometer.

2.2 Deubiquitination Assay

1. Assay buffer: 50-mM sodium phosphate, pH 7.4, 100-mM NaCl, 5-mM dithiothreitol (DTT), 0.05% Brij35, and 0.2-mg/mL bovine serum albumin or ovalbumin.
2. Deubiquitinase enzyme; here we used Usp2cc.
3. Ub.
4. Ub–protein conjugate substrate; here we used Ub-GB1 (protein G B1 domain).
5. Atto532-tUI or Atto532-tUI2.
6. Spectrofluorometer (or fluorescence microplate reader) with a temperature-controlled sample chamber.
7. Quartz fluorometer microcuvette (e.g., Starna Cells, catalog no. 16.45F-Q-3) or microplate for fluorescence measurements.
8. Benchtop microcentrifuge.

2.3 Programs for Data Analysis

1. Microsoft Excel.
2. GraphPad Prism.

3 Methods

3.1 Preparation of Atto532-tUI

1. Prepare the 6× His-tagged tUI by expression in *E. coli* transformed with the tUI plasmid (Addgene plasmid #122661) and purification as described previously [9] (*see* **Note 2**).
2. Generate a stock concentration of tUI between 200 μM and 1 mM in storage buffer.
3. Add anhydrous DMSO to preweighed, dry Atto532-maleimide (*see* **Notes 3** and **4**) to yield a stock solution of 25-mM Atto532-maleimide. Vortex the mixture at room temperature to completely dissolve it.
4. Mix tUI and Atto532-maleimide in labeling buffer at a 1:1.3 molar ratio; see the example in Table 1.
5. Incubate this labeling reaction mixture at 25 °C for 2 h, protected from light.
6. After the 2-h incubation, add 10-mM β-mercaptoethanol to the reaction mixture and incubate it at room temperature for 1 h to consume unreacted Atto532-maleimide.
7. After the 1-h incubation, remove an aliquot ("BEFORE" purification) containing a few μg of protein to use for analysis by SDS-PAGE to confirm the formation of Atto532-labeled tUI and to evaluate the ratio of bound dye to residual, free dye. Do not boil the sample; boiling quenches the fluorescence of Atto532 (*see* **Note 5**).
8. Resuspend the Ni-NTA resin and transfer the equivalent of 50 μL of packed resin to a 1.5-mL microcentrifuge tube. As described below (**steps 13–16**), the Ni-NTA resin is used to capture the total tUI (i.e., both unlabeled and Atto532-labeled) to remove unconjugated Atto532 dye (*see* **Note 6**).
9. Allow the resin to settle for 5 min and discard the supernatant (i.e., the resin storage buffer that contains 30% ethanol).

Table 1
Reaction mixture used to label tUI with Atto532-maleimide

	200-μM tUI	25-mM Atto532-maleimide	Labeling Buffer	H_2O	Total
Vol (μL)	200	1	80	119	400

10. Resuspend the resin with the addition of 300 μL of the binding buffer.
11. Allow 5 min for the resin to settle to the bottom of the tube and discard the supernatant.
12. Repeat **steps 10–11** two more times.
13. Transfer the Atto532-tUI reaction mixture to the tube with the Ni-NTA resin, and incubate it for 10 min in the shaking incubator at room temperature. Adjust the shaking rate to be just enough to keep the resin suspended; it is best not to shake the tube too vigorously to prevent the resin from sticking to the upper part of the tube wall or the tube cap.
14. After shaking, leave the tube for 5 min to let the resin settle to the bottom.
15. Remove the supernatant. The 6×His-tagged Atto532-tUI (and any remaining unlabeled tUI) will remain bound to the resin.
16. Add 300 μL of wash buffer and incubate at room temperature for 5 min in the shaking incubator.
17. Leave the tube without shaking for 5 min, and then remove the supernatant.
18. Repeat **steps 16–17** another nine times to remove unconjugated Atto532 dye thoroughly.
19. Before adding the elution buffer to the tube with the Ni-NTA resin, measure the fluorescence (excitation, 532 nm; emission, 553 nm) of the last supernatant to check if unconjugated Atto532 has been completely removed. If Atto532 dye remains present in the supernatant, repeat **steps 16** and **17** until the fluorescence intensity is negligible. Removing the unconjugated Atto532 dye from the sample is critical to preparing an Atto532-tUI stock that will show the full 3.7-fold fluorescence intensity increase in response to Ub binding.
20. Add 200 μL of the elution buffer and incubate in the shaking incubator at room temperature for 5 min.
21. Leave the tube without shaking for 5 min to allow the resin to settle.
22. Remove the supernatant and save it in a new tube. The supernatant has Atto532-tUI eluted from the resin.
23. Repeat **steps 20–22** two more times to recover all the Atto532-tUI from the Ni-NTA resin.
24. Measure the absorbance at 280 nm (A_{280}) and 540 nm (A_{540}; the dye absorption maximum of Atto532-tUI) of the eluted samples to determine the concentration and degree of labeling (DOL). To minimize the consumption of the samples, a Nanodrop spectrophotometer can be used for these measurements.

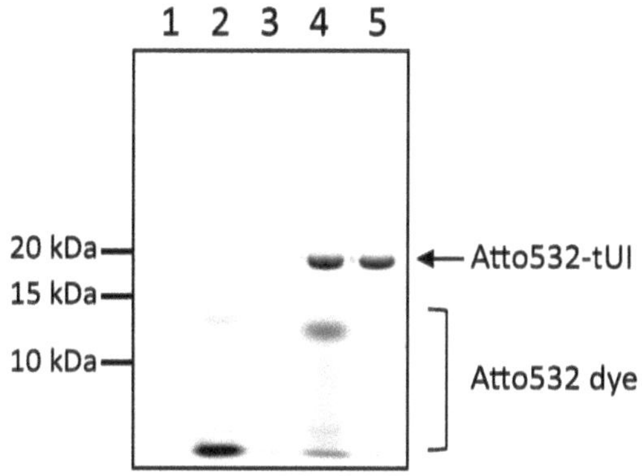

Fig. 3 Labeling of tUI with Atto532-maleimide. The labeling reactants and products were separated by SDS-PAGE. The samples loaded are SDS-PAGE sample buffer alone (lane 1), Atto532 dye (lane 2), unlabeled tUI (lane 3), Atto532-tUI reaction mixture before the purification (lane 4), and Atto532-tUI after the purification (lane 5). Fluorescence was detected using a Typhoon FLA 9500 laser scanner with 532-nm excitation

$$\text{Atto532–tUI concentration (M)} = \frac{A_{280} - A_{540} \times 0.09}{18480} \quad (1)$$

$$\text{DOL} = \frac{A_{540} \times 18480}{(A_{280} - A_{540} \times 0.09) \times 115000} \quad (2)$$

25. Analyze 0.5 to 1 μg of the eluted Atto532-tUI sample(s) by 15% SDS-PAGE to confirm the success of the labeling and removal of the free dye. Designate this sample as "AFTER" purification. Do not boil the sample (*see* **Note 5**).
26. Compare the BEFORE and AFTER samples from **steps** 7 and **25**, respectively, by detecting Atto532 fluorescence in the samples separated by 15% SDS-PAGE. If the 15% SDS-PAGE gel is sandwiched in glass, you do not need to remove the gel to detect the Atto532 fluorescence; remove it if the gel is between plastic plates. Do not stain the gel with Coomassie blue before imaging the fluorescence. To see whether unreacted Atto532 dye has been completely removed (*see* Fig. 3), detect the fluorescence using a laser-based scanner (e.g., GE Typhoon FLA 9500) or a camera-based gel documentation system equipped with a light box and appropriate filters. The excitation and emission wavelengths should be set to 532 nm and 540–600 nm, respectively.

3.2 Real-Time Deubiquitination Assay with Atto532-tUI

3.2.1 Standard Curve

The standard curve is used to convert fluorescence intensity changes of Atto532-tUI samples to concentrations of free Ub.

1. Label eight 0.5-mL tubes (e.g., STD1–STD8) as in Table 2.
2. Make the master mix in a 1.5-mL tube as in Table 2; this will be used for both the standard curve and the deubiquitination assays (*see* **Note 7**).

Table 2
Example of sample compositions for a standard curve

	Final Ub (nM)	10-nM Ub (μL)	100-nM Ub (μL)	1000-nM Ub (μL)	Master mix[a] (μL)	PBS (μL)	Total (μL)
STD1	0	–	–	–	27.2	32.8	60
STD2	1	6.0	–	–	27.2	26.8	60
STD3	2	12.0	–	–	27.2	20.8	60
STD4	4	–	2.4	–	27.2	30.4	60
STD5	8	–	4.8	–	27.2	28.0	60
STD6	16	–	9.6	–	27.2	23.2	60
STD7	32	–	–	1.9	27.2	30.9	60
STD8	64	–	–	3.8	27.2	29.0	60

[a]The master mix solution is designed to give a final reaction containing 100-nM Atto532-tUI, 50-mM sodium phosphate, pH 7.4, 0.05% Brij35, 1-mM DTT, and 0.2-mg/mL bovine serum albumin or ovalbumin. Important: Use the same batch of the master mix for both the standard curve assays and the deubiquitination assays

3. Add PBS (volumes listed in Table 2) to each tube.
4. Add 27.2 μL of the master mix shown in Table 2 to each tube.
5. Add Ub to each tube (*see* **Note 8**).
6. Add the STD1 sample to a fluorometer cuvette and put the cuvette in the fluorometer (*see* **Note 9**).
7. Measure the fluorescence intensity of the first standard (STD1) five times. Averaged values will be used to generate a standard curve (*see* **Note 10**).
8. Remove the first standard (STD1) and clean the cuvette.
9. Repeat **steps 6–8** for the other Ub standards (STD2–STD8).
10. Use the fluorescence intensities from the measurements to generate a standard curve in the Microsoft Excel program (*see* **Note 11**).
11. Use the scatter function in Excel charts to generate a data graph.
12. Display the equation of the standard curve on the graph (Fig. 4a).
13. Use the equation to convert the fluorescence intensities to concentrations of free Ub.

3.2.2 Deubiquitination Assay (See **Notes 11** and **12**)

Usp2cc and Ub-GB1 are presented as a DUB and a substrate, respectively, to exemplify the application of this assay. Usp2cc is the catalytic core of the DUB Usp2. Ub-GB1 is a mimic of a Ub–protein conjugate in which the Ub C-terminus is extended by GB1. The kinetic parameters, K_m and k_{cat}, of Usp2cc cleavage of Ub-GB1 were obtained from this assay at 25 °C (*see* **Note 13**). The basic

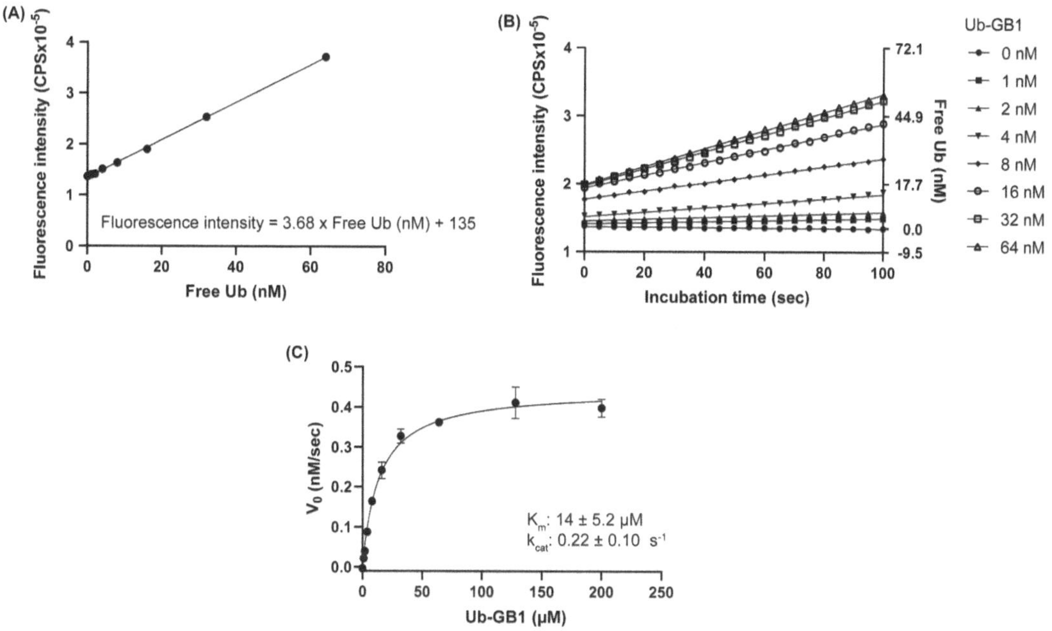

Fig. 4 Deubiquitination monitored with Atto532-tUI2. (**a**) A standard curve is established to convert the increases in Atto532-tUI2 fluorescence to the corresponding free Ub concentrations. (**b**) Incubation of Usp2cc with different concentrations of Ub-GB1 substrate shows increasing fluorescence (left Y-axis) that corresponds to free Ub (right Y-axis) release over time. (**c**) The initial velocities determined from the slopes of the lines in (**b**) are used to fit parameters of Michaelis–Menten kinetics. The assay is performed in duplicate; the points and error bars represent the means and standard deviations, respectively

format of this assay can be applied to quantify activities of other DUBs and (poly)Ub conjugate substrates (*see* **Note 14**), and, when desired, other reaction conditions can be used.

1. Label ten 0.5-mL tubes S1–S10.
2. Add PBS to the tube as outlined in Table 3.
3. Add 27.2 μL of the master mix (Table 3) to each tube.
4. Add Ub-GB1, the substrate, to the corresponding tube.
5. Add 2.4 μL of the DUB to the S1 tube and mix by gently pipetting up and down five times. The S1 tube does not contain any substrate.
6. Add the S1 sample to the fluorometer cuvette and put it in the fluorometer.
7. Measure the fluorescence intensity of the S1 sample every 2 s for 5 min to obtain a baseline.
8. Remove the S1 sample from the cuvette.
9. Add 2.4 μL of the DUB to the S2 sample and mix by gently pipetting up and down five times.
10. Add the S2 sample to a clean cuvette.

Table 3
Deubiquitination assay

	(μM) Final Ub-GB1	(nM) Final DUB	(μL) 10-μM Ub-GB1	(μL) 100-μM Ub-GB1	(μL) 564-μM Ub-GB1	(μL) 50-nM DUB	(μL) Master mix[a]	(μL) PBS	(μL) Total
S1	0	2	0.0	0.0	0.0	2.4	27.2	30.4	60
S2	1	2	6.0	0.0	0.0	2.4	27.2	24.4	60
S3	2	2	12	0.0	0.0	2.4	27.2	18.4	60
S4	4	2	0.0	2.4	0.0	2.4	27.2	28.0	60
S5	8	2	0.0	4.8	0.0	2.4	27.2	25.6	60
S6	16	2	0.0	9.6	0.0	2.4	27.2	20.8	60
S7	32	2	0.0	0.0	3.4	2.4	27.2	27.0	60
S8	64	2	0.0	0.0	6.8	2.4	27.2	23.6	60
S9	128	2	0.0	0.0	13.6	2.4	27.2	16.8	60
S10	256	2	0.0	0.0	21.3	2.4	27.2	9.1	60

[a]The master mix solution contains 100-nM Atto532-tUI, 50-mM sodium phosphate, pH 7.4, 0.05% Brij35, 1-mM DTT, and 0.2-mg/mL bovine serum albumin or ovalbumin. Important: Use the same batch of the master mix for both the standard curve assays and the deubiquitination assays

11. Measure the fluorescence intensity of the S2 sample every 2 s for 5 min to get an initial velocity.
12. Remove the S2 sample from the cuvette.
13. Repeat **steps 9–12** for the remaining samples.
14. Complete at least one additional set of assays to evaluate the precision of the data (*see* **Notes 10** and **11**).
15. Calculate initial velocities for each substrate (here, Ub-GB1) concentration (Fig. 4b).
16. Convert the fluorescence intensities (e.g., Fig. 4b) to the product concentration (i.e., free Ub) using the equation in Fig. 4a obtained from the standard curve (Subheading 3.2.1).
17. Enter the data into the GraphPad Prism program and select the Michaelis–Menten equation to determine best-fit values for K_m and k_{cat} (Fig. 4c).

4 Notes

1. Decolorization of the imidazole stock used in the buffers is recommended by treatment with activated charcoal (e.g., Norit CA) to remove fluorescent impurities that can elevate background fluorescence. We find that it is convenient to prepare a

1-M imidazole stock solution. The solution is heated to 50–60 °C in a water bath, the activated charcoal (0.1 g per liter) is added, and the heating is continued while the mixture is occasionally swirled to keep the charcoal suspended. After 5–10 min, the mixture is cooled to room temperature, the charcoal is removed with a 0.22-micron filter, and the imidazole stock is stored at 4 °C. Note also that EDTA should not be used in any of the solutions because it can deplete the Zn^{2+} ion that is an integral component of the $IsoT^{BuZ}$ domain in tUI.

2. Atto532-maleimide reacts with the thiol group of C573 in $UQ1^{UBA}$ of tUI (Fig. 1). Therefore, additional thiol-containing compounds, such as the commonly used reducing agent β-mercaptoethanol or dithiothreitol (DTT), should be avoided in the labeling reaction. TCEP can be used to protect tUI thiols from oxidation as it is not reactive with maleimides. Note that tUI has four cysteines, but only C573 reacts readily with maleimide because the other cysteines are protected by coordination with a tightly bound Zn^{2+} ion.
3. Solid Atto532-maleimide is stored at −20 °C. To avoid contamination with water when the chemical is weighed, allow the container to warm to room temperature before opening it; use an anhydrous solvent to dissolve the Atto532-maleimide. The stock solutions should be stored in small (and ideally, single-use) aliquots to minimize exposure to water. The aliquots can be stored for at least several months at −20 °C. DMF can be used as the solvent for the Atto532-maleimide stock as an alternative to DMSO.
4. Atto532 dye solutions should be protected from light.
5. The BEFORE and AFTER samples are used to analyze if the purification of Atto532-tUI successfully removed unreacted Atto532-maleimide. Bromophenol blue, commonly used as a tracking dye in SDS-PAGE sample buffer, fluoresces and can obscure unconjugated Atto532 that runs near the dye front in an SDS-PAGE gel. Therefore, SDS-PAGE sample buffer made without bromophenol blue is used with the BEFORE and AFTER samples for SDS-PAGE analysis.
6. The nominal binding capacity of the Ni-NTA resin used here is 50 mg of recombinant protein per 1-mL resin. The amount of Ni-NTA resin used to recover the Atto532-tUI product from the dye-labeling reaction should be equal to or higher than the total amount of Atto532-tUI and tUI. However, to minimize losses due to nonspecific binding and unnecessary dilution of the eluted Atto532-tUI, avoid using more than a five- to tenfold excess of resin.
7. The fluorescence intensity change from Atto532-tUI is proportional to the free Ub concentration; thus, it is important to use the same master mix for all samples in the experiment (i.e.,

all samples used for the standard curve as well as the DUB reactions).

8. Accurate concentrations of Ub stocks are critical for this standard curve. Measurement of absorbance at 280 nm is recommended to determine the Ub concentration. Ub has a low extinction coefficient (1490 M^{-1} cm^{-1}) at 280 nm; therefore, the Ub standard needs to be devoid of 280-nm absorbing contaminants for reliable quantitation and should be sufficiently concentrated for a precise absorbance reading (typically, 0.1–1 A.U.). We recommend the preparation of a 0.1–1-mM Ub stock. After determining the stock concentration, lower concentrations are made by serial dilution.
9. The free Ub concentration, i.e., the product of the DUB activity, is determined by the measurement of the increase of Atto532-tUI fluorescence intensity in the reaction solution. Because bubbles in the sample solution can decrease the measured fluorescence intensity of the sample, it is important to ensure that there are no bubbles in the light path. The nonionic detergent (Brij35) included in the reaction reduces the surface tension of the solution, thereby helping to prevent the formation of trapped bubbles on the walls of the cuvette or microplate wells. The Brij35 also minimizes nonspecific interactions of the proteins in the assays.
10. The excitation and emission slit widths can be adjusted to increase the fluorescence signal-to-noise ratio. With 100-nM Atto532-tUI in 45-μL samples, we use 5-nm slits. Under these conditions, the experimental error typically is affected more by pipetting accuracy than the fluorescence measurement. However, the sensitivity observed will depend upon the particular instrument being used and could, in principle, limit the precision of the assays.
11. Replicates of the assays should be done for statistical evaluation of the results.
12. The availability of substrates often is a practical limitation in DUB assays. Therefore, to minimize consumption of substrates, the assay procedure described here uses a fluorometer microcuvette that requires only 45 μL of sample. If other types of cuvettes or microwell plates are used, the volumes of the reactions in Tables 2 and 3 should be scaled accordingly.
13. If samples are not 25 °C, equilibrate to 25 °C before preparing the reaction solutions.
14. Free (e.g., unanchored) polyUb cannot be used as substrates in DUB assays monitored with Atto532-tUI because the proximal end Ub of the substrate would be bound by Atto532-tUI. However, unanchored polyUb substrates can be used if the C-terminal glycylglycine (GG) of the proximal Ub is modified to preclude tight binding by Atto532-tUI. Thus, because

DUB cleavage would be required to expose a Ub C-terminus for binding by Atto532-tUI, either polyUbΔGG or polyUb conjugated to a small adduct (e.g., polyUb with Ub-D77 in the proximal position) can be used as a substrate.

Acknowledgments

This work was supported in part by National Institutes of Health Grants R01GM115997 and R21GM135818 to R.E. Cohen. Y.-S. Choi was supported by the South Dakota Biomedical Research Infrastructure Network (SD BRIN) through an IDeA award from NIH-NIGMS (grant no. P20GM103443). The Usp2cc assays were supported by WestCore at BHSU.

References

1. Mevissen TET, Komander D (2017) Mechanisms of deubiquitinase specificity and regulation. Annu Rev Biochem 86:159–192. https://doi.org/10.1146/annurev-biochem-061516-044916
2. Clague MJ, Urbé S, Komander D (2019) Breaking the chains: deubiquitylating enzyme specificity begets function. Nat Rev Mol Cell Biol 20:338–352. https://doi.org/10.1038/s41580-019-0099-1
3. Harrigan JA, Jacq X, Martin NM et al (2018) Deubiquitylating enzymes and drug discovery: emerging opportunities. Nat Rev Drug Discov 17:57–78. https://doi.org/10.1038/nrd.2017.152
4. Dharadhar S, Kim RQ, Uckelmann M et al (2019) Quantitative analysis of USP activity in vitro. Methods Enzymol 618:281–319. https://doi.org/10.1016/bs.mie.2018.12.023
5. Wiener R, DiBello AT, Lombardi PM et al (2013) E2 ubiquitin-conjugating enzymes regulate the deubiquitinating activity of OTUB1. Nat Struct Mol Biol 20:1033–1039. https://doi.org/10.1038/nsmb.2655
6. Faesen AC, Luna-Vargas MPA, Geurink PP et al (2011) The differential modulation of USP activity by internal regulatory domains, interactors and eight ubiquitin chain types. Chem Biol 18:1550–1561. https://doi.org/10.1016/j.chembiol.2011.10.017
7. Deol KK, Crowe SO, Du J et al (2020) Proteasome-bound UCH37/UCHL5 debranches ubiquitin chains to promote degradation. Mol Cell 80:796–809.e9. https://doi.org/10.1016/j.molcel.2020.10.017
8. Song A, Hazlett Z, Abeykoon D et al (2021) Branched ubiquitin chain binding and deubiquitination by UCH37 facilitate proteasome clearance of stress-induced inclusions. eLife 10:e72798. https://doi.org/10.7554/eLife.72798
9. Choi Y-S, Bollinger SA, Prada LF et al (2019) High-affinity free ubiquitin sensors for quantifying ubiquitin homeostasis and deubiquitination. Nat Methods 16:771–777. https://doi.org/10.1038/s41592-019-0469-9
10. Zhang D, Raasi S, Fushman D (2008) Affinity makes the difference: nonselective interaction of the UBA domain of ubiquilin-1 with monomeric ubiquitin and polyubiquitin chains. J Mol Biol 377:162–180. https://doi.org/10.1016/j.jmb.2007.12.029
11. Reyes-Turcu FE, Horton JR, Mullally JE et al (2006) The ubiquitin binding domain ZnF UBP recognizes the C-terminal diglycine motif of unanchored ubiquitin. Cell 124:1197–1208. https://doi.org/10.1016/j.cell.2006.02.038

Chapter 16

A Method to Monitor Activity of SARS-CoV-2 Nsp3 from Cells

Christopher Rainville, David E. Sterner, and Kumar Suresh

Abstract

SARS-CoV-2 protease Nsp3 is a therapeutic target for developing anti-SARS-CoV-2 drugs. Nsp3 is a large multi-spanning membrane protein, and its characterization in vitro has been challenging. Here we describe an in vitro assay to characterize the biochemical activity of full-length Nsp3 isolated from cells. The assay can be used to evaluate Nsp3 inhibitors.

Key words SARS-CoV-2, Nsp3, PLpro, Deubiquitinase, deISGylase, ISG15-CHOP2, Ubiquitin, Ub-Rh110

1 Introduction

In addition to hijacking cellular deubiquitinases (DUBs), DNA viruses and RNA viruses encode their own deubiquitinase enzymes that can reverse ubiquitination-mediated host cell processes and suppress the innate antiviral immune response of the host [1]. SARS-CoV-1 and SARS-CoV-2 are coronaviruses with RNA genomes, and they encode a deubiquitinase called papain-like protease (PLpro) within their Nsp3 proteins [2–4]. Along with the main protease Nsp5 (3CLpro), Nsp3 cleaves the viral polyproteins to generate key viral proteins. Studies of PLpro from SARS-CoV-1 and SARS-CoV-2 have shown that these two enzymes, although sharing 83% sequence identity, have different substrate specificities [5]. Whereas SARS-CoV-1 PLpro preferentially targets ubiquitin chains, PLpro from SARS-CoV-2 cleaves the ubiquitin-like modifier ISG15 (deISGylation) [5–18]. Notably, the PLpros of SARS-CoV and SARS-CoV-2 antagonize multiple components of the type I interferon (IFN) pathway [12, 19–22].

Since Nsp3 is essential for viral replication, it has been considered an attractive therapeutic target for developing antiviral drugs. Nsp3 is a large, multi-pass transmembrane protein, and most of the Nsp3 biochemical studies have used the much smaller PLpro

Julie A. Maupin-Furlow and Mariola J. Edelmann (eds.), *Deubiquitinases: Methods and Protocols*,
Methods in Molecular Biology, vol. 2591, https://doi.org/10.1007/978-1-0716-2803-4_16,

domain expressed and purified from recombinant *Escherichia coli*. The biochemical activity of full-length Nsp3 from SARS-CoV-2 and many functions of Nsp3 have not been fully characterized. Here we describe a method to transiently express, purify, and determine the biochemical activity of SARS-CoV-2 full-length Nsp3 using HEK293T cells.

Expression, purification, and protease assays were performed with full-length Nsp3 and PLpro domain to compare the two enzymes, and they displayed similarity with regard to substrate specificity and the effect of a known PLpro inhibitor. Initially, the full-length SARS-CoV-2 Nsp3 and PLpro domain were expressed with N-terminal hemagglutinin (HA-) tags and purified, as demonstrated by Western blotting with anti-HA (Fig. 1a) and anti-Nsp3 (Fig. 1b) antibodies. Purified full-length HA-Nsp3 and HA-PLpro show strong deISGylase activity using the ISG15-CHOP2 reporter (Fig. 2). Full-length HA-Nsp3 did not show appreciable activity using ubiquitin-Rh110 as a substrate (data not shown), consistent with SARS-CoV-2 Nsp3 being a much weaker deubiquitinase enzyme. Recombinant PLpro (5 nM) was used as a positive control, and it showed strong activity with ISG15-CHOP2 (Fig. 3). Additionally, GRL0617, a small molecule inhibitor previously shown to inhibit the recombinant PLpro enzyme, is shown to inhibit full-length Nsp3 with an IC_{50} of 3.14 μM (Fig. 4). Thus, the method described herein can be used to characterize inhibitors of full-length SARS-CoV-2 Nsp3.

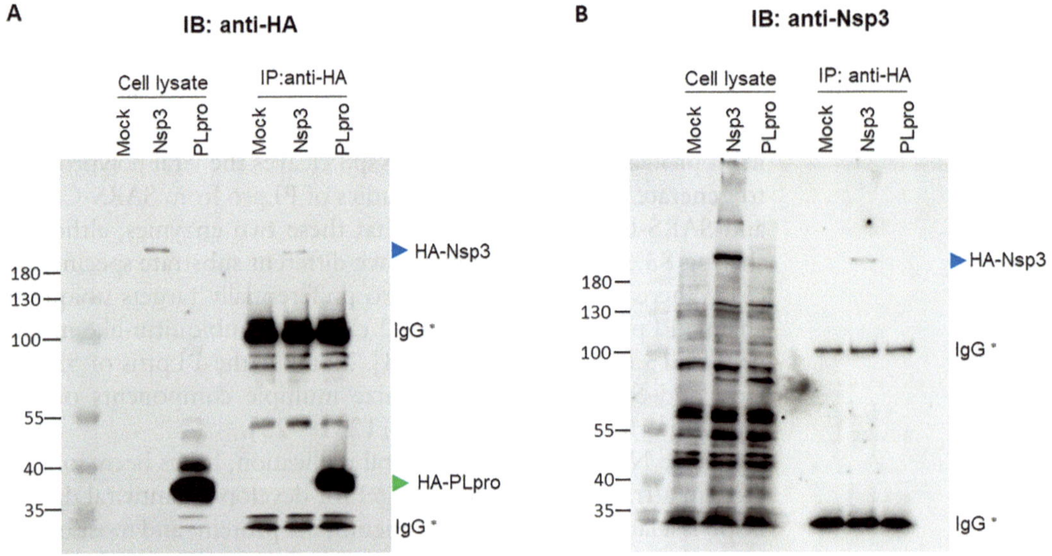

Fig. 1 Western blot of SARS-CoV-2 Nsp3 and PLpro transiently expressed with N-terminal HA tags in HEK293T cells. Whole cell lysate and pull-down samples were probed with anti-HA antibody (**a**) or anti-Nsp3 antibody (**b**). The anti-Nsp3 antibody does not detect HA-PLpro, as the antibody is raised against an epitope that does not include the PLpro domain. Asterisks mark IgG heavy and light chains from the anti-HA antibody

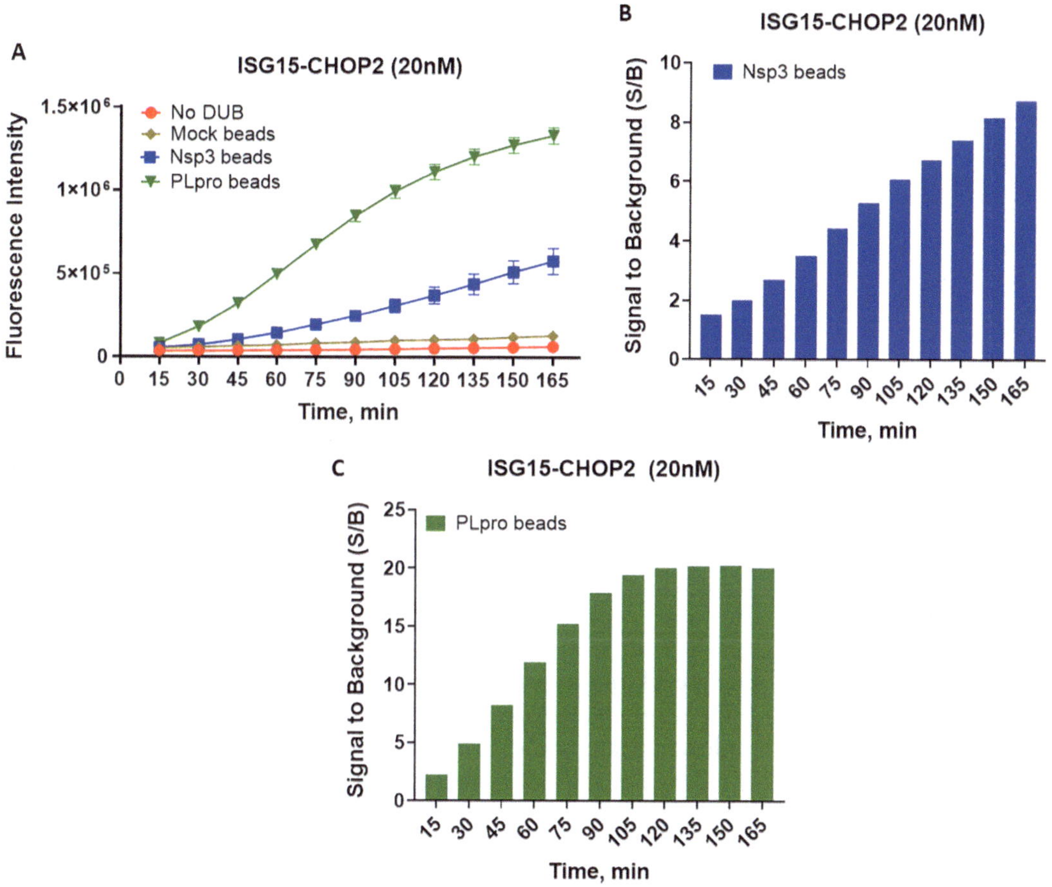

Fig. 2 Biochemical assay of SARS-CoV-2 Nsp3 and PLpro beads with ISG15-CHOP2 as substrate. (**a**) Time-dependent increase in deISGylase activity of Nsp3 and PLpro pull-down beads using 20-nM ISG15-CHOP2 reporter. (**b**) Signal to the background ratio of Nsp3 beads. (**c**) Signal to the background ratio of PLpro beads

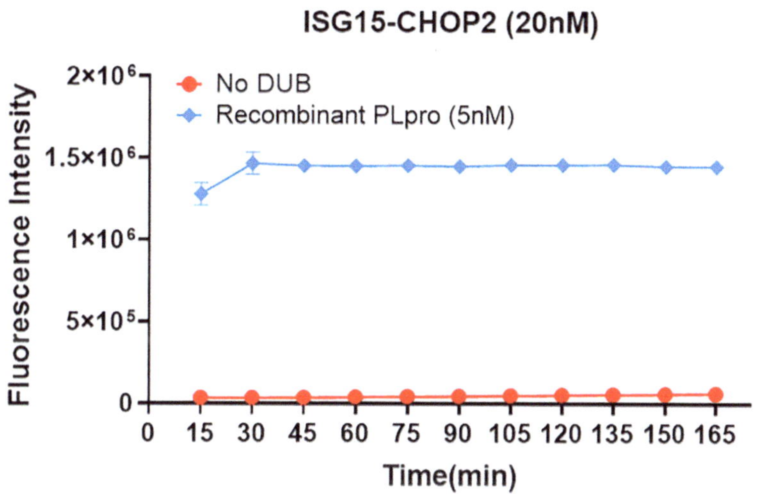

Fig. 3 The activity of recombinant SARS-CoV-2 PLpro. The recombinant SARS-CoV-2 PLpro (5 nM) activity was determined in assays using 20-nM ISG15-CHOP2

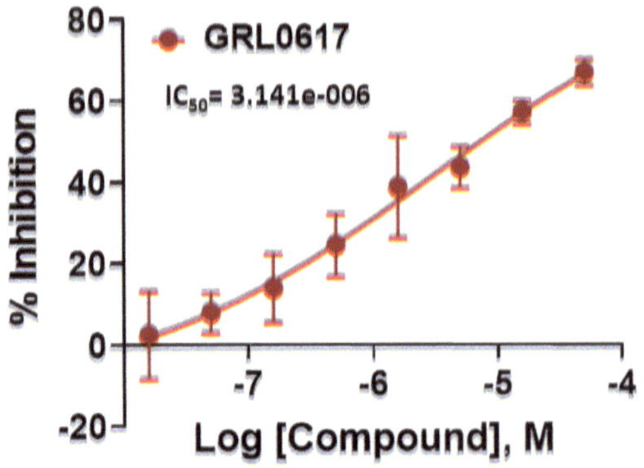

Fig. 4 Inhibition of SARS-CoV-2 Nsp3 by GRL0617. Nsp3 beads were preincubated with various doses of GRL0617 for 30 min, and inhibition of deISGylase activity of Nsp3 was monitored using the ISG15-CHOP2 reporter. Percent inhibition was calculated and plotted using GraphPad Prism 9.0

2 Materials

2.1 Transfection

Prepare all solutions using ultrapure, deionized water and analytical grade reagents. Prepare and store all reagents at room temperature (unless indicated otherwise). All waste materials should be disposed of following waste disposal regulations.

1. HEK293T (human embryonic kidney) cell line.
2. Dulbecco's Modified Eagle Medium (DMEM).
3. Fetal bovine serum (FBS).
4. Penicillin-streptomycin solution, 100×.
5. Complete cell culture medium: DMEM, 10% FBS, 1× penicillin-streptomycin.
6. 175-cm^2 (T175) cell culture flasks.
7. 100-mm cell culture dishes.
8. Tissue culture hood with laminar flow.
9. Water-jacketed CO_2 incubator for incubation of cultured cells.
10. Microscope for observing and counting cells (100× total magnification).
11. 1× phosphate-buffered saline (PBS): 10-mM Na_2HPO_4, 1.8-mM KH_2PO_4, pH 7.4, 137-mM NaCl.
12. TrypLE Express cell-dissociation reagent (Thermo Fisher Cat#12604-021).
13. Hemocytometer.
14. Microcentrifuge tubes, 1.5 mL.

15. Lipofectamine 3000 transfection reagent (Thermo Fisher, Cat#130182).
16. Opti-MEM medium (Thermo Fisher, Cat#31985-062).
17. Plasmid DNA (pEF empty vector, pEF-HA-SARS-CoV-2 Nsp3, pEF-HA-SARS-CoV-2 PLpro) (*see* **Note 1**).

2.2 Cell Lysis and Immunoprecipitation

1. Cell lysis buffer: 50-mM Tris–HCl, pH 7.5, 150-mM NaCl, 10% glycerol, 1% NP-40, 1-mM phenylmethylsulfonyl fluoride (PMSF) (*see* **Note 2**; to avoid cysteine protease inhibition, no protease inhibitor cocktail was added).
2. Cell scraper.
3. Vortex mixer.
4. Microcentrifuge.
5. Bradford protein assay kit (e.g., Bio-Rad Protein Assay Dye Reagent Concentrate).
6. Bovine serum albumin (BSA), 2-mg/mL stock (for preparing protein quantification standards).
7. Spectrophotometer (for quantification of protein in lysates).
8. Mouse anti-HA antibody (e.g., BioLegend Purified anti-HA.11 Epitope Tag Antibody, Clone 16B12).
9. Protein G Sepharose beads (e.g., Cytiva Protein G Sepharose 4 Fast Flow).
10. Rotating mixer (e.g., Labquake tube rotator, 8 rpm).

2.3 Enzymatic Assays

1. ISG15-CHOP2 reporter system (LifeSensors, Cat#PR1105).
2. CHOP2 substrate (LifeSensors, Cat#PR1105).
3. Ubiquitin-rhodamine110 resuspended in 100% DMSO to 50-μM stock (LifeSensors, Cat#SI230).
4. Recombinant SARS-CoV-2 papain-like protease (LifeSensors, Cat#DB604).
5. Assay buffer: 20-mM Tris–HCl, pH 8.0, 0.05% CHAPS, 2-mM β-mercaptoethanol.
6. 384-well polypropylene plate (for compound dilution).
7. 384-well shallow well plate, black (for plate reading).
8. 16-channel repeater pipette.
9. Microplate reader (e.g., BMG Labtech CLARIOstar). Settings: Ex 482-16, Dichroic-Auto 500, Em 530-40, flashes = 5, plates read kinetically every 5 min for 3 h.

2.4 Western Blotting

1. SDS-PAGE gels or system for casting gels (with 30% acrylamide/bis-acrylamide solution, 37.5:1).
2. SDS-PAGE protein gel and electrophoresis system (e.g., Bio-Rad Mini-PROTEAN Tetra).
3. SDS-PAGE running buffer: 25-mM Tris base, 192-mM glycine, 0.1% SDS.
4. SDS-PAGE sample loading buffer, 6×: 125-mM Tris–HCl, pH 6.8, 10% SDS, 50% glycerol, 0.06% bromophenol blue, 1% β-mercaptoethanol. Store aliquots at −20 °C.
5. Prestained protein molecular weight marker (for Western blot) (e.g., Thermo Fisher PageRuler Prestained Protein Ladder, 10–180 kDa).
6. Electroblotting system (e.g., Bio-Rad Trans-Blot Turbo Western transfer system).
7. Western blot transfer buffer appropriate for the electroblotting system used (e.g., 25-mM Tris base, 192-mM glycine, 0.1% SDS, 20% ethanol, pH 8.3 for Bio-Rad Trans-Blot Turbo).
8. 0.2-μm PVDF transfer membrane.
9. Filter paper for electroblotting.
10. 10× Tris-buffered saline (TBS): 250-mM Tris–HCl, pH 8.0, 1.37-M NaCl.
11. TBST: 1× TBS containing 0.1% Tween-20.
12. Blocking solution: 5% nonfat dry milk in TBST. Store at 4 °C.
13. Western blot incubation boxes.
14. Mouse anti-HA antibody (e.g., BioLegend Purified anti-HA.11 Epitope Tag Antibody, Clone 16B12).
15. Anti-SARS-CoV-2 (COVID-19) nsp3 antibody (rabbit) (GeneTex, Cat#GTX135614).
16. Anti-rabbit IgG-HRP conjugate (e.g., Jackson ImmunoResearch Peroxidase-conjugated AffiniPure Goat Anti-Rabbit IgG (H + L)).
17. Anti-mouse IgG-HRP conjugate (e.g., Jackson ImmunoResearch Peroxidase-conjugated AffiniPure Goat Anti-Mouse IgG (H + L)).
18. Enhanced chemiluminescence (ECL) detection reagent (e.g., MilliporeSigma Immobilon Western Chemiluminescent HRP Substrate).
19. Chemiluminescent Western blot imaging system (e.g., LI-COR Odyssey).
20. Western blot stripping buffer: 200-mM glycine, 3.5-mM SDS, 1% Tween-20, pH 2.2.

3 Methods

3.1 Transfection of HEK293T Cells

Perform all procedures at room temperature unless otherwise specified.

1. Grow HEK293T cells in a T175 flask to ~70–80% confluency using complete cell culture medium. On the day of transfection, remove the medium, and rinse one time with sterile 1× PBS. Remove PBS, add 1–2-mL TrypLE Express reagent (with trypsin-like activity), and incubate for 1–2 min at 37 °C (*see* **Note 3**). Examine the cells with a microscope to ensure that all the cells are dissociated from the plate. Add 8–9-mL complete cell culture medium and gently resuspend cells to obtain a uniform suspension. Count the cells using a hemocytometer.
2. Plate 6×10^6 cells into 10-cm dishes in a total of 10-mL complete culture medium. Allow the cells to attach for 3–4 h at 37 °C.
3. Prepare the DNA–lipid complex following the manufacturer's recommended protocol for Lipofectamine 3000 reagent. Add 12.5 μg of plasmid DNA to 500 μL of Opti-MEM in a sterile 1.5-mL microcentrifuge tube. Add 25 μL of Lipofectamine 3000 reagent and mix well by inverting the tube four to five times. Incubate for 15 min at room temperature.
4. In a separate sterile 1.5-mL tube, add 25 μL of Lipofectamine 3000 reagent to 500-μL Opti-MEM. Mix well and incubate for 15 min at room temperature.
5. Combine the two solutions and gently mix well by inverting the tube four to five times. Incubate for 15 min at room temperature.
6. Add the 1.0-mL DNA–lipid complex to cells in a 10-cm dish dropwise and mix by gentle rocking.
7. Incubate the cells at 37 °C, 5% CO_2 for 24 h.

3.2 Cell Lysis and Immunoprecipitation

Perform all procedures on ice or at 4 °C unless otherwise specified.

1. Remove the dishes containing cells from the incubator and place the dishes on ice.
2. Gently aspirate the medium and gently rinse the cells with ice-cold 1× PBS one time (*see* **Note 3**).
3. Add 500-μL ice-cold lysis buffer and scrape the cells using a cell scraper.
4. Transfer the cell lysates into 1.5-mL tubes.
5. Incubate the lysates on ice for 20 min with occasional mixing by vortexing (*see* **Note 4**).

6. Centrifuge the lysate at 16,000 × g for 15 min to pellet cell debris.
7. Prepare Protein G Sepharose beads by washing with ice-cold 1× TBS (*see* **Note 5**).
8. Transfer the cell lysate to a new 1.5-mL tube and add 200 μL of Protein G Sepharose beads. Incubate the samples on a rotating mixer (low speed, e.g., 8 rpm) for 30 min at 4 °C (*see* **Note 6**).
9. Centrifuge at 2000 × g for 4 min at 4 °C.
10. Transfer the lysate to a fresh 1.5-mL tube. Discard the beads.
11. Determine protein concentration by Bradford protein assay using BSA as a standard. Measure absorbance at 595 nm.
12. Transfer 1 mg of cell lysate to a fresh 1.5-mL tube. Add 5-μg anti-HA antibody. Incubate on a rotating mixer for 1 h at 4 °C.
13. Wash the Protein G Sepharose beads three times using lysis buffer. To wash, centrifuge the slurry for 4 min at 2000 × g and discard the supernatant. Resuspend the beads in lysis buffer as a 1:1 slurry.
14. Add 200 μL of Protein G Sepharose beads (1:1 slurry in lysis buffer) to the 1-mg cell lysate in the 1.5-mL tube. Incubate on a rotating mixer for 4 h at 4 °C.
15. Centrifuge at 2000 × g for 4 min. Discard the supernatant.
16. Resuspend the beads in 1-mL assay buffer.
17. Store the beads on ice for immediate enzyme assay.

3.3 Enzyme Assay

1. Dilute the compounds for testing by diluting the 10-mM compound stocks in 100% DMSO to 1.05 mM in DMSO. Subsequently, prepare half-log serial dilutions of the compounds in DMSO for 8-point dose–responses. Example of compound dilution plate layout is shown in Fig. 5.
2. Transfer 0.5-μL DMSO or the serially diluted compounds to black shallow well assay plates with 384 wells. Example of assay plate layout is shown in Fig. 6.
3. Add 5 μL per well of Protein G Sepharose bead slurry containing immunoprecipitated enzymes to the 384-well plate containing DMSO or the compound dilutions.
4. Add 5 μL per well of 2× ISG15-CHOP2 (2× = 40 nM) or ubiquitin-rhodamine110 (2× = 200 nM) substrate solution prepared in assay buffer to corresponding wells. Final concentrations are 20-nM ISG15-CHOP2 reporter and 20-μM QXL substrate. Recombinant PLpro (5 nM for ISG15-CHOP2 and 100 nM for Ub-Rh110) is used as a positive control (*see* **Note 7**).

	1	2	4	Diluted Stock Concentration (mM)	Final Concentration in 10.5 µl (µM)
A	DMSO, 50µl	DMSO, 50µl	Add 5.25 µl of 10mM GRL-0617 to 44.75 DMSO	1.05	50
B	DMSO, 50µl	DMSO, 50µl	Add 19 µl from well A4 to 41 µl DMSO	0.332	15.8
C	DMSO, 50µl	DMSO, 50µl	Add 19 µl from well B4 to 41µl DMSO	0.105	5
D	DMSO, 50µl	DMSO, 50µl	Add 19 µl from well C4 to 41 µl DMSO	0.0332	1.6
E	DMSO, 50µl	DMSO, 50µl	Add 19 µl from well D4 to 41 µl DMSO	0.0105	0.5
F	DMSO, 50µl	DMSO, 50µl	Add 19 µl from well E4 to 41 µl DMSO	0.00332	0.16
G	DMSO, 50µl	DMSO, 50µl	Add 19 µl from well F4 to 41 µl DMSO	0.00105	0.05
H	DMSO, 50µl	DMSO, 50µl	Add 19 µl from well G4 to 41 µl DMSO	0.000332	0.016

Fig. 5 Example of compound dilution plate layout

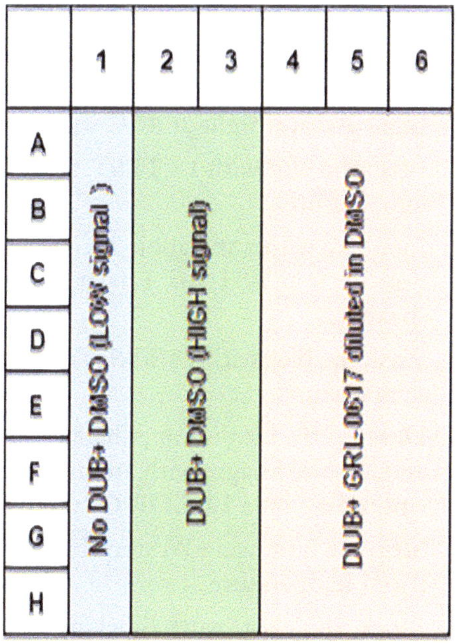

Fig. 6 Example of assay plate layout

5. Read plate using a microplate reader kinetically every 5 min for 3 h.

6. Export the data as Excel file and perform data analysis. Percent inhibition is calculated using the formula Percent Inhibition = 100 × [1 − (X − LOW)/(HIGH − LOW)], where X is the signal at a given concentration of inhibitor, LOW is the signal with no DUB added (100% inhibition), and HIGH is the

signal with DUB in presence of DMSO (0% inhibition). Plot the percent inhibition data using GraphPad Prism software, and calculate IC_{50} using the equation log (inhibitor) vs. response—Variable slope (four parameters).

3.4 Western Blotting

1. Place 30-μL immunoprecipitated Protein G Sepharose beads and 30-μg whole cell lysates (input) in 1.5-mL tubes.
2. Add 2× SDS sample buffer containing 10% β-mercaptoethanol.
3. Incubate at 37 °C for 5 min (*see* **Note 8**).
4. Centrifuge at 16,000 × g for 4 min at room temperature.
5. Load supernatants and inputs on an 8% SDS-PAGE gel.
6. Run the gel at 150 V until the dye front moves to the bottom of the gel.
7. Transfer the proteins in the gel to a PVDF membrane for 30 min at 0.4 A/25 V using an electroblotting system.
8. Block the membrane in 5% nonfat dry milk in 1× TBST (blocking buffer) for 1 h at room temperature with gentle shaking.
9. Discard the blocking buffer and add 10 mL of fresh blocking buffer. Add anti-Nsp3 antibody at 1:1000 dilution prepared in blocking buffer.
10. Incubate overnight at 4 °C with gentle shaking.
11. Wash the blot with 1× TBST, three times, 10 min each at room temperature.
12. Incubate with anti-rabbit IgG-HRP conjugate at 1:1000 dilution in 1× TBST for 1 h at room temperature with gentle shaking.
13. Wash the blot with 1× TBST, three times, 10 min each at room temperature.
14. Develop the blot using enhanced chemiluminescence solution, and acquire images with an imaging system (e.g., 2-min acquisition time on a LI-COR Odyssey imager).
15. Strip the blot using Western blot stripping buffer for 30 min at room temperature.
16. Block the blot with blocking buffer for 30 min at room temperature.
17. Add anti-HA antibody (1:1000 dilution in blocking buffer) and incubate for 2–3 h at room temperature with gentle shaking.
18. Wash the blot with 1× TBST, three times, 10 min each at room temperature.
19. Incubate with anti-mouse IgG-HRP conjugate at 1:1000 dilution in 1× TBST for 1 h at room temperature.

20. Wash the blot with 1× TBST, three times, 10 min each at room temperature.
21. Develop and image the blot as above, using enhanced chemiluminescence solution and an imager.

4 Notes

1. The plasmids used in this study (pEF-HA-PLpro, pEF-HA-Nsp3, and pEF empty vector) were based on the mammalian expression vector pEF-HA, the original version of which was constructed by Eugenia Spanopoulou [23]. The coding sequences used for PLpro and Nsp3 (which contains PLpro) were as described by Kim et al. [24], based on the genomic sequences reported for SARS-CoV-2 strains HKU-SZ-005b [25] and Wuhan-Hu-1 [26]. The subcloned sequences encode replicase polyprotein 1ab residues 1564–1878 for PLpro and 819–2763 for Nsp3, respectively (https://www.uniprot.org/uniprot/P0DTD1#P0DTD1-1), with a stop codon added at the end of each. For subcloning PLpro, the DNA insert was produced by PCR with primers that allowed the generation of XbaI and NotI overhangs after appropriate restriction digestion. Because of the presence of an XbaI restriction site in the second half of the Nsp3 coding sequence, the Nsp3 insert was produced as two separate DNA fragments via PCR. The PCR primers were designed to allow the generation of XbaI/KpnI (first half) and KpnI/NotI (second half) inserts for three-way ligation with the vector DNA. To make plasmid constructs expressing the N-terminally HA-tagged SARS-CoV-2 PLpro and Nsp3, the pEF-HA-Ub plasmid construct was digested with restriction enzymes XbaI and NotI to remove the ubiquitin coding sequence, followed by ligation with PLpro or Nsp3 inserts as described above. The resulting constructs encoded two residues (AS) between the start codon and HA tag, and the linker after the tag was SLSR for PLpro and SLSRMPTLYKKVGM for Nsp3. To generate the control plasmid, QuikChange site-directed mutagenesis [27] was performed on pEF-HA-Ub to replace the HA-Ub coding sequence with only an XbaI site between the upstream XhoI and downstream NotI sites (CTCGAGTCTAGAGCGGCCGC), producing the pEF empty vector.
2. Nsp3 and PLpro are cysteine protease enzymes. Thus, adding a protease inhibitor cocktail containing covalent cysteine protease inhibitors in the lysis buffer may reduce the activity of Nsp3 and PLpro.
3. HEK293T cells are susceptible to dislodging from the culture dishes and should be handled with the utmost care. Removing

media and washing with 1× PBS should be done gently to avoid cells detaching from the culture dish.

4. Vortex the cell lysates for 5 s and place the tubes on ice. Repeat vortexing every 2–4 min.
5. For pipetting Protein G Sepharose beads, use 200-μL pipette tips with several millimeters cut off to avoid tip clogging and facilitate bead transfer.
6. Pre-clearing the cell lysates with Protein G Sepharose beads will help reduce the nonspecific and background deubiquitinase activity of the beads after the anti-HA pull-down.
7. When performing ISG15-CHOP2 assays, work as quickly as possible, since the PLpro ISG15-CHOP2 kinetics are rapid. Signal saturation can be avoided by adjusting the plate reader parameters, reducing the amount of ISG15-CHOP2 substrate (10 nM or lower), or lowering the volume of the beads added.
8. Do not heat the cell lysates and the pull-down samples to 95 °C prior to loading onto SDS-PAGE gels. Nsp3, being a multi-pass transmembrane protein, can form aggregates if samples are heated to these high temperatures.

Acknowledgments

We thank Dr. Michael Mattern for his help in preparing the manuscript. This work was supported by Progenra Inc.

References

1. Proulx J, Borgmann K, Park IW (2021) Role of virally-encoded deubiquitinating enzymes in regulation of the virus life cycle. Int J Mol Sci 22(9):4438. https://doi.org/10.3390/ijms22094438
2. Imbert I, Snijder EJ, Dimitrova M, Guillemot JC, Lecine P, Canard B (2008) The SARS-Coronavirus PLnc domain of nsp3 as a replication/transcription scaffolding protein. Virus Res 133(2):136–148. https://doi.org/10.1016/j.virusres.2007.11.017
3. Lei J, Kusov Y, Hilgenfeld R (2018) Nsp3 of coronaviruses: structures and functions of a large multi-domain protein. Antivir Res 149:58–74. https://doi.org/10.1016/j.antiviral.2017.11.001
4. Angeletti S, Benvenuto D, Bianchi M, Giovanetti M, Pascarella S, Ciccozzi M (2020) COVID-2019: the role of the nsp2 and nsp3 in its pathogenesis. J Med Virol 92(6):584–588. https://doi.org/10.1002/jmv.25719
5. Shin D, Mukherjee R, Grewe D, Bojkova D, Baek K, Bhattacharya A, Schulz L, Widera M, Mehdipour AR, Tascher G, Geurink PP, Wilhelm A, van der Heden van Noort GJ, Ovaa H, Muller S, Knobeloch KP, Rajalingam K, Schulman BA, Cinatl J, Hummer G, Ciesek S, Dikic I (2020) Papain-like protease regulates SARS-CoV-2 viral spread and innate immunity. Nature 587(7835):657–662. https://doi.org/10.1038/s41586-020-2601-5
6. Barretto N, Jukneliene D, Ratia K, Chen Z, Mesecar AD, Baker SC (2005) The papain-like protease of severe acute respiratory syndrome coronavirus has deubiquitinating activity. J Virol 79(24):15189–15198. https://doi.org/10.1128/JVI.79.24.15189-15198.2005
7. Lindner HA, Fotouhi-Ardakani N, Lytvyn V, Lachance P, Sulea T, Menard R (2005) The papain-like protease from the severe acute respiratory syndrome coronavirus is a deubiquitinating enzyme. J Virol 79(24):

15199–15208. https://doi.org/10.1128/JVI.79.24.15199-15208.2005

8. Barretto N, Jukneliene D, Ratia K, Chen Z, Mesecar AD, Baker SC (2006) Deubiquitinating activity of the SARS-CoV papain-like protease. Adv Exp Med Biol 581:37–41. https://doi.org/10.1007/978-0-387-33012-9_5
9. Ratia K, Saikatendu KS, Santarsiero BD, Barretto N, Baker SC, Stevens RC, Mesecar AD (2006) Severe acute respiratory syndrome coronavirus papain-like protease: structure of a viral deubiquitinating enzyme. Proc Natl Acad Sci U S A 103(15):5717–5722. https://doi.org/10.1073/pnas.0510851103
10. Lindner HA, Lytvyn V, Qi H, Lachance P, Ziomek E, Menard R (2007) Selectivity in ISG15 and ubiquitin recognition by the SARS coronavirus papain-like protease. Arch Biochem Biophys 466(1):8–14. https://doi.org/10.1016/j.abb.2007.07.006
11. Baez-Santos YM, Mielech AM, Deng X, Baker S, Mesecar AD (2014) Catalytic function and substrate specificity of the papain-like protease domain of nsp3 from the Middle East respiratory syndrome coronavirus. J Virol 88(21):12511–12527. https://doi.org/10.1128/JVI.01294-14
12. Chen X, Yang X, Zheng Y, Yang Y, Xing Y, Chen Z (2014) SARS coronavirus papain-like protease inhibits the type I interferon signaling pathway through interaction with the STING-TRAF3-TBK1 complex. Protein Cell 5(5):369–381. https://doi.org/10.1007/s13238-014-0026-3
13. Mielech AM, Kilianski A, Baez-Santos YM, Mesecar AD, Baker SC (2014) MERS-CoV papain-like protease has deISGylating and deubiquitinating activities. Virology 450–451:64–70. https://doi.org/10.1016/j.virol.2013.11.040
14. Ratia K, Kilianski A, Baez-Santos YM, Baker SC, Mesecar A (2014) Structural basis for the ubiquitin-linkage specificity and deISGylating activity of SARS-CoV papain-like protease. PLoS Pathog 10(5):e1004113. https://doi.org/10.1371/journal.ppat.1004113
15. Baez-Santos YM, St John SE, Mesecar AD (2015) The SARS-coronavirus papain-like protease: structure, function and inhibition by designed antiviral compounds. Antivir Res 115:21–38. https://doi.org/10.1016/j.antiviral.2014.12.015
16. Clasman JR, Everett RK, Srinivasan K, Mesecar AD (2020) Decoupling deISGylating and deubiquitinating activities of the MERS virus papain-like protease. Antivir Res 174:104661. https://doi.org/10.1016/j.antiviral.2019.104661
17. Klemm T, Ebert G, Calleja DJ, Allison CC, Richardson LW, Bernardini JP, Lu BG, Kuchel NW, Grohmann C, Shibata Y, Gan ZY, Cooney JP, Doerflinger M, Au AE, Blackmore TR, van der Heden van Noort GJ, Geurink PP, Ovaa H, Newman J, Riboldi-Tunnicliffe A, Czabotar PE, Mitchell JP, Feltham R, Lechtenberg BC, Lowes KN, Dewson G, Pellegrini M, Lessene G, Komander D (2020) Mechanism and inhibition of the papain-like protease, PLpro, of SARS-CoV-2. EMBO J 39(18):e106275. https://doi.org/10.15252/embj.2020106275
18. Wang L, Hu W, Fan C (2020) Structural and biochemical characterization of SARS-CoV papain-like protease 2. Protein Sci 29(5):1228–1241. https://doi.org/10.1002/pro.3857
19. Li SW, Wang CY, Jou YJ, Huang SH, Hsiao LH, Wan L, Lin YJ, Kung SH, Lin CW (2016) SARS coronavirus papain-like protease inhibits the TLR7 signaling pathway through removing Lys63-linked polyubiquitination of TRAF3 and TRAF6. Int J Mol Sci 17(5):678. https://doi.org/10.3390/ijms17050678
20. Frieman M, Ratia K, Johnston RE, Mesecar AD, Baric RS (2009) Severe acute respiratory syndrome coronavirus papain-like protease ubiquitin-like domain and catalytic domain regulate antagonism of IRF3 and NF-kappaB signaling. J Virol 83(13):6689–6705. https://doi.org/10.1128/JVI.02220-08
21. Matthews K, Schafer A, Pham A, Frieman M (2014) The SARS coronavirus papain like protease can inhibit IRF3 at a post activation step that requires deubiquitination activity. Virol J 11:209. https://doi.org/10.1186/s12985-014-0209-9
22. Moustaqil M, Ollivier E, Chiu HP, Van Tol S, Rudolffi-Soto P, Stevens C, Bhumkar A, Hunter DJB, Freiberg AN, Jacques D, Lee B, Sierecki E, Gambin Y (2021) SARS-CoV-2 proteases PLpro and 3CLpro cleave IRF3 and critical modulators of inflammatory pathways (NLRP12 and TAB1): implications for disease presentation across species. Emerg Microbes Infect 10(1):178–195. https://doi.org/10.1080/22221751.2020.1870414
23. Li Y, Gazdoiu S, Pan ZQ, Fuchs SY (2004) Stability of homologue of Slimb F-box protein is regulated by availability of its substrate. J Biol Chem 279(12):11074–11080. https://doi.org/10.1074/jbc.M312301200
24. Kim DK, Knapp JJ, Kuang D, Chawla A, Cassonnet P, Lee H, Sheykhkarimli D, Samavarchi-Tehrani P, Abdouni H, Rayhan A, Li R, Pogoutse O, Coyaud E, van der Werf S, Demeret C, Gingras AC, Taipale M, Raught B,

Jacob Y, Roth FP (2020) A comprehensive, flexible collection of SARS-CoV-2 coding regions. G3 (Bethesda) 10(9):3399–3402. https://doi.org/10.1534/g3.120.401554

25. Chan JF, Kok KH, Zhu Z, Chu H, To KK, Yuan S, Yuen KY (2020) Genomic characterization of the 2019 novel human-pathogenic coronavirus isolated from a patient with atypical pneumonia after visiting Wuhan. Emerg Microbes Infect 9(1):221–236. https://doi.org/10.1080/22221751.2020.1719902
26. Wu F, Zhao S, Yu B, Chen YM, Wang W, Song ZG, Hu Y, Tao ZW, Tian JH, Pei YY, Yuan ML, Zhang YL, Dai FH, Liu Y, Wang QM, Zheng JJ, Xu L, Holmes EC, Zhang YZ (2020) A new coronavirus associated with human respiratory disease in China. Nature 579(7798):265–269. https://doi.org/10.1038/s41586-020-2008-3
27. Wang W, Malcolm BA (1999) Two-stage PCR protocol allowing introduction of multiple mutations, deletions and insertions using QuikChange Site-Directed Mutagenesis. BioTechniques 26(4):680–682. https://doi.org/10.2144/99264st03

Chapter 17

Use of Deubiquitinase Fusion Proteins to Characterize Endocytic Trafficking in Yeast

Nathaniel L. Hepowit, Aeva S. Denise, and Jason A. MacGurn

Abstract

Ubiquitin modification is known to regulate endocytic trafficking of many different types of cargo in eukaryotic cells, but it can be challenging to determine what role, if any, ubiquitin plays in the trafficking of a novel or uncharacterized endocytic cargo. Here, we describe a useful approach that leverages fusion to deubiquitinase (DUB) catalytic domains to explore the role ubiquitin plays in endocytic trafficking. This approach can be applied to the analysis of many different endocytic cargos in different cell types, and it can also be used to study linkage specificity in endocytic trafficking. Several different trafficking assays are described to illustrate the broad utility of this "DUB fusion" approach.

Key words DUB fusion, Endocytosis, Mup1, Can1, Flow cytometry, Fluorescence microscopy, Spot plate dilution assay

1 Introduction

In eukaryotic cells, ubiquitin modification of integral membrane proteins at the plasma membrane (PM) often results in capture and sorting into endocytic vesicles, thus removing ubiquitin-modified "cargo" from the cell surface [1–5]. Despite this central role in specifying endocytic cargo for clearance from the PM, demonstrating a role for ubiquitin in the regulation of specific cargo can be challenging—especially when there is no strong precedent for ubiquitin modification of a cargo of interest. Additionally, ubiquitin modification can be complex and difficult to characterize, given that conjugation can occur on multiple lysine residues redundantly, and complex polymer structure (e.g., linkage-type and branching) can impact regulatory outcomes [6–8]. Given these complexities, there is a need for tools and approaches that test the regulatory function of ubiquitin conjugation to specific endocytic cargo.

One useful tool for characterizing the role of ubiquitin in trafficking involves fusing specific cargo to catalytic domains of

Julie A. Maupin-Furlow and Mariola J. Edelmann (eds.), *Deubiquitinases: Methods and Protocols*,
Methods in Molecular Biology, vol. 2591, https://doi.org/10.1007/978-1-0716-2803-4_17,

deubiquitinases (DUBs). This approach was first developed in yeast by Robert Piper and colleagues [9] and has been used to characterize ubiquitin-dependent endocytosis [10] as well as other trafficking events such as sorting on endosomes [11, 12] and SNARE recycling [13]. The DUB fusion approach has also been used to identify regulatory functions of ubiquitin in cultured mammalian cells [10, 14]. Furthermore, linkage-specific DUB domains have been used to uncover roles for specific ubiquitin polymer linkage types in trafficking [10]. Importantly, these experiments rely on cargo fusion to catalytic dead DUB variants as critical controls.

Here, we describe methods for using DUB fusion domains in three different endocytic assays. The first assay uses fluorescence microscopy to analyze the effects of DUB fusion on stimulated endocytosis of the methionine transporter MUP1. The second assay uses flow cytometry to analyze the same trafficking event. The third assay analyzes endocytosis indirectly by measuring sensitivity to canavanine, a toxic amino acid analog that is taken up only by surface-localized Can1, an arginine transporter. Each of these different assays illustrates the utility of the DUB fusion strategy for investigating the role of ubiquitin in endocytic trafficking.

2 Materials

1. Oligonucleotides (*see* Table 1).
2. DUB fusion Longtine plasmids (*see* Table 2).
3. *Saccharomyces cerevisiae* strains: SEY6210, NHY379, and BY4741 (*see* Table 3 for genotypic description).
4. PCR reaction reagents:
   (a) DUB fusion PCR mix (99 μL): 71 μL nuclease-free water, 20 μL 5× HF buffer (New England Biolabs), 2 μL 10 mM dNTP mix, 10-μM oligonucleotides (2.5 μL forward and 2.5 μL reverse), and 1-μL Phusion polymerase.
   (b) Diagnostic PCR mix (24 μL): 15.5 μL nuclease-free water, 5 μL 5× HF buffer (New England Biolabs), 0.5 μL 10 mM dNTP mix, 20 μM oligonucleotides (1.25 μL forward and 1.25 μL reverse), and 0.5 μL Phusion polymerase.
5. 25× TAE stock solution (1 L): 121 g Tris–HCl base, 28.5 mL glacial acetic acid, and 50-mL EDTA (pH 8.0) in dH_2O up to 1 L.
6. Agarose gel (1% gel): Dissolve 0.5 g agarose in 50-mL 1× TAE buffer by microwaving for 2 min. Add 2 μL GreenGlo (DNA dye) and allow gel to solidify in a molding tray.
7. Thermal cycler.
8. DNA agarose gel electrophoresis equipment.

Table 1
Oligonucleotides to generate DUB fusion cassettes and for diagnostic colony screening

Primer	Description	Sequence
ojam525	Mup1 C-terminal tagging FWD	TATAATCGAACATTACAAAAGTGAACAAGAAAAA TCGCTGCGGATCCC CGGGTTAATTAA
ojam526	Mup1 C-terminal tagging REV	TATAAGAATCGAGATGAGATGGTAAGTACC TTTTTGGTTAGCGCGTTG GCCGATTCATTA
ojam643	Mup1 C-terminal tag diagnostic FWD	GGCGATGTCGATACACAAGA
ojam644	Mup1 C-terminal tag diagnostic REV	TGTTCCGCCAACTCAGCTTA
ojam1147	Can1 C-terminal tagging FWD	ACCAAAGACTTTTTGGGACAAATTTTGGAATGTTG TAGCACGGATCCC CGGGTTAATTAA
ojam1148	Can1 C-terminal tagging REV	ATGGCGTGGAAATGTGATCAAAGGTAATAAAACG TCATATGAATTCGA GCTCGTTTAAAC
ojam1149	Can1 C-terminal tag diagnostic FWD	CCGGCTTGGCTTATTATGCG
ojam1150	Can1 C-terminal tag diagnostic REV	GTCATCGGAGTCGTTATCAG

Table 2
Plasmids with DUB fusion Longtine cassettes and their corresponding primer pair for PCR amplification

DUB fusion Longtine plasmid		Primer pair for cassette amplification	
Code	**Description of cassette**	**Forward**	**Reverse**
LT4	*MUP1-GFP::kanMX6* (no DUB control)	ojam525	ojam526
pJAM1641	*MUP1-GFP-UL36-3XHA::kanMX6* (*see* **Note 2**)	ojam525	ojam526
pJAM1642	*MUP1-GFP-UL36 cat. dead-3XHA::kanMX6* (*see* **Note 2**)	ojam525	ojam526
pBW1571	*MUP1-pHluorin::kanMX6* (no DUB control) (*see* **Note 3**)	ojam525	ojam526
pJAM1570	*MUP1-pHluorin-UL36-3XHA::kanMX6*	ojam525	ojam526
pJAM1571	*MUP1-pHluorin-UL36 cat. dead-3XHA::kanMX6*	ojam525	ojam526
pJAM800	*CAN1-UL36-3XHA::kanMX6*	ojam1147	ojam1148
pJAM801	*CAN1-UL36 cat. dead -3XHA::kanMX6*	ojam1147	ojam1148
pJAM1572	*Can1-AMSH-3XHA::kanMX6* (*see* **Note 4**)	ojam1147	ojam1148
pJAM1573	*Can1-AMSH cat. dead -3XHA::kanMX6*	ojam1147	ojam1148

Table 3
Genotypic description of yeast strains for DUB fusion cassette transformation

Strain	Genotype
SEY6210	*MATα leu2-3,112 ura3-52 his3-Δ200 trp1-Δ901 suc2-Δ9 lys2-801; GAL*
NHY379	SEY6210 *VPH1::mCherry::kanMX6* (*see* **Note 5**)
BY4741	*MATa his3Δ1 leu2Δ0 met15Δ0 ura3Δ0*

9. PCR product purification kit: E.Z.N.A. Gel Extraction Kit (Omega).
10. Salmon sperm DNA (Sigma).
11. Yeast transformation solution (900-μL 50% polyethylene glycol in TE buffer [10 mM Tris–HCl, 1 mM EDTA, pH 7.5], 100 μL 1 M lithium acetate (LiAc, pH 7.5), 10 μL 1 M Tris–HCl (pH 8.0), 2 μL 0.5-M EDTA (pH 8.0)).
12. Smash and Grab (S&G) buffer (100 mL): 2 mL Triton X-100, 5 mL 20% SDS, 2 mL 5 M NaCl, 1 mL 1 M Tris–HCl (pH 8.0), 200 μL 0.5 M EDTA (pH 8.0), 89.8 mL dH_2O.
13. Synthetic media stock solutions (filter-sterilized): 20× quadruple dropout stock (200 mg adenine, 200 mg arginine, 2300-mg lysine, 3000 mg threonine, and 300 mg tyrosine dissolved in 500 mL dH_2O), 100× histidine (0.5 g histidine dissolved in 250 mL dH_2O), 100× leucine (1.5 g histidine dissolved in 250 mL dH_2O), 100× tryptophan (0.5 g tryptophan dissolved in 250 mL dH_2O), 100× uracil (0.5 g dissolved in 250 mL dH_2O), 1000× methionine (10 g methionine in 500 mL dH_2O), 50% glucose (500 g glucose dissolved in dH_2O up to 1 L final volume), 10× yeast nitrogen base (YNB) without amino acids and with ammonium sulfate (67 g YNB dissolved in dH_2O up to 1 L final volume).
14. Yeast culture media:
    (a) Yeast peptone dextrose (YPD, 1 L) broth: Dissolve 50 g YPD (Fisher) in 1 L dH_2O. Autoclave.
    (b) YPD agar with G418 (1 L): Dissolve 50-g YPD (Fisher) and 15 g agar (high-gel strength) in 1 L dH_2O. Autoclave. Cool to around 50 °C and add 1 mL of 200 mg/mL filter-sterilized G418. Pour into petri dishes (30 mL per dish) and let the agar solidify at room temperature. Store plates at 4 °C until use (*see* **Note 1**).
    (c) Methionine-dropout synthetic media (500 mL): Mix 50 mL 10× YNB, 20 mL 50% glucose, 25 mL 20× quadruple dropout stock, 5 mL 10× histidine, 5 mL 10× leucine, 5 mL 10× tryptophan, and 5 mL uracil, and adjust to 1 L with dH_2O.

(d) Canavanine plate (1 L): 1-L dH_2O, 6.7 g YNB (without amino acids), 0.7 g complete amino acid mix, 15 g agar. Autoclave. Cool to around 50 °C and add 80 μL of 25 mg/mL filter-sterilized canavanine. Pour into petri dishes (30 mL per dish) and let the agar solidify at room temperature. Store plates at 4 °C until use (*see* **Note 1**).

3 Methods

3.1 Generating DUB Fusion Knock-in Cassettes

1. Mix 1 μL of DUB fusion plasmid (10–20 ng) and 99 μL of DUB fusion PCR mix in a 0.2 mL PCR tube (*see* Table 2 for template and primer set).
2. Amplify the DUB fusion cassette in a PCR machine using the thermal cycle setting: 98 °C (30 s); 40 cycles of 98 °C (10 s), 57 °C (20 s), and 72 °C (1 min and 30 s); and 72 °C (3 min).
3. To confirm the presence of amplicons, separate 5 μL PCR product (from 100 μL PCR product), mix with 1 μL 6× DNA loading dye, and run on 1% agarose gel in 1× TAE buffer at 100 volts for 30 min.
4. Visualize the DNA agarose gel by UV transillumination or by using a Gel Doc EZ Imager (Bio-Rad) at SafeStain or ethidium bromide imaging setting.
5. Purify the remaining 95 μL PCR product using a DNA purification kit as recommended by the supplier.
6. Store the purified DNA cassette at −20 °C until use.

3.2 Transformation of DUB Fusion Knock-in Cassette

1. A day before transformation, inoculate a 5 mL YPD broth with *S. cerevisiae* (SEY6210 or BY4741) cells, and let it grow overnight at 30 °C in a shaker incubator (*see* Table 3 for genotypic description of parental yeast strains).
2. Next morning, subculture cells at starting 0.1 OD_{600} unit in 10 mL YPD broth and grow cells to log phase (~0.5 to 0.8 OD_{600} unit).
3. Harvest cells by centrifugation at 2500 × *g* for 1 min. Wash cells with 10 mL sterile dH_2O and collect cells by centrifugation at 2500 × *g* for 1 min.
4. Resuspend cells in 1 mL 0.1-M LiAc (in TE buffer, pH 8.0) by pipetting in and out and transfer into a 1.5 mL microfuge tube. Harvest cells by centrifugation at 2500 × *g* for 10 s and discard the supernatant. Resuspend cells in 100 μL 0.1 M LiAc (in TE buffer, pH 8.0).
5. Heat salmon sperm DNA at 98 °C for 5 min and rapidly cool on ice.
6. Mix the following in a 0.2 mL PCR tube by pipetting in and out: 4 μL salmon sperm DNA, 2 μL knock-in DNA (PCR-amplified DUB fusion cassette), 10 μL cells (from #4,

SEY6210 cells for Mup1-pHluorin-UL36 fusion strains, SEY6210 and BY4741 cells for Can1 fusion strains), and 60 μL yeast transformation solution.

7. Incubate the PCR tube at 42 °C for 45 min using a thermal cycler.
8. Transfer the transformation mixture into 5 mL YPD broth and let cells grow overnight (12–18 h) at 30 °C with agitation.
9. Transfer 0.5 mL of the culture into a 1.5 mL microfuge tube. Harvest cells by centrifugation (2500 × *g* for 10 s) and plate cells onto YPD with G418. Incubate plate at 30 °C for 2 days.

3.3 Screening of DUB Fusion Strains

1. Re-streak growing colonies onto fresh YPD with G418 plate to obtain a pure culture.
2. Add a loopful of cells into a microfuge tube with 200 μL S&G buffer. Include parent strain as a negative control.
3. Add 200 μL glass beads (0.5-mm dia.) and 200 μL cold phenol and vortex for 15 min.
4. Add 200 μL dH_2O and spin down at top speed for 5 min at 4 °C.
5. Remove 5 μL supernatant and add it into 95 μL dH_2O. Store DNA at −20 °C until use.
6. Add 1 μL DNA into 24 μL diagnostic PCR mix (*see* Table 1 for diagnostic primer pair).
7. Amplify the DUB fusion cassette in a PCR machine using the thermal cycle setting: 98 °C (30 s); 40 cycles of 98 °C (10 s), 57 °C (20 s), and 72 °C (1 min and 30 s); and 72 °C (3 min).
8. Mix 5 μL PCR product and 1 μL 6× DNA loading dye, and run on 1% agarose gel (with GreenGlo) in 1× TAE buffer at 100 volts for 30 min.
9. Visualize the DNA agarose gel by UV transillumination or by using a Gel Doc EZ Imager (Bio-Rad) at SafeStain or ethidium bromide imaging setting.

3.4 Analyzing the Endocytic Trafficking of Mup1-GFP DUB Fusions by Fluorescence Microscopy

1. Grow pure cultures of Mup1-GFP, Mup1-GFP-UL36, and Mup1-GFP-UL36 C40S (cat. dead) strains overnight (14–16 h) in 10 mL –Met SM (*see* **Note 6**).
2. Subculture in fresh 10 mL –Met SM with a starting OD_{600} of 0.1. Grow cells to 0.4 OD_{600} unit at 26 °C with agitation (OD_{600} of yeast doubles approximately every 2 h).
3. Turn on the fluorescence microscope (DeltaVision Elite Imaging Station) and select an appropriate objective (*see* **Note 7**). Make sure that the objective is lowered prior to initializing the system.

4. Open the Softworx software. Click the file and initialize the 'acquire resolve' function.. Wait for around 10 min, while the fluorescence microscope is self-calibrating.
5. On the Softworx window, activate the Alexa filter wheel setting (excitation, emission, and eyepiece filter wheel).
6. Open the Design/Run Experiment window and set the Sectioning setting as follows:
   Check Z sectioning.
   Image Scan Sequence: Wavelength then Z.
   Focus point when the scan starts: Middle of Sample.
   Optical section setting (μM): 0.20 (*see* **Note 8**).
   Number of optical sections: 20.
   Sample thickness (μM): 4.00.
7. Open the Channels window.
8. Set the following excitation/emission filters: FITC (green, 50% transmission), A594 (red, 100% transmission), and POL (DIC, 32% transmission). Click the "Refresh exposure condition" button.
9. Revolve the filter wheel of the microscope to set the light transmission through the eyepiece to POL (excitation, −50/28; emission, −50/0; polychroic, Quad; %T, 10%).
10. Set exposure to 0.0800 and image size to 1024 × 1024.
11. When culture (in #2) reaches 0.4 OD_{600} unit, induce the endocytic trafficking of Mup1-GFP from the plasma membrane into the vacuole (enlarged lysosome) by adding 10 μL of 20 mg/mL methionine (10× Met). Vortex culture for 3 s to mix.
12. Take 1-mL culture (at 0, 20, and 40 min after adding Met) and concentrate cells by centrifugation at 2500 × *g* for 1 min.
13. Take 1.5 μL of cell suspension, drop onto a coverslip, and lay a glass slide on top to immobilize the cells.
14. Visualize the cells under 100× objective (oil immersion) through the eyepiece of the microscope.
15. Acquire images and deconvolve to remove off-focus fluorescence signal (*see* **Note 9**).
16. Overlay the images at different channels, crop to desired frame size, and add the scale bar using Fiji (Fig. 1a).
17. Randomly select 30 cells (at least) (*see* **Note 10**), and measure the fluorescence signal (with background subtraction) at the plasma membrane and in the vacuole using Fiji.
18. Compute the ratio of fluorescence signal at the plasma membrane and in the vacuole using Microsoft Excel.
19. Plot (scatter dot plot) using GraphPad Prism 9.0 (Fig. 1b).

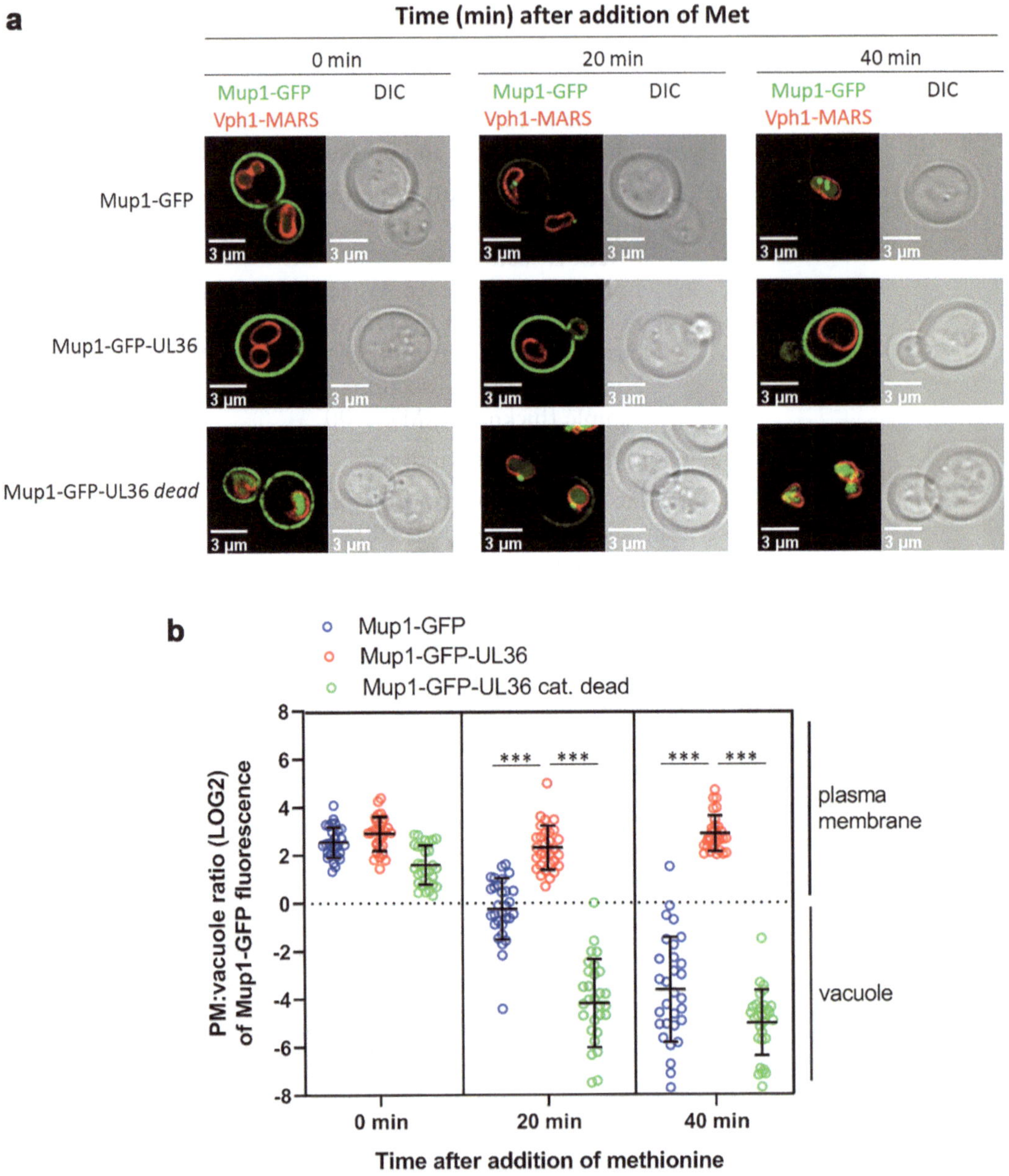

Fig. 1 (**a**) Representative differential interference contrast (gray) and fluorescence microscopy images of yeast cells with endogenously expressed Mup1-GFP and Mup1-GFP-DUB fusions (green) and Vph1-MARS (red, vacuole marker) obtained at incremental time points after the addition of methionine. Mup1 trafficking and localization in live yeast cells at mid-log phase were visualized using a DeltaVision Elite Imaging system (Olympus IX-71 inverted microscope; Olympus 100× oil objective (1.4 NA); DVElite sCMOS camera, GE Healthcare). Images in different channels (FITC [excitation 475/28 nm, emission 523/36 nm] and A594 [excitation 575/25, emission 632/60]) were resolved and deconvolved using softWoRx 7.0.0, and channels were merged and analyzed using Fiji as previously described [16]. (**b**) The Mup1 mean fluorescence intensity ratio at the plasma membrane and the vacuole was measured on 30 randomly selected cells from each time point using Fiji. The dot plot was generated using Prism 9.0. ***$p < 0.0001$

3.5 Analyzing the Endocytic Trafficking of Mup1-pHluorin by Flow Cytometry

1. Grow pure cultures of Mup1-pHl, Mup1-pHl-UL36, and Mup1-pHl-UL36 C40S (catalytically dead) strains overnight (~14–16 h) in 10 mL –Met SM in triplicates.
2. Subculture in fresh 10 mL –Met SM with a starting OD_{600} of 0.1. Grow cells to 0.4 OD_{600} unit at 26 °C with agitation (OD_{600} of yeast doubles approximately every 2 h).
3. Open the BD CSampler Plus software on the computer attached to the flow cytometer.
4. Turn on the flow cytometer (BD Accuri C6 Plus), prepare all required solutions, and flush-clean the lines and column as recommended.
5. When cultures (in #2) reach 0.4 OD_{600} unit, induce the endocytic trafficking of Mup1-GFP from the plasma membrane into the vacuole (enlarged lysosome) by adding 10 μL of 20 mg/mL methionine (10× Met). Vortex culture for 3 s to mix.
6. Quickly aliquot cultures using a multi-well pipette into a flat-bottom 96-well plate (280 μL per well) as indicated in Fig. 2a.
7. Eject the sample platform of the flow cytometer and place the 96-well plate.
8. Open the Auto Collect window and set the run setting as follows:

   Plate type: 96-well plate: flat bottom.
   Provide the filename of the experimental setup.
   Click "Select All" on the 96-well view.
   Run with limits: 10,000 events.
   Fluidics: medium.
   Click "Run Horizontally."
   Click "Run SIP Clean After Samples."
9. Click the "OPEN RUN DISPLAY" button.
10. Click the "AUTORUN" button.
11. When the instrument reads all samples, open the "Analyze" window.
12. Select the histogram plot icon and set the plot's X-axis to FITC-A.
13. Select the A1 well of the 96-well view and set the gating bar on the histogram plot to approximately 90%.
14. Open the "Statistics" window.
15. Copy the % of all values, expressed as %cells with Mup1-pHl fluorescence signal, into a Microsoft Excel spreadsheet.
16. Normalize the raw data of each row using the fluorescence reading on column 1 as time zero.
17. Plot the line graph (Fig. 2b) (*see* **Note 11**).

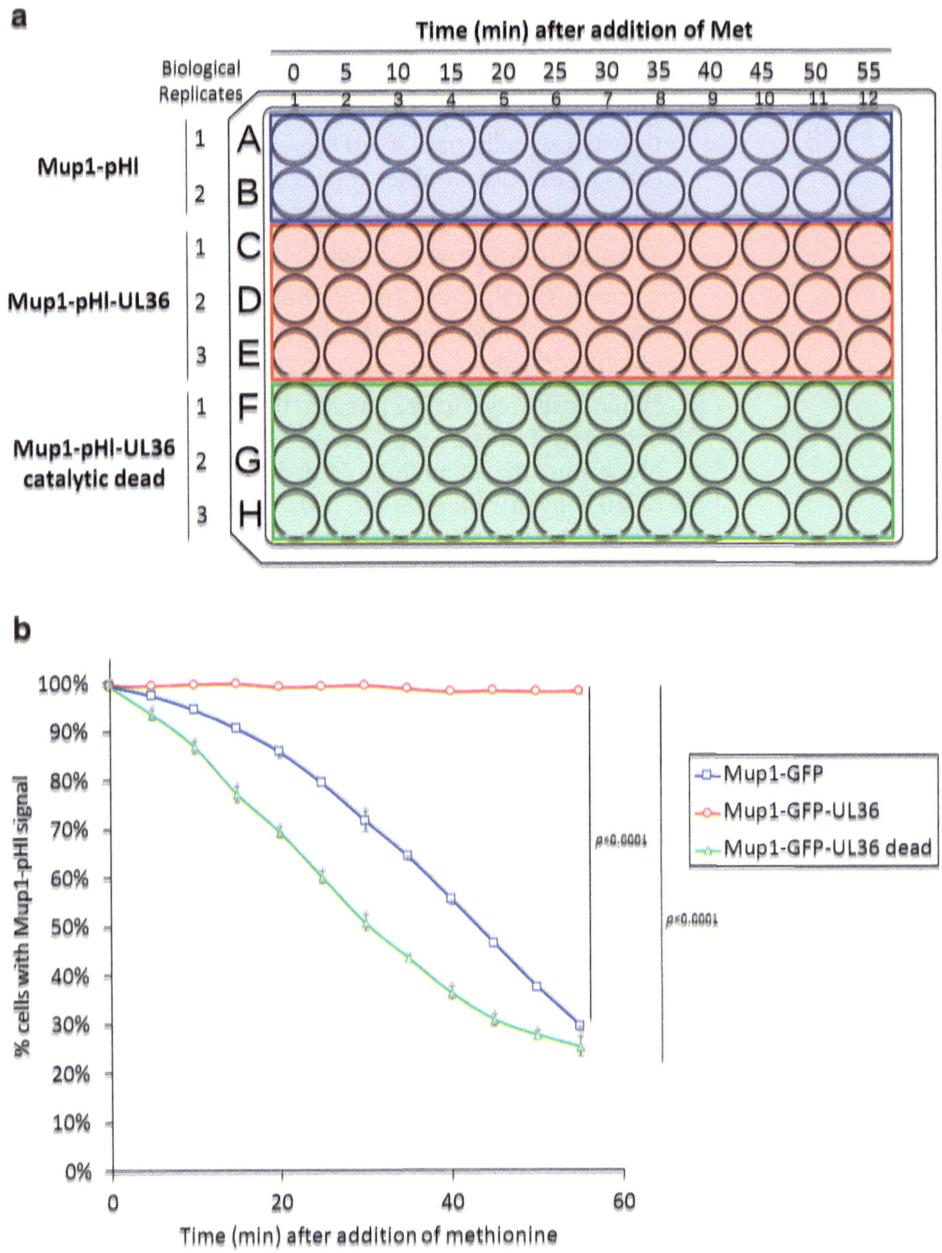

Fig. 2 (**a**) A 96-well template for a flow cytometry experimental design. Each well contains 280 μL of yeast culture at 0.4 OD_{600} unit expressing endogenous Mup1-pHluorin and Mup1-pHluorin-DUB fusions. (**b**) Percentage of the cell population (10,000 events) expressing endogenous Mup1-pHluorin and Mup1-pHluorin-DUB fusions. Cell fluorescence through the FITC filter was determined using a BD Accuri C6 Plus flow cytometer as previously described [11, 17]. Line graph was generated using Microsoft Excel. $^{***}p < 0.0001$

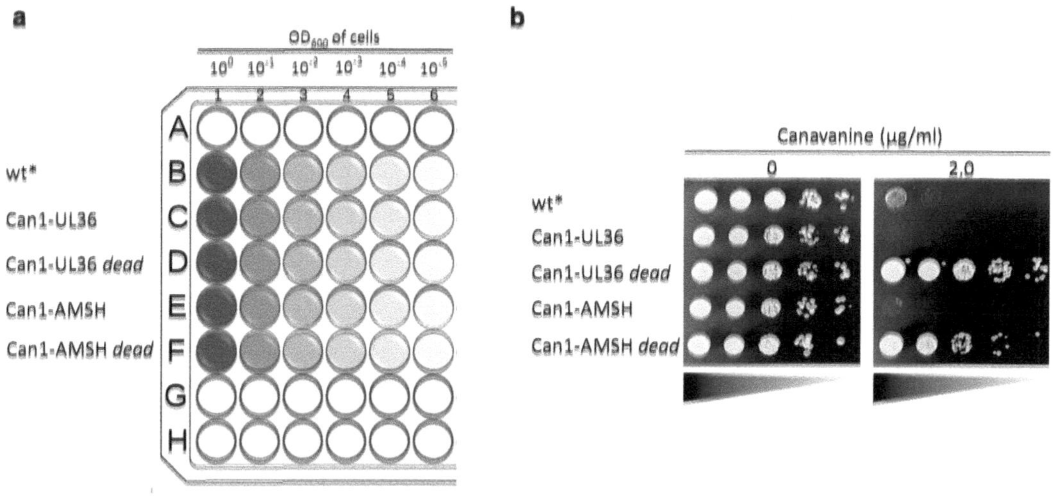

Fig. 3 (**a**) A 96-well template for spot plate dilution assay experimental design. Each well contains 250-μL cell suspension serially diluted to 10^0, 10^1,10^2,10^3,10^4, and 10^5 OD_{600} unit. Spot plate inoculation from the 96-well plate onto the synthetic media plates was done using a flame-sterilized 48-well replica plater (Sigma). (**b**) Yeast cells expressing endogenous Can1-DUB fusions were spot-inoculated on synthetic complete dextrose media with or without canavanine (toxic arginine analog). Cells were serially diluted to 10^0, 10^1,10^2,10^3,10^4, and 10^5 OD_{600} units on a 96-well plate and were replica-inoculated onto agar plates using a 48-well replica plater (Sigma) as previously described [18]. Photograph images were obtained after incubation of plates at 26 °C for 3 days

3.6 Analysis of Can1 PM Activity by Measurement of Growth in the Presence of Canavanine

1. Grow pure cultures of BY4741, BY4741 Can1-AMSH, and BY4741 Can1-AMSH cat. dead strains overnight (~14–16 h) in 10 mL –Met SM (*see* **Note 12**).
2. Serially dilute cells to 10^0, 10^1,10^2,10^3,10^4, and 10^5 OD_{600} units in a 96-well plate (250 μL per well), as shown in Fig. 3a.
3. Using a flame-sterilized frogger, spot plate the cell suspensions onto SM agar plate with or without canavanine. Let the liquid spots dry on the plate.
4. Incubate the plates at 26 °C for 3 days.
5. Obtain the image of plates with black background (Fig. 3b).

4 Notes

1. For optimal efficiency of strain selection, yeast media plates should be used within 60 days of preparation.
2. UL36 is a broad-specificity DUB from the type 1 herpesvirus. Cargo is fused to residues 15–260 of UL36, and the catalytic dead variant harbors the C40S substitution [9].
3. Superecliptic pHluorin was chosen for these experiments since it exhibits a strong fluorescence signal quenched in the acidic environment of yeast vacuoles [15].

4. AMSH is a K63-specific DUB [10]. The catalytic dead variant harbors the E280A and D309A substitutions.
5. Vph1 is a marker of the vacuolar membrane, and using a yeast strain expressing Vph1-RFP-MARS allows for visualization of the limiting membrane of the vacuole. Thus, this marker is useful for quantifying how much a given cargo is localized to the vacuole. Markers of other compartments (e.g., endosomes, PM) could also be used.
6. Mup1 is a methionine transporter in yeast from the major facilitator superfamily of transporters. It is highly expressed at the PM in yeast grown in the absence of methionine, and stimulation of yeast with methionine triggers its rapid internalization by endocytosis followed by trafficking to the vacuole for degradation.
7. We used a DeltaVision Elite Imaging system (GE) with a 100× immersion oil objective for the analysis shown.
8. For yeast imaging, optical sections can be taken at a distance as small as 0.15 μm. However, if the distance between sections decreases, the total number of sections acquired must be increased to achieve the same sample thickness.
9. For deconvolution, it is critical to input a point spread function specific to the objective used to collect the images. The objective manufacturer should supply a file containing the point spread function.
10. We find that quantification of 30 cells is typically sufficient to determine if a significant difference exists between two different samples (based on an ANOVA test). However, it may be necessary to quantify fluorescence distribution in additional cells in some cases.
11. In our experience, analysis of pHluorin-fused cargo by flow cytometry works well when cargo is highly expressed. In contrast, for cargo expressed at low levels, a fluorescence signal is insufficient for this assay to provide a valid dynamic range, limiting the assay's sensitivity.
12. Can1 is an arginine transporter in yeast from the major facilitator superfamily of transporters. It is expressed at low levels at the PM, and its activity at the PM confers sensitivity to canavanine, a toxic analog of arginine. Thus, perturbations that block Can1 endocytic turnover result in hypersensitivity to canavanine.

Acknowledgments

This work was supported by NIH grant R35GM144112 (to JAM).

References

1. Léon S, Haguenauer-Tsapis R (2009) Ubiquitin ligase adaptors: regulators of ubiquitylation and endocytosis of plasma membrane proteins. Exp Cell Res 315:1574–1583
2. MacGurn JA, Hsu PC, Emr SD (2012) Ubiquitin and membrane protein turnover: from cradle to grave. Annu Rev Biochem 81:231–259
3. MacGurn JA (2014) Garbage on, garbage off: new insights into plasma membrane protein quality control. Curr Opin Cell Biol 29C:92–98
4. Di Fiore PP, von Zastrow M (2014) Endocytosis, signaling, and beyond. Cold Spring Harb Perspect Biol 6:a016865
5. Henry AG, White IJ, Marsh M, von Zastrow M, Hislop JN (2011) The role of ubiquitination in lysosomal trafficking of δ-opioid receptors. Traffic 12:170–184
6. Haakonsen DL, Rape M (2019) Branching out: improved signaling by heterotypic ubiquitin chains. Trends Cell Biol 29: 704–716
7. Cappadocia L, Lima CD (2018) Ubiquitin-like protein conjugation: structures, chemistry, and mechanism. Chem Rev 118:889–918
8. Mevissen TET, Komander D (2017) Mechanisms of Deubiquitinase specificity and regulation. Annu Rev Biochem 86:159–192
9. Stringer DK, Piper RC (2011) A single ubiquitin is sufficient for cargo protein entry into MVBs in the absence of ESCRT ubiquitination. J Cell Biol 192:229–242
10. Huang F, Zeng X, Kim W, Balasubramani M, Fortian A, Gygi SP, Yates NA, Sorkin A (2013) Lysine 63-linked polyubiquitination is required for EGF receptor degradation. Proc Natl Acad Sci U S A 110:15722–15727
11. Lee S, Tumolo JM, Ehlinger AC, Jernigan KK, Qualls-Histed SJ, Hsu PC, McDonald WH, Chazin WJ, MacGurn JA (2017) Ubiquitin turnover and endocytic trafficking in yeast are regulated by Ser57 phosphorylation of ubiquitin. elife 6:e29176
12. MacDonald C, Buchkovich NJ, Stringer DK, Emr SD, Piper RC (2012) Cargo ubiquitination is essential for multivesicular body intralumenal vesicle formation. EMBO Rep 13:331–338
13. Xu P, Hankins HM, MacDonald C, Erlinger SJ, Frazier MN, Diab NS, Piper RC, Jackson LP, MacGurn JA, Graham TR (2017) COPI mediates recycling of an exocytic SNARE by recognition of a ubiquitin sorting signal. elife 6: e28342
14. Nielsen CP, Jernigan KK, Diggins NL, Webb DJ, MacGurn JA (2019) USP9X deubiquitylates DVL2 to regulate WNT pathway specification. Cell Rep 28:1074–1089.e5
15. Prosser DC, Whitworth K, Wendland B (2010) Quantitative analysis of endocytosis with cytoplasmic pHluorin chimeras. Traffic 11:1141–1150
16. Tumolo JM, Hepowit NL, Joshi SS, MacGurn JA (2020) A Snf1-related nutrient-responsive kinase antagonizes endocytosis in yeast. PLoS Genet 16:e1008677
17. Hepowit NL, Macedo JKA, Young LEA, Liu K, Sun RC, MacGurn JA, Dickson RC (2021) Enhancing lifespan of budding yeast by pharmacological lowering of amino acid pools. Aging (Albany NY) 13:7846–7871
18. Hepowit NL, Pereira KN, Tumolo JM, Chazin WJ, MacGurn JA (2020) Identification of ubiquitin Ser57 kinases regulating the oxidative stress response in yeast. elife 9:e58155

Index

Julie A. Maupin-Furlow and Mariola J. Edelmann (eds.), *Deubiquitinases: Methods and Protocols*, Methods in Molecular Biology, vol. 2591, https://doi.org/10.1007/978-1-0716-2803-4,

MIX
Papier aus verantwortungsvollen Quellen
Paper from responsible sources
FSC® C105338
FSC
www.fsc.org

If you have any concerns about our products,
you can contact us on
ProductSafety@springernature.com

In case Publisher is established outside the EU,
the EU authorized representative is:
Springer Nature Customer Service Center GmbH
Europaplatz 3, 69115 Heidelberg, Germany

Printed by Libri Plureos GmbH
in Hamburg, Germany